Pearson Education
AP* Test Prep Series

AP BIOLOGY

Fred W. Holtzclaw
Theresa Knapp Holtzclaw

To accompany:

Pearson's CAMPBELL BIOLOGY Programs

PEARSON

Boston Columbus Indianapolis New York San Francisco Upper Saddle River Amsterdam
Cape Town Dubai London Madrid Milan Munich Paris Montreal Toronto Delhi
Mexico City São Paulo Sydney Hong Kong Seoul Singapore Taipei Tokyo

Vice President/Editor-in-Chief: *Beth Wilbur*
Senior Acquisitions Editor: *Josh Frost*
Senior Editorial Manager: *Ginnie Simione Jutson*
Senior Supplements Project Editor: *Susan Berge*
Assistant Editor: *Katherine Harrison-Adcock*
Managing Editor, Production: *Michael Early*
Production Project Manager: *Jane Brundage*
Production Management and Composition: *S4Carlisle Publishing Services*
Cover Production: *Seventeenth Street Studios*
Manufacturing Buyer: *Michael Penne*
Executive Marketing Manager: *Lauren Harp*
Text and Cover printer: *Edwards Brothers Malloy*
Cover Photo Credit: *"Succulent I" ©2005 Amy Lamb, www.amylamb.com*

This AP* Test Prep Workbook replaces
ISBNs 0-13-137553-9/978-0-13-137553-6
and 0-13-135749-2/978-0-13-135749-5.

6 7 8 9 10—EBM—16 15 14

ISBN-10:   0-321-85663-5
ISBN-13: 978-0-321-85663-0

www.PearsonSchool.com/Advanced

Brief Contents

About Your Pearson AP* Guide

Pearson Education is the leading education solution provider worldwide. With operations on every continent, we make it our business to understand the changing needs of students at every level, from kindergarten to college. We think that makes us especially qualified to offer this series of AP* Test Prep workbooks, tied to some of our best-selling programs.

Our reasoning is that as you study for your course, you're preparing along the way for the AP Exam. If you can connect the material in the book directly to the exam, it makes the material that much more relevant, and enables you to focus your time most efficiently. And that's a good thing!

The AP Exam is an important milestone in your education. A high score means you're in a better position for college acceptance, and possibly puts you a step ahead with college credits. Our goal is to provide you with the tools you need to succeed.

Good luck!

Revisions to This Edition

Part I: *Introduction to the AP Biology Examination*

- The introduction is now aligned with the new Curriculum Framework (CF) that launches in the 2012–2013 school year.
- An outline of the new CF is included and its organization is explained.
- The science practices that will be tested are introduced.
- A revised topic correlation shows how concepts in *Campbell Biology* correspond to the new CF.
- The description of the exam and the testing hints are revised to reflect the changes in the course beginning with the 2013 exam.

Part II: *A Review of Topics with Sample Questions*

- You Must Know boxes have been edited to reflect the change in emphasis of the new CF.
- New boxes titled What's Important to Know? are scattered throughout the content areas to remind students of the types of questions they might be asked. These often focus on science practices, which is an emphasis of the new course.
- Content that is no longer relevant to the exam has been removed, or notes have been added to make it explicit whether material reflects illustrative examples or is required content.
- New test questions have been added to each topic to reflect changes that will be seen on the exam beginning in 2013. This includes questions that require interpretation of data or application of knowledge.

Part III: *The Laboratory*

- The 12 classic labs in the 2001 AP Biology Laboratory Manual have been modified or replaced. The College Board has released a new group of laboratory investigations, and so this section has been heavily revised to reflect these changes.
- The new CF asks students to be able to apply mathematics to a variety of topics. A sheet of formulas will be supplied to them on the exam. This is included in this revised edition, along with a number of tutorials and problems that take students through sample mathematical applications.
- New boxes titled Science Practices: Can You. . . have been added to focus students on making connections and applying science practices.

Part IV: *Sample Test*

- The format of the exam changed in 2012–2013. The sample test more accurately reflects the types of questions students may encounter. It will be very important for students to practice with these types of questions throughout the year.
- Grid-in questions are included in the sample test. Students will be expected to have a calculator for the new exam and provide numerical responses with a grid-in system.
- Free-response questions will be of varying lengths. The sample test follows this new format.

Part V: *Answers and Explanations*

- The format of this section remains similar to past editions, but some explanations focus on science practices and applications.

Part I

Introduction to the AP Biology Examination

This section gives an overview of the Advanced Placement* program and the AP Biology Examination. Part I introduces the types of questions you will encounter on the exam, explains the procedures used to grade the exam, and provides helpful test-taking strategies. A correlation chart shows where in *Campbell Biology*, Ninth Edition, by Reece et al., you will find key information that commonly appears on the AP Biology Examination. Review Part I carefully before trying the sample test items in Part II and Part III.

The Advanced Placement* Program

Probably you are reading this book for a couple of reasons. You may be a student in an Advanced Placement (AP) Biology class, and you have some questions about how the whole AP Program works and how it can benefit you. Also, perhaps, you will be taking an AP Biology Examination, and you want to find out more about it. This book will help you in several important ways. The first part of this book introduces you to the AP Biology course and the AP Biology Exam. You'll learn helpful details about the different question formats—multiple-choice and free-response—that you'll encounter on the exam. In addition, you'll find many test-taking strategies that will help you prepare for the exam. A correlation chart at the end of Part I shows how to use your textbook, *Campbell Biology*, Ninth Edition, by Reece et al., to find the information you'll need to know to score well on the AP Biology Exam. By the way, this chart is useful, too, in helping you to identify any extraneous material that won't be tested. Part II of this book provides an extensive content review correlated to each unit of your textbook, along with sample multiple-choice and free-response questions. Finally, in Part III, you will find a full-length sample test. This will help you practice taking the exam under real-life testing conditions. The more familiar you are with the AP Biology Exam ahead of time, the more comfortable you'll be on testing day.

The AP Program is sponsored by the College Board, a nonprofit organization that oversees college admissions examinations. The AP Program offers thirty-three college-level courses to qualified high school students. If you receive a grade of 3 or higher on an AP exam, you may be eligible for college credit, depending on the policies of the institution you plan to attend. Over 3,000 colleges and universities around the world grant credit to students who have performed well on AP exams. Some institutions grant sophomore status to incoming first-year students who have demonstrated mastery of several AP subjects. You can check the policies of specific institutions on the College Board's website (www.collegeboard.com). In addition, the College Board confers a number of AP Scholar Awards on students who score 3 or higher on three or more AP exams. Additional awards are available to students who receive very high grades on four or five AP exams.

Why Take an AP Course?

You may be taking an AP course simply because you like challenging yourself and you are thirsty for knowledge. Another reason may be that you know that colleges look favorably on applicants who have AP courses on their secondary school transcripts. AP classes involve rigorous, detailed lessons, a lot of homework, and numerous tests. College admissions officers may see your willingness to take these courses as evidence of your work ethic and commitment to your education. Because AP course work is more difficult than average high school work, many admissions officers evaluate AP grades on a higher academic level. For example, if you receive a B in an AP class, it might carry the same weight as an A in a regular-level high school class.

Your AP Biology course prepares you for many of the skills you will need to succeed in college. For example, your teacher may assign a major research paper or require you to perform several challenging laboratory exercises using proper scientific protocol. AP Biology teachers routinely give substantial reading assignments, and students learn how to take detailed lecture notes and participate vigorously in class discussions. The AP Biology course will challenge you to gather and consider information in new—and sometimes unfamiliar—ways. You can feel good knowing that your ability to use these methods and skills will give you a leg up as you enter college.

Each college or university decides whether or not to grant college credit for an AP course, and each bases this decision on what it considers satisfactory grades on AP exams. Depending on what college you attend and what area of study you pursue, your decision to take the AP Biology Exam could save you tuition money. You can contact schools directly to find out their guidelines for accepting AP credits, or use the College Board's online feature, "AP Credit Policy Info."

Taking an AP Examination

The AP Biology Exam is given annually in May. Your AP teacher or school guidance counselor can give you information on how to register for an AP exam. Remember, the deadline for registration and payment of exam fees is usually in March, two months before the actual exam date in May. The cost of the exam is subject to change and can differ depending on the number of exams taken. However, in 2012 a single exam costs $87. For students who can show financial need, the College Board will reduce the price by $26, and your school might also waive its regular rebate of $8, so the lowest possible total price is $53. Moreover, schools in some states are willing to pay the exam fee for the student. If you feel you may qualify for reduced rates, ask your school administrators for more information.

The exams are scored in June. In mid-July the results will be available online for you, your high school, and any colleges or universities you indicated on your answer sheet. If you want to know your score as early as possible, you can get it (for an additional charge of $8) beginning July 1 by calling the College Board at (888) 308-0013. On the phone, you'll be asked to give your AP number, your social security number, your birth date, and a credit card number.

If you decide that you want your score sent to additional colleges and universities, you can fill out the appropriate information on your AP Grade Report (which you will receive by mail in July) and return it to the College Board. There is an additional charge of $15 for each additional school that will receive your AP score.

On the other hand, if a feeling of disaster prevents you from sleeping on the nights following the exam, you could choose to withhold or cancel your grade. (Withholding is temporary, whereas canceling is permanent.) Grade withholding carries a $10 charge per college or university, whereas canceling carries no fee. You'll need to write to (or e-mail) the College Board and include your name, address, gender, birth date, AP number, the date of the exam, the name of the exam, a check for the exact amount due, and the name, city, and state of the college(s) from which you want the score withheld. You should check the College Board website for the deadline for withholding your score, but it's usually in mid-July. It is strongly suggested that you *do not* cancel your scores, since you won't know your score until mid-July. Instead, relax and try to assume that the glass is half full. At this point, you have nothing to lose and a lot to gain.

If you would like to get back your free-response booklet for a post-exam review, you can send another check for $7 to the College Board. You'll need to do this by mid-September. Finally, if you have serious doubts about the accuracy of your score for the multiple-choice section, the College Board will rescore it for an additional $25.

AP Biology: Course Goals

Beginning with the 2012–2013 school year, the AP Biology course has been revised, and the May examination reflects a change in emphasis. Although you may have friends who have taken the AP exam in earlier years, their course and yours are not the same. We will help you be prepared for the current format.

The new course and exam focuses on enduring, conceptual understandings and the content that supports them. There will be less factual recall and more emphasis on inquiry-based learning of essential concepts. The goal is to help you develop the reasoning skills necessary to be prepared for the study of advanced topics in college science courses.

The development of advanced inquiry and reasoning skills will involve an increased emphasis on science practices. You should be able to design a plan for collecting data, analyzing data, making predictions, applying mathematical routines, and connecting concepts across domains (levels of organization). Throughout this guide, we will make references to these science practices. Let's be clear: You will still need to master content! And then, you will apply your newfound knowledge.

In the revised AP Biology course, content, inquiry, and reasoning are equally important. The exam will be based on a series of Learning Objectives (LOs), and each LO combines content and science practices. Here is an abbreviated list of the Science Practices from pages 97–102 of the College Board's publication *Course and Exam Description 2012*. You will find an expanded list of the Science Practices on pages 97–102 of the guide.

Science Practices for AP Biology

Source: AP Biology—Course and Exam Description. © 2012. The College Board. www.collegeboard.org. Reproduced with permission.

Science Practice 1: The student can use representations and models to communicate scientific phenomena and solve scientific problems.
Science Practice 2: The student can use mathematics appropriately.
Science Practice 3: The student can engage in scientific questioning to extend thinking or to guide investigations within the context of the AP course.
Science Practice 4: The student can plan and implement data collection strategies appropriate to a particular scientific question.
Science Practice 5: The student can perform data analysis and evaluation of evidence.

Science Practice 6: The student can work with scientific explanations and theories.

Science Practice 7: The student is able to connect and relate knowledge across various scales, concepts, and representations in and across domains.

Overview of the Course

Source: AP Biology—Course and Exam Description. © 2012. The College Board. www.collegeboard.org. Reproduced with permission.

The key concepts and related content of the revised AP Biology course and exam are organized around four underlying principles called **Big Ideas**. These encompass the core scientific principles, theories, and processes governing living organisms and biological systems. Each Big Idea has several **Enduring Understandings**, which are the fundamental concepts you should retain from your course.

Each Enduring Understanding has statements of **Essential Knowledge** that you should know. All of the details in the outline are required elements of the course and may be included in the AP Biology Exam. Finally, your teacher may elect to help you learn about biology this year by selecting from many possible **Illustrative Examples**. This is what will make each course unique, and allow your teacher to teach a rich course where they may select unique examples of their own choosing. For this reason, this guide may cover content that your teacher has not have selected. You will need to pick and choose through these areas, realizing that there are many ways a teacher may elect to teach this course.

We have taken the Curriculum Framework (all 91 pages) and summarized it in an outline (4 pages). There are four **Big Ideas**. Under each Big Idea, the A, B, C, and so on represent **Enduring Understandings**. The numbers under A, B, C, etc., are **Essential Knowledges**. The **Illustrative Examples** are not represented, as they vary from course to course. The outline that follows is also correlated to *Campbell Biology,* Ninth Edition, but if you do not use this text, the outline will still be helpful to you as a summary.

It is important for you to know that this outline gives you an idea of the topics that will be covered on the AP exam in May, but the new course will also emphasize Science Practices linked to this content. Facts will not be sufficient! You will need to work with the Science Practices all year long to do well on the exam.

Correlation between the AP Biology Curriculum Framework 2012–2013 and CAMPBELL BIOLOGY 9e AP* Edition

Summary Outline of Topics in Curriculum Framework	Correlation to *Campbell Biology 9E AP* Edition*
Big Idea 1: The process of evolution drives the diversity and unity of life	
A. Evolution: Change over Time	
1. Natural selection as a mechanism	22.2, 23.2, 51.3, 51.4
2. Natural selection acts on phenotypes	23.1, 23.4, 51.3, 51.4

     3. Evolutionary change is random   23.3, 51.3, 51.4

     4. Evolution is supported by scientific evidence   22.3, 25.2, 51.3, 51.4

B. Descent from Common Ancestry

     1   Many essential processes and features are widely   25.1, 25.3
        conserved

     2. Phylogenetic trees and cladograms   26.1–26.3

C. Evolution Continues in a Changing Environment

     1. Speciation and extinction (rates/adaptive radiation)   24.3, 24.2, 24.4, 25.2, 25.4

     2. Role of reproductive isolation   24.1

     3. Populations continue to evolve (e.g., drug resistance)   22.3, 24.2

D. Natural Processes and the Origin of Living Systems

     1. Hypotheses and evidence of these origins   4.1, 25.1, 25.3

     2. Scientific evidence from many disciplines   26.6

Big Idea 2: Biological systems utilize free energy and molecular building blocks to grow, to reproduce, and to maintain dynamic homeostasis

A. Role of Free Energy and Matter for Life Processes

     1. Constant input of free energy is required   8.2, 40.1, 40.2, 40.3, 40.4, 55.1

        a. Energy pathways, ecosystem effects   51.3, 53.3, 53.4, 55.2, 55.3

        b. Laws of thermodynamics/coupled reactions/exergonic,   8.1, 8.2, 8.3
           endergonic

     2. Role of autotrophs and heterotrophs in free energy   10.1
        capture

        a. Light reactions/chemiosmosis/Calvin cycle   10.2, 10.3

        b. Glycolysis, pyruvate oxidation, Krebs cycle, etc.   9.1, 9.2, 9.3, 9.4, 9.5

     3. Matter and exchange with environment

        a. Role of carbon, nitrogen, and phosphorus in organic   4.1, 4.2, 4.3
           compounds

        b. Properties of water   3.1, 3.2

        c. Surface area/volume ratios and exchange   6.2

        d. Role of apoptosis   11.5

B. Cell Maintenance of Internal Environment

     1. Cell membrane structure and selective permeability   7.1, 7.2

     2. Transport processes across membranes   7.3, 7.4, 7.5

     3. Compartmentalization   6.2, 6.3, 6.4, 6.5

C. Role of Feedback Mechanisms in Homeostasis

     1. Positive and negative feedback and examples   40.2, 45.2

     2. How organisms respond to changes in external   40.3   *(continued)*
        environment

D. **Growth and Homeostasis Are Influenced by Environmental Changes**

  1. Biotic and abiotic factors/effect on cells → ecosystems   52.2, 53.1, 53.2, 53.3, 53.4, 53.5, 54.1, 54.2, 54.3, 54.4, 54.5, 55.1, 55.2, 55.3, 55.4

  2. Homeostatic mechanisms reflect evolution   40.2, 40.3

  3. Effect of homeostatic disruption at various levels   40.2, 40.3, 56.1

  4. Plant and animal defense systems against infection   39.5, 43.1, 43.2, 43.3, 43.4

E. **Temporal Regulation and Coordination to Maintain Homeostasis**

  1. Regulation of timing and coordination in development

    a. Cell differentiation   11.5, 18.2, 18.4, 25.5, 47.3

    b. Homeotic genes/induction   18.3, 18.4

    c. Gene expression/microRNAs   18.3

  2. Mechanism of control/coordination of timing

    a. Plants: photoperiodism/tropisms/germination   38.1, 39.1, 39.2, 39.3

    b. Animals   24.1

    c. Fungi/protists/bacteria   11.1

  3. Mechanisms that time and coordinate behavior

    a. Innate behaviors/learning   51.1, 51.2, 51.4

    b. Plant/animal behaviors   39.2, 39.3, 51.1, 51.4

    c. Cooperative behaviors   54.1

Big Idea 3: Living systems store, retrieve, transmit, and respond to information essential to life processes

A. **Heritable Information**

  1. DNA and RNA

    a. Structure and function   5.5, 16.1

    b. Replication   16.1, 16.2

    c. Role of RNA and its processing   17.1, 17.2, 17.3, 17.4

    d. Prokaryotic/viral differences   19.2, 27.1

    e. Manipulation of DNA   20.1, 20.2

  2. Cell cycle, Mitosis and Meiosis   12.1, 12.2, 12.3, 13.1, 13.2, 13.3

  3. Chromosomal basis of inheritance   14.1, 14.2, 14.3, 14.4

  4. Non-Mendelian inheritance   15.1, 15.2, 15.3, 15.5

B. **Cellular and Molecular Mechanisms of Gene Expression**

  1. Gene regulation and differential gene expression   18.1, 18.2, 18.3

  2. Signal transmission and gene expression   11.1, 11.4, 18.4, 45.1, 45.2

C. **Genetic Variation Can Result from Imperfect Processing**

  1. Changes in genotype → phenotype changes   15.4, 16.2, 17.5, 21.2, 23.4

  2. Mechanisms to increase variation   27.2, 13.4

  3. Role of viruses   19.1, 19.2

D. How Cells Transmit and Receive Signals

1. Processes reflect evolutionary history	11.1, 45.2
2. Local and long-distance signaling	11.1, 11.2, 45.1, 45.2, 45.3
3. Signal transduction pathways	11.2, 11.3
4. Effect of changes in pathways	11.1, 11.2, 11.3, 11.4

E. How Information Transmission Results in Changes

1. Organisms exchange information which changes behavior	51.1
2. Role of the animal nervous system	48.1, 48.2, 48.3, 48.4
a. Neurons/synapses/signaling	48.1, 48.2, 48.3, 48.4
b. Mammalian brain	49.2

Big Idea 4: Biological systems interact, and these systems and their interactions possess complex properties

A. Interactions within Biological Systems

1. Properties of biological molecules and their components; monomers and polymers	5.1, 5.2, 5.3, 5.4, 5.5
2. Cellular organelles and their role in cell processes	6.2, 6.3, 6.4, 6.5
3. Role of interaction between stimuli and gene expression in specialization	18.4
4. Complex properties are due to interaction of constituent parts	48.4
5. Populations and organisms interact in communities	
a. Ecological field data	53.1, 53.2, 53.3, 54.1, 54.2, 54.4
b. Growth curves, demographics	53.1, 53.2, 53.3, 53.5, 53.6
6. Interactions and energy flow in the environment	55.1, 55.3
a. Human impact on ecosystems	55.4, 55.5

B. Competition and Cooperation

1. Interaction and structure/function of molecules	5.4, 8.4, 8.5
a. Enzymes and their action	8.4, 8.5
2. Interactions within organisms and use of energy/matter	6.4, 6.5, 6.6
a. Compartments (e.g., digestion, excretion, circulation)	40.1
3. Interactions of populations and effect on species distribution/abundance	54.1
4. Change in ecosystem distribution over time	56.1, 56.4, 25.4

C. Diversity Affects Interactions within the Environment

1. Variation in molecular units provides cells with a wide range of functions (e.g., chlorophylls, antibodies, allelic variants)	5.1, 5.2, 5.3, 5.4, 5.5, 21.5
2. Gene expression of genotype is influenced by environment	14.3
3. Variation affects cell structure and function	21.5, 23.4, 26.4
4. Variation affects population dynamics	23.1, 23.2, 23.3
5. Diversity affects ecosystem stability	14.3, 23.2, 54.2, 56.1

Understanding the AP Biology Examination

The AP Biology Exam takes three hours and includes both a 90-minute multiple-choice section and a 90-minute free-response (essay) section that begins with a 10-minute reading period. The multiple-choice section will be one-half of your exam grade, and the free-response section will account for the other half. Both sections include questions that assess students' understanding of the Big Ideas, Enduring Understandings, and Essential Knowledge statements. The exam probably looks like many other tests you've taken. At the core of the examination are questions designed to measure your knowledge and understanding of modern biology. You should be prepared to recall basic facts and concepts, to apply scientific facts and concepts to particular problems, to synthesize facts and concepts, and to demonstrate reasoning and analytical skills by organizing written answers to broad questions.

The AP Biology Exam is very challenging. When you sit down to take the test, you are expected not only to be fluent in the areas of biology that you find fascinating (the ones that probably inspired you to take a special interest in the subject originally), but also to have an intimate knowledge of topics you don't find interesting at all. Whatever those topics might be—DNA replication, the dizzying details of gene expression or the immune system—you need to be comfortable with and knowledgeable about all of the AP Biology topics.

Section I, Part A: Multiple-Choice Questions

The College Board has finalized the exam for 2013. There will be 63 standard multiple-choice questions and 6 questions that involve mathematical calculation for which you will grid in a response. You will have 90 minutes to complete Section I. Each of the multiple-choice questions will be directly paired with a Learning Objective from the Curriculum Framework. The questions will require both an understanding of important concepts and biological processes, and then the ability to apply information that is given to you in the question. Because of this, the stem of the questions may be longer than many multiple-choice questions you have seen before. Read the stem carefully, study the accompanying charts and figures, and then work methodically through the possible responses. Because each question may require a deep conceptual understanding, rather than the recall of facts, it will take you longer to answer this type of question. Compared to AP exam questions prior to 2013, the questions are more involved, so the number of questions has been reduced from 100 to 63 multiple-choice and 6 grid-in questions. This portion of the exam is followed by a 5–10 minute break—the only official break during the examination.

The directions for the multiple-choice section of the test are straightforward and similar to the following:

Directions: Each of the questions or incomplete statements below is followed by four suggested answers or completions. Select the answer that is best in each case.

Here is a sample item from page 152 of the College Board's *Course and Exam Description 2012.*

A human kidney filters about 200 liters of blood each day. Approximately two liters of liquid and nutrient waste are excreted as urine. The remaining fluid and dissolved substances are reabsorbed and continue to circulate throughout the body. Antidiuretic hormone (ADH) is secreted in response to reduced plasma volume. ADH targets the collecting ducts in the kidney, stimulating the insertion of aquaporins into their plasma membranes and an increased reabsorption of water.

If ADH secretion is inhibited, which of the following would initially result?

(A) The number of aquaporins would increase in response to the inhibition of ADH.

(B) The person would decrease oral water intake to compensate for the inhibition of ADH.

(C) Blood filtration would increase to compensate for the lack of aquaporins.

(D) The person would produce greater amounts of dilute urine.

Essential Knowledge	3.D.3: Signal transduction pathways link signal reception with cellular response.
Science Practice	1.5: The student can re-express key elements of natural phenomena across multiple representations in the domain.
Learning Objective	3.36: The student is able to describe a model that expresses the key elements of signal transduction pathways by which a signal is converted to a cellular response.

Your course may not have included a study of kidney function, but all the information you need to answer this item is given in the stem. You are told what ADH does and how it is controlled, and given the role of aquaporins. It is expected that you can use this information to select the correct answer. If you are certain you know the answer, fill in the corresponding oval on the answer sheet. However, what if you're not certain? The next step is to see if you can eliminate one or more of the choices.

> **TIP FROM THE READERS**
> In the past, there was a penalty for guessing, but that has been eliminated. *Answer every question!*

Let's look again at the question. Choice *A* suggests a response when ADH is inhibited. You must know that this means there is less ADH. The stem tells you that ADH "stimulates the insertion of aquaporins" so you will eliminate this choice. Continue to methodically analyze the other choices, and try to eliminate another choice based on your understanding of the information that is given in the stem. Confidently mark your answer sheet and continue. Remember, answer *every* question!

Lab-Based or Experimental Questions

Another type of question you will see in Section I is the lab-based or experimental question. These questions either present you with a set of data in graph (or other) form, or they describe an experiment and ask you to make predictions, select appropriate data, form hypotheses, analyze data mathematically, and perform other science practices. These questions often occur in groups that use the same data set.

It is known that plant cells require oxygen in order to obtain ATP. Those who work with plants have long known that it is possible to quickly kill a plant by overwatering. The graph below shows the results of a study of the effect of soil air spaces on plant growth.

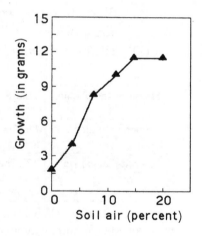

1. The data from the above graph show that the plant
   (A) grows fastest when the soil is 5–10% air.
   (B) grows fastest when the soil is 15–20% air.
   (C) grows at the same rate regardless of the soil air percentage.
   (D) grows most slowly when the soil is 5–10% air.

The correct choice is *A*. The graph shows the line with the greatest slope (the highest degree of change over the shortest amount of time) between the percentages 5 and 10. During this time, the plant grows by about 9 – 5 = 4 grams. Just to be sure, check the amount this plant grows when the soil is 15–20% air. At the start, when the air was 15% air, the plant weighed 12 grams. At the end, when the soil is 20% air, the plant weight is the same—12 grams. Virtually no growth occurred during this time.

Clearly this question requires you to be able to interpret a graph, but at this point in your biology education you should be quite capable of doing that. In order to brush up on the various ways that graphs present information, you might review Appendix B of the AP Biology Investigation Manual.

2. It is seen from the graph that plant growth is negatively affected by decreased soil oxygen. Which of the following statements bests justifies the reason for this effect?
   (A) Plant root cells require oxygen because no photosynthesis occurs underground.
   (B) Plant stem and leaf cells are able to do photosynthesis at a higher rate when there is more oxygen in the soil.
   (C) Water in the soil eliminates oxygen from the soil spaces, so the root cells are unable to produce ATP
   (D) The rate of cellular respiration is decreased under conditions of high soil moisture.

The correct choice is *C*. This question requires you to recall the reactants and products of photosynthesis and cellular respiration. *All* plant cells do cellular respiration, consuming O_2 in order to produce ATP. Choice *C* justifies (explains) why plant growth is low when there is low soil O_2. Choice *D* restates the information from the graph. While it is a true statement, it does not justify the reason for this correlation.

Section I, Part B: Grid-In Questions

These questions will require you to calculate an answer for the question, and enter it in a grid in that section on your answer sheet, as shown on the next page. Be sure to practice gridding responses correctly prior to the exam. By the way, you will be able to use a four-function calculator during the exam, as well as a formula sheet (Appendix B of the Investigation Manual), because the emphasis here is on your ability to apply mathematical techniques. To set your mind at ease, the electronic scoring for these responses is set so there is a range of correct scores, to allow for variations in rounding.

The acceptable answer is in the range of 6030–6156, depending on rounding. This question requires that you be able to use the Hardy-Weinberg equation to calculate allelic frequencies. Be sure to remember to carefully follow any instructions that are given for rounding, such as "round to the nearest tenth." Study the grid pattern on the next page, and you will see that you are able to indicate negative numbers, fractions, and the location of decimal points.

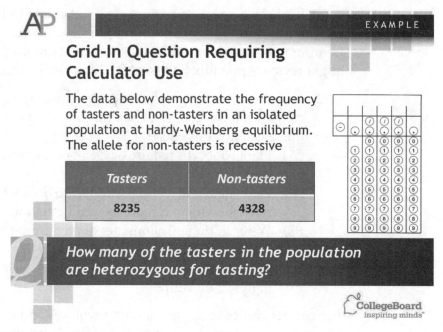

Grid-In Question Requiring Calculator Use

The data below demonstrate the frequency of tasters and non-tasters in an isolated population at Hardy-Weinberg equilibrium. The allele for non-tasters is recessive

Tasters	Non-tasters
8235	4328

How many of the tasters in the population are heterozygous for tasting?

CollegeBoard
inspiring minds

Source: AP Biology—Course and Exam Description. © 2012. The College Board. www.collegeboard.org. Reproduced with permission.

Integer answer 502	Integer answer 502	Decimal answer −4.13	Fraction answer −2/10

Source: AP Biology—Course and Exam Description. © 2012. The College Board. www.collegeboard.org. Reproduced with permission.

Section II: Free-Response Questions

At the beginning of Section II of the AP Biology Exam, you will be given a 10-minute time period where you may read the questions and plan and outline your responses before you may begin writing. You should use this time to carefully consider each question. Underline key words or phrases such as "Select *two*" or "*Explain* and *predict*" or "Using an example from *each* domain." Begin an outline of your response; include key words. If a graph will be required, decide what type is appropriate, and what will go on each axis. After the reading period, you will be given a response booklet in which to write your essays.

Section II of the AP Biology Exam will have 6 free-response questions. This section contains two types of free-response questions, 2 long and 4 short. The long questions will consist of several sections or related tasks, and be evaluated on a 10-point scale. You should allow approximately 20 minutes for a long free-response question. This is the format that was used on AP Biology exams prior to 2013.

There will be 4 short-response questions that call for brief responses. The instructions may say something like "In a sentence or two. . . ." Heed these parameters to use your time wisely. The short free-response questions should be completed in about 6 minutes or so. For all free-response questions, you must write your answers out, and cannot use an outline form.

BIOLOGY

Section II

Time—80 minutes

Directions: Answer all questions.

Answers must be in essay form. Outline form is NOT acceptable. Labeled diagrams may be used to supplement discussion, but in no case will a diagram alone suffice. It is important that you read each question completely before you begin to write.

1. Water comprises roughly 70% of the human body; cells are roughly 70–95% water, and water covers about three-quarters of the Earth's surface.

   (a) **Describe** the major physical properties of water that make it unique from other liquids.
   (b) **Explain** the properties of water that enable it to travel up through the roots and stems of plants to reach the leaves.
   (c) **Explain** why the temperature of the oceans can remain relatively stable and support vast quantities of both plant and animal life, when air temperature fluctuates so significantly throughout the year.

   Like many free-response questions on the AP Biology Exam, this sample is broken into three distinct parts. Each contains a clear directive. In fact, they are printed in boldface to help you focus on exactly how you should answer the question. First you will need to explain the uniqueness of water by describing its major physical properties. (In your response to this first part of the question, you might wish to include a labeled diagram of the structure of water, complete with electrons and bonds.) Then you must explain the properties of water that allow it to travel from root to leaf. Finally, you should explain the reason(s) why ocean water temperature remains stable and supports plant and animal life—even in the face of great air temperature variations. Of course, limiting your answer by addressing exactly what the question asks will make writing the essay easier for you and earn you a higher score. Always take the time to determine precisely what is being asked before you begin to formulate a concrete thesis and focus on writing your relevant supporting paragraphs.

Grading Procedures for the AP Biology Examination

The raw scores of the AP Biology Examination are converted to the following 5-point scale:

> 5—Extremely Well Qualified
> 4—Well Qualified
> 3—Qualified
> 2—Possibly Qualified
> 1—No Recommendation

Some colleges give undergraduate course credit to students who achieve scores of 3 or better on AP exams. Other colleges require students to achieve scores of 4 or 5. You may check the policy for individual colleges on the College Board website (www.collegeboard.com). Below is a breakdown of how the grading of the AP Biology Exam works.

Section I: Multiple-Choice Questions

The multiple-choice section of the exam is worth 50% of your total score. The raw score of Section I is determined by crediting one point for each correctly answered question. Therefore, it is important to answer each question. *There is no penalty for guessing!*

Section II: Free-Response Questions

Section II counts for 50% of your examination grade. The free-response section is scored by several hundred faculty consultants, including high school teachers and college instructors from all over the country who work in a central location to grade the essays. This period of scoring exams is called the "Reading." To ensure that scoring of all exams is consistent, grading rubrics, or standards, are developed and then faculty consultants are trained in their application. Because of this intense training, group discussion, and supervision, your essay should receive the same score regardless of who reads it. Ongoing internal checks during the Reading ensure this. Each of your essays will be evaluated by a faculty consultant trained to score that single response.

Your answers to the free-response questions must be presented in essay form. Outlines or unlabeled and unexplained diagrams are not given credit. Your performance on any single essay is evaluated independently of the other essays. Do not assume that information provided in one question will be considered during the grading of another essay. You should repeat information from question to question if it is necessary to illustrate your point.

Test-Taking Strategies for the AP Biology Examination

Here are a few tips for preparing yourself in the weeks leading up to the examination.

▌ The earlier you start studying for the AP Biology Exam, the better. Some students use this AP Biology prep book along with their textbook throughout the course, taking notes in the margin to supplement their teacher's lectures. You should definitely begin serious preparation for the test at least one month in advance.

▌ Each chapter in Part II is correlated to *Campbell Biology*, Ninth Edition, by Reece et al. For each topic, review your lecture notes, study the figures in your text that explain key concepts, and then make your way through the corresponding section of Part II of this book. If possible, retake your unit test on the topic, and also answer the questions in this guide for each unit. This will help you identify topics that will require further study. Do not try to reread your text; use it as a tool for those topics that need further study. You can use the correlation guide at the end of Part I of this book to link AP Biology topics to your textbook. Pace yourself!

AP Review: Lab Essays

The College Board has published a manual with 13 recommended inquiry labs. It is expected that you will complete two labs for each Big Idea during the course of the school year. Your teacher may elect to use some of the labs from the College Board's *AP Biology Investigative Labs,* or others of his/her choosing. All the laboratories will emphasize the science practices, and you will become more proficient in working and thinking like a scientist only by engaging in this type of work. These labs will often run over multiple days or even weeks, and will include components where you learn new techniques, and then use them to answer your own experimental questions.

Here is a list of the labs in the College Board's *AP Biology Investigative Labs.* An asterisk is placed by those labs that use techniques or cover concepts many teachers are familiar with from the "old" *AP Biology Laboratory Manual.*

Source: AP Biology—Investigative Labs: An Inquiry-Based Approach. © 2012. The College Board. www.collegeboard.org. Reproduced with permission.

Big Idea 1: Evolution

Investigation 1: Artificial Selection
Investigation 2: Mathematical Modeling: Hardy-Weinberg
Investigation 3: Comparing DNA Sequences to Understand Evolutionary Relationships with BLAST

Big Idea 2: Cellular Processes: Energy and Communication

Investigation 4: Diffusion and Osmosis*
Investigation 5: Photosynthesis*
Investigation 6: Cellular Respiration*

Big Idea 3: Genetics and Information Transfer

> Investigation 7: Cell Division: Mitosis and Meiosis*
> Investigation 8: Biotechnology: Bacterial Transformation*
> Investigation 9: Biotechnology: Restriction Enzyme Analysis of DNA*

Big Idea 4: Interactions

> Investigation 10: Energy Dynamics
> Investigation 11: Transpiration*
> Investigation 12: Fruit Fly Behavior*
> Investigation 13: Enzyme Activity*

It is likely that at least one essay on the exam will be based on an AP laboratory, though it may not be one that you have done. Not to fear! Since the questions are based on objectives for the course, your teacher will have prepared you to work with data from a variety of sources, so you will simply need to apply your expertise in scientific thinking acquired over the course to a novel situation. If you have done a similar experiment, this may be helpful—but be sure you read carefully and do not anticipate what is being asked.

You may be asked to "design an experiment to determine . . ." If you performed a lab in your AP class that would answer this question, it is fine to describe this lab. There are a number of items that would generally be included in your design, so consider including each of these:

Section	Question Type	Number of Questions	Timing
I	Part A: Multiple Choice	63	90 minutes
	Part B: Grid-In	6	
II	Long Free Response	2	80 minutes + 10-minute reading period
	Short Free Response	6	

Source: AP Biology—Course and Exam Description. © 2012. The College Board. www.collegeboard.org. Reproduced with permission.

- **State a hypothesis** as an "**If** . . . (conditions), **then** . . . (results)" statement. Your hypothesis must be testable.
- **Identify the variable factor** for the experiment (e.g., temperature).
- **Identify a control**. You must explain the control for the experiment.
- **Hold all other variables constant**. Explain how you would do this.
- **Manipulate the variable** (e.g., one group at 10°C, one at 20°C, and one at 30°C).
- **Measure the results** (e.g., cm grown, mass increase in grams).
- **Discuss results expected** as related to hypothesis.
- **Replication or verification**. The experiment must be repeated or large sample sizes must be used.

If appropriate, you could also consider using statistical analysis of data (see Chi-square analysis, Lab 7) and review of the literature.

Graphing Data

It is likely that you will be asked to graph data. You will need to consider the type of graph that is appropriate for your data. Bar graphs are used when data points are discrete, that is, not related to each other, such as the number of girls in AP Biology vs. the number of boys in AP Biology. Line graphs are used when the data are continuous, such as the change in an individual's height at each birthday. Consider if there is a data point at 0 on the graph. Be sure to extend your line to 0 if there is, but do not take the line to 0 if there is no measurement for that data point. Also,

▍ Label the graph with a descriptive title.

▍ Label the *x*- and *y*-axes. Be sure you know which variable is independent and which is dependent.

▍ Keep all measurement units constant. Each division on the graph must be a unit equal to all the others.

▍ If you are asked to draw a line to predict what would be shown if some change occurred, be sure that you include a legend for the second line.

Sample Tests

When you are ready to check your preparation, take the sample exam in Part III of this book. Keep track of your time, and try to simulate test conditions. Circle the items you get wrong, or could not answer, and keep a list of the subject matter of those questions. Then analyze the list to look for patterns—are you having a hard time with questions on animal physiology or the process of photosynthesis specifically? Spend the next week or so studying the topics in which you are weak. Spend the final days before the test looking through this guide, your class notes, and your textbook to fill in any remaining gaps.

The Day of the Exam

If you have followed this suggested study plan, you should feel well prepared by test day. Plan your schedule so that you get two very good nights of uninterrupted sleep before exam day. The night before the exam, relax, think positive thoughts, and focus on getting a good night's rest. Below is a brief list of basic tips and strategies to think about before you arrive at the exam site.

1. **Arrive early!** It's a good idea to arrive at the exam site 30 minutes before the start time. On the day of the exam, make sure that you eat a good, nutritious meal. These tips may sound corny or obvious, but your body must be in peak form in order for your brain to perform well. Remember, you are going to need ATP to fuel brain cells at peak efficiency for more than three hours.

2. **Bring a photo ID.** (It's essential if you are taking the exam at a school other than your own.) Carrying a driver's license or a student ID card will allow you to prove your identity.

3. **Bring at least two sharpened #2 pencils** for the multiple-choice section. Also, bring a clean pencil eraser with you. Many pencils today have cheap erasers that smudge. Invest in a good eraser. The machine that scores Section I of the

exam recognizes only marks made by a #2 pencil. Poorly erased responses are often misscored.

4. **Bring two black ballpoint pens** for the free-response portion of the test. Felt-tip pens run and pencils and inks of other colors are harder to read.

5. **Bring a watch** with you to the exam. Most testing rooms do have clocks. Still, having your own watch makes it easy to keep close track of your own pace. Watches with calculators or alarms are not permitted in the exam room.

6. **Bring a four-function calculator** (or your school may arrange to have them available).

Several other items that are forbidden from the testing room are books, notes, laptops, beepers, cameras, and portable listening or recording devices. If you must bring a cellular phone with you, be prepared to turn it off and to give it to the test proctor until you are finished with your exam. For a complete list of what not to bring, see the College Board website.

Educational Testing Service prohibits the objects listed above in the interest of fairness to all test-takers. Similarly, the test administrators are very clear and very serious about what types of conduct are not allowed during the examination. Below is a list of actions to avoid at all costs, since each can result in your immediate dismissal from the exam room.

▌ Do not consult any outside materials during the three hours of the exam period. Remember, the break is technically part of the exam—you are not free to review any materials at that time either.

▌ Do not speak during the exam. If you have a question for the test proctor, raise your hand to get the proctor's attention.

▌ When you are told to stop working on a section of the exam, you must stop immediately.

▌ Do not open your exam booklet before the test begins.

▌ Never tear a page out of your test booklet or try to remove the exam from the test room.

▌ Do not behave disruptively—even if you're distressed about a difficult test question or because you've run out of time. Stay calm and make no unnecessary noise.

Section I: Strategies for Multiple-Choice Questions

Obviously, having a firm grasp of biology is, of course, the key to doing well on the AP Biology Examination. In addition, being well-informed about the exam itself increases your chances of achieving a high score. Below is a list of strategies that you can use to increase your comfort, your confidence, and your chances of excelling on the multiple-choice section of the exam.

▌ Become as familiar as possible with the format of Section I. The more comfortable you are with the multiple-choice format and with the kinds of questions you'll encounter, the easier the exam will be. Remember, Part II and Part III of this book provide you with invaluable practice on the kinds of multiple-choice questions you will encounter on the AP Biology Exam.

- Every question you answer correctly is a point, so pacing is important. If you have done all the suggested practice tests, you should have a good sense of how to pace yourself. You will have 90 minutes to answer 63 multiple-choice and 6 grid-in questions (about 78 seconds per question). Keep track of time!

- Some of the questions will require calculations. If you encounter a question that will require extra time, leave it blank and make a note. Your goal should be to reach the end of the test, picking up all the points from questions you can answer easily.

- The test is organized with three types of questions: standard multiple choice, lab sets, and grid-ins. Lab sets are generally the most tedious. When a data table or graph is presented, proceed directly to the related questions. Determine what information is needed to answer the questions, and then return to the data table or graph and seek the information. Sometimes, although the data appear daunting, the questions are actually very easy.

- Make a light mark in your test booklet next to any questions you can't answer. Return to these questions after you reach the end of Section I. Sometimes questions that appear later in the test will refresh your memory on a particular topic, and you will be able to answer one or more of those earlier questions.

- Always read the entire question carefully, and underline key words or ideas. You might wish to double underline words such as NOT or EXCEPT in multiple-choice questions.

- Read each and every one of the answer choices carefully before you make your final selection.

- Use the process of elimination to help you arrive at the correct answer. Even if you are quite sure of an answer, cross out the letters of incorrect choices in your test booklet as you eliminate them. This cuts down on the incorrect choices and allows you to narrow the remaining choices even further.

- Become completely familiar with the instructions for the multiple-choice questions before you take the exam. By knowing the instructions cold, you'll save yourself the time of reading them carefully on exam day.

- If you finish early, you should go back over as many items as possible to catch any careless errors.

Section II: Strategies for Free-Response Questions

Below is a list of strategies that you can use to increase your chances of excelling on the free-response section of the exam.

- You will have a 10-minute period to review the essay questions before you receive a response book. During this time, you should organize your thoughts and outline your essays in the space provided. After the preparation time, you should record your answer on the pages provided for each question response book, and this is where you should record each answer. You have about 22 minutes to spend on each 10 point essay, and you should gauge your time for the shoft essays based on the instructions.

- *Read the question; then read the question again!* Be sure you answer the question that is asked and that you address each part of the question. As you read a question, underline any directive words (usually the first word in an essay) that indicate how you should answer and focus the material in your essay. Some of the most frequently used directives on the AP Biology Exam are listed below, along with descriptions of what you need to do in your writing to answer the question.

- *Analyze* (show relationships between events; explain)
- *Compare* (discuss similarities and differences)
- *Contrast* (discuss points of difference or divergence between two or more things)
- *Describe* (give a detailed account)
- *Design* (create an experiment and convey its ideas)
- *Explain* (clarify; tell the meaning)
- *Predict* (tell what you expect to happen when conditions change)
- *Justify* (explain why a response is reasonable)

▌ Reread the question as many times as necessary to make sure that you will cover each aspect of the topic. Free-response questions frequently have several parts, so you will need to take this into account as you outline your ideas.

▌ Write an essay! As the exam states clearly in the directions to free-response questions, a diagram or graph by itself is never an acceptable way to answer a free-response question. However, you should think about whether you could use a labeled diagram or graph to develop your written answer in some useful way.

▌ The essay you craft for this exam is not the same type of essay you should write for an English course. Yes, it should be well-organized; however, introductory sentences and conclusions are absolutely not necessary. Readers are interested in what you know and how well you express your knowledge. Spend your time packing the essay with the biological information you have worked so hard to learn.

▌ If the question has several parts, answer the parts in the sequence given. Use a letter or some other indication for each part, so that the faculty consultant does not overlook a section of your response.

▌ If you are asked to perform a calculation, be sure to show the steps used to arrive at your answer. You have heard this before: Show your work!

▌ If you cannot remember a specific term, describe the structure or process.

▌ Define any scientific term that you use that is directly related to your response. For example, if you discuss hydrogen bonding and how it relates to properties of water, be sure to explain what hydrogen bonds are, and then describe or define adhesion, cohesion, and so on.

▌ Your handwriting can affect your results. Although faculty consultants make every attempt to read each essay, sometimes it is impossible to decipher messy handwriting. When your handwriting is poor, the reader may lose concentration or patience and miss an important word or phrase.

▌ Don't leave any part of any essay blank. Every point made is worth approximately twice as much as each multiple-choice point.

■ If time allows, proofread your essays. Don't worry about crossing out material—readers understand that your responses are first drafts and that you are writing down ideas under the pressure of time.

The success of your free-response essays will depend a great deal on how clearly and extensively you answer the questions posed. Of course, the structure of your essays will depend entirely on your knowledge of the subjects at hand. Take a look at an example of a free-response question below.

2. In cancer, the cell's reproductive machinery experiences a loss of control that makes cancer cells reproduce continually, and eventually form a tumor.

   (a) **Describe** three DNA-related cellular events that could lead to the loss of cell division control that contributes to cancer.
   (b) **Describe** why tumors are detrimental to the body.
   (c) **Predict** two cell processes that could be targeted to find a cure for cancer.

In order to answer this question, you should isolate exactly what it is that you must answer. You may want to underline the relevant information in the question to remind yourself of your focus:

2. In cancer, the cell's reproductive machinery experiences a <u>loss of control</u> that makes cancer cells reproduce continually, and eventually form a tumor.

   (a) <u>**Describe**</u> <u>three</u> DNA-related cellular events that could lead to the <u>loss of cell division control</u> that contributes to cancer.
   (b) <u>**Describe**</u> <u>why</u> tumors are detrimental to the body.
   (c) <u>**Predict**</u> <u>two</u> cell processes that could be targeted to find a cure for cancer.

In order to answer the first part of the question, you'll need to identify three appropriate DNA-related cellular events.

1. A mutation or change in the original DNA sequence
2. Errors in DNA replication that go undetected by the cell's proofreading devices
3. A translocation

Under each of the three events, you should list any and all details you remember about those events to use in your description. When you flesh out these details, you'll need to clearly connect them to the concept of the loss of cell division control leading to cancer.

To answer the second part of the question, you'll need to list as many reasons as you can think of as to why tumors are harmful to the body. These might include cancerous cells' ability to metastasize; tumors' ability to occur almost anywhere in the body; their tendency to block the flow of blood when they grow near blood vessels; disruption of the natural function of any organ in the body; and endangering of homeostasis. Of course, after you list reasons, you'll need to add details to each item in your list.

To answer the final part of the question, you'll need to consider first what causes cancer. Then you must think creatively in order to predict reasonable approaches to dealing with each specific cause.

Part II of this book contains a review of everything that you learned in your textbook that could be on the AP Biology test. Many questions will be posed along the way so that you can get used to being tested on the concepts in the way that the College Board will test you. In Part III, inquiry labs have been reviewed, including sample test questions to check your understanding. In Part IV, there is a practice test for you to try on your own. Part V is where you will find answers and explanations to all the questions asked in this book.

AP Biology Concepts at a Glance

Source: AP Biology—Investigative Labs: An Inquiry-Based Approach. © 2012. The College Board. www.collegeboard.org. Reproduced with permission.

BIG IDEA 1: The process of evolution drives the diversity and unity of life.	
Enduring understanding 1.A: Change in the genetic makeup of a population over time is evolution.	**Essential knowledge 1.A.1:** Natural selection is a major mechanism of evolution.
	Essential knowledge 1.A.2: Natural selection acts on phenotypic variations in populations.
	Essential knowledge 1.A.3: Evolutionary change is also driven by random processes.
	Essential knowledge 1.A.4: Biological evolution is supported by scientific evidence from many disciplines, including mathematics.
Enduring understanding 1.B: Organisms are linked by lines of descent from common ancestry.	**Essential knowledge 1.B.1:** Organisms share many conserved core processes and features that evolved and are widely distributed among organisms today.
	Essential knowledge 1.B.2: Phylogenetic trees and cladograms are graphical representations (models) of evolutionary history that can be tested.
Enduring understanding 1.C: Life continues to evolve within a changing environment.	**Essential knowledge 1.C.1:** Speciation and extinction have occurred throughout the Earth's history.
	Essential knowledge 1.C.2: Speciation may occur when two populations become reproductively isolated from each other.
	Essential knowledge 1.C.3: Populations of organisms continue to evolve.
Enduring understanding 1.D: The origin of living systems is explained by natural processes.	**Essential knowledge 1.D.1:** There are several hypotheses about the natural origin of life on Earth, each with supporting scientific evidence.
	Essential knowledge 1.D.2: Scientific evidence from many different disciplines supports models of the origin of life.
BIG IDEA 2: Biological systems utilize free energy and molecular building blocks to grow, to reproduce, and to maintain dynamic homeostasis.	
Enduring understanding 2.A: Growth, reproduction, and maintenance of the organization of living systems require free energy and matter.	**Essential knowledge 2.A.1:** All living systems require constant input of free energy.

	Essential knowledge 2.A.2: Organisms capture and store free energy for use in biological processes.
	Essential knowledge 2.A.3: Organisms must exchange matter with the environment to grow, reproduce, and maintain organization.
Enduring understanding 2.B: Growth, reproduction, and dynamic homeostasis require that cells create and maintain internal environments that are different from their external environments.	Essential knowledge 2.B.1: Cell membranes are selectively permeable due to their structure.
	Essential knowledge 2.B.2: Growth and dynamic homeostasis are maintained by the constant movement of molecules across membranes.
	Essential knowledge 2.B.3: Eukaryotic cells maintain internal membranes that partition the cell into specialized regions.
Enduring understanding 2.C: Organisms use feedback mechanisms to regulate growth and reproduction, and to maintain dynamic homeostasis.	Essential knowledge 2.C.1: Organisms use feedback mechanisms to maintain their internal environments and respond to external environmental changes.
	Essential knowledge 2.C.2: Organisms respond to changes in their external environments.
Enduring understanding 2.D: Growth and dynamic homeostasis of a biological system are influenced by changes in the system's environment.	Essential knowledge 2.D.1: All biological systems from cells and organisms to populations, communities, and ecosystems are affected by complex biotic and abiotic interactions involving exchange of matter and free energy.
	Essential knowledge 2.D.2: Homeostatic mechanisms reflect both common ancestry and divergence due to adaptation in different environments.
	Essential knowledge 2.D.3: Biological systems are affected by disruptions to their dynamic homeostasis.
	Essential knowledge 2.D.4: Plants and animals have a variety of chemical defenses against infections that affect dynamic homeostasis.

Enduring understanding 2.E: Many biological processes involved in growth, reproduction, and dynamic homeostasis include temporal regulation and coordination.	**Essential knowledge 2.E.1:** Timing and coordination of specific events are necessary for the normal development of an organism, and these events are regulated by a variety of mechanisms.
	Essential knowledge 2.E.2: Timing and coordination of physiological events are regulated by multiple mechanisms.
	Essential knowledge 2.E.3: Timing and coordination of behavior are regulated by various mechanisms and are important in natural selection.
BIG IDEA 3: Living systems store, retrieve, transmit, and respond to information essential to life processes.	
Enduring understanding 3.A: Heritable information provides for continuity of life.	**Essential knowledge 3.A.1:** DNA, and in some cases RNA, is the primary source of heritable information.
	Essential knowledge 3.A.2: In eukaryotes, heritable information is passed to the next generation via processes that include the cell cycle and mitosis or meiosis plus fertilization.
	Essential knowledge 3.A.3: The chromosomal basis of inheritance provides an understanding of the pattern of passage (transmission) of genes from parent to offspring.
	Essential knowledge 3.A.4: The inheritance pattern of many traits cannot be explained by simple Mendelian genetics.
Enduring understanding 3.B: Expression of genetic information involves cellular and molecular mechanisms.	**Essential knowledge 3.B.1:** Gene regulation results in differential gene expression, leading to cell specialization.
	Essential knowledge 3.B.2: A variety of intercellular and intracellular signal transmissions mediate gene expression.
Enduring understanding 3.C: The processing of genetic information is imperfect and is a source of genetic variation.	**Essential knowledge 3.C.1:** Changes in genotype can result in changes in phenotype.
	Essential knowledge 3.C.2: Biological systems have multiple processes that increase genetic variation.
	Essential knowledge 3.C.3: Viral replication results in genetic variation, and viral infection can introduce genetic variation into the hosts.

Enduring understanding 3.D: Cells communicate by generating, transmitting, and receiving chemical signals.	**Essential knowledge 3.D.1:** Cell communication processes share common features that reflect a shared evolutionary history.
	Essential knowledge 3.D.2: Cells communicate with each other through direct contact with other cells or from a distance via chemical signaling.
	Essential knowledge 3.D.3: Signal transduction pathways link signal reception with cellular response.
	Essential knowledge 3.D.4: Changes in signal transduction pathways can alter cellular response.
Enduring understanding 3.E: Transmission of information results in changes within and between biological systems.	**Essential knowledge 3.E.1:** Individuals can act on information and communicate it to others.
	Essential knowledge 3.E.2: Animals have nervous systems that detect external and internal signals, transmit and integrate information, and produce responses.
BIG IDEA 4: Biological systems interact, and these systems and their interactions possess complex properties.	
Enduring understanding 4.A: Interactions within biological systems lead to complex properties.	**Essential knowledge 4.A.1:** The subcomponents of biological molecules and their sequence determine the properties of that molecule.
	Essential knowledge 4.A.2: The structure and function of subcellular components, and their interactions, provide essential cellular processes.
	Essential knowledge 4.A.3: Interactions between external stimuli and regulated gene expression result in specialization of cells, tissues, and organs.
	Essential knowledge 4.A.4: Organisms exhibit complex properties due to interactions between their constituent parts.
	Essential knowledge 4.A.5: Communities are composed of populations of organisms that interact in complex ways.
	Essential knowledge 4.A.6: Interactions among living systems and with their environment result in the movement of matter and energy.

Enduring understanding 4.B: Competition and cooperation are important aspects of biological systems.	**Essential knowledge 4.B.1:** Interactions between molecules affect their structure and function.
	Essential knowledge 4.B.2: Cooperative interactions within organisms promote efficiency in the use of energy and matter.
	Essential knowledge 4.B.3: Interactions between and within populations influence patterns of species distribution and abundance.
	Essential knowledge 4.B.4: Distribution of local and global ecosystems changes over time.
Enduring understanding 4.C: Naturally occurring diversity among and between components within biological systems affects interactions with the environment.	**Essential knowledge 4.C.1:** Variation in molecular units provides cells with a wider range of functions.
	Essential knowledge 4.C.2: Environmental factors influence the expression of the genotype in an organism.
	Essential knowledge 4.C.3: The level of variation in a population affects population dynamics.
	Essential knowledge 4.C.4: The diversity of species within an ecosystem may influence the stability of the ecosystem.

SCIENCE PRACTICES FOR AP BIOLOGY
SCIENCE PRACTICE 1: The student can use representations and models to communicate scientific phenomena and solve scientific problems.
1.1 The student can *create representations and models* of natural or man-made phenomena and systems in the domain.
1.2 The student can *describe representations and models* of natural or man-made phenomena and systems in the domain.
1.3 The student can *refine representations and models* of natural or man-made phenomena and systems in the domain.
1.4 The student can *use representations and models* to analyze situations or solve problems qualitatively and quantitatively.
1.5 The student can *reexpress key elements* of natural phenomena across multiple representations in the domain.
SCIENCE PRACTICE 2: The student can use mathematics appropriately.
2.1 The student can *justify the selection of a mathematical routine* to solve problems.
2.2 The student can *apply mathematical routines* to quantities that describe natural phenomena.

2.3 The student can *estimate numerically* quantities that describe natural phenomena.
SCIENCE PRACTICE 3: The student can engage in scientific questioning to extend thinking or to guide investigations within the context of the AP course.
3.1 The student can *pose scientific questions.*
3.2 The student can *refine scientific questions.*
3.3 The student can *evaluate scientific questions.*
SCIENCE PRACTICE 4: The student can plan and implement data collection strategies appropriate to a particular scientific question.
4.1 The student can *justify the selection of the kind of data* needed to answer a particular scientific question.
4.2 The student can *design a plan* for collecting data to answer a particular scientific question.
4.3 The student can *collect data* to answer a particular scientific question.
4.4 The student can *evaluate sources of data* to answer a particular scientific question.
SCIENCE PRACTICE 5: The student can perform data analysis and evaluation of evidence.
5.1 The student can *analyze data* to identify patterns or relationships.
5.2 The student can *refine observations and measurements* based on data analysis.
5.3 The student can *evaluate the evidence provided by data sets* in relation to a particular scientific question.
SCIENCE PRACTICE 6: The student can work with scientific explanations and theories.
6.1 The student can *justify claims with evidence.*
6.2 The student can *construct explanations of phenomena based on evidence* produced through scientific practices.
6.3 The student can *articulate the reasons that scientific explanations and theories are refined or replaced.*
6.4 The student can *make claims and predictions about natural phenomena* based on scientific theories and models.
6.5 The student can *evaluate alternative scientific explanations.*
SCIENCE PRACTICE 7: The student is able to connect and relate knowledge across various scales, concepts, and representations in and across domains.
7.1 The student can *connect phenomena and models* across spatial and temporal scales.
7.2 The student can *connect concepts* in and across domain(s) to generalize or extrapolate in and/or across enduring understandings and/or big ideas.

Part II

A Review of Topics with Sample Questions

Part II is keyed to *Campbell Biology*, Ninth Edition, by Reece et al. It gives an overview of important information and provides sample multiple-choice and free-response questions. The necessary content is included in the bulleted information, but you will need to rely on sample questions and your teacher's instruction to reinforce the science practices. Answers and explanations can be found in Part V. Use the summary of key concepts section at the end of each chapter in your textbook before attempting the practice questions. Be sure to review the answers thoroughly to prepare yourself for the range of questions you will encounter on the AP Biology Examination.

The Chemistry of Life

Chapter 2: The Chemical Context of Life

WHAT'S IMPORTANT TO KNOW?
This chapter is considered prior knowledge for the AP Biology Examination. However, you will need to know this information to proceed with the required topics, so we include what is most important in this area.

YOU MUST KNOW

- The three subatomic particles and their significance.
- The types of chemical bonds, how they form, and their relative strengths.

Concept 2.1 Matter consists of chemical elements in pure form and in combinations called compounds

▮ **Matter** is anything that takes up space and has mass.

▮ An **element** is a substance that cannot be broken down to other substances by chemical reactions. *Examples*: gold, copper, carbon, and oxygen.

▮ A **compound** is a substance consisting of two or more elements combined in a fixed ratio. *Examples*: water (H_2O) and table salt (NaCl).

▮ **C, H, O, N** make up 96% of living matter. About 25 of the 92 natural elements are known to be essential to life.

▮ **Trace elements** are those required by an organism in only minute quantities. *Examples*: iron and iodine.

Concept 2.2 An element's properties depend on the structure of its atoms

▮ **Atoms** are the smallest unit of an element that still retains the property of the element. Atoms are made up of neutrons, protons, and electrons.

▮ **Protons** are positively charged particles. They are found in the nucleus and determine the element.

▮ **Electrons** are negatively charged particles that are found in electron shells around the nucleus. They determine the chemical properties and reactivity of the element.

▮ **Neutrons** are particles with no charge. They are found in the nucleus. Their number can vary in the same element, resulting in isotopes.

- **Isotopes** are forms of an element with differing numbers of neutrons. Example: ^{12}C and ^{14}C are isotopes of carbon. Both have 6 protons, but ^{12}C has 6 neutrons whereas ^{14}C has 8 neutrons.
- The **atomic number** is the number of protons an element possesses. This number is unique to every element. (See Figure 1.1.)
- The **mass number** of an element is the sum of its protons and neutrons.

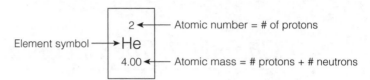

Figure 1.1   An element of the periodic table

Concept 2.3   *The formation and function of molecules depend on chemical bonding between atoms*

- **Chemical bonds** are defined as interactions between the valence electrons of different atoms. Atoms are held together by chemical bonds to form molecules.
- A **covalent bond** occurs when valence electrons are shared by two atoms.

  - **Nonpolar covalent bonds** occur when the electrons being shared are shared equally between the two atoms. *Examples*: O=O, H−H.
  - Atoms vary in their *electronegativity*, a tendency to attract electrons of a covalent bond. Oxygen is strongly electronegative.
  - In **polar covalent bonds**, one atom has greater electronegativity than the other, resulting in an unequal sharing of the electrons. *Example*: Refer to Figure 1.2 and note that within each molecule of H_2O the electrons are shared unequally, resulting in the region of the oxygen atom being slightly negative, whereas the regions about the hydrogen atoms are slightly positive.

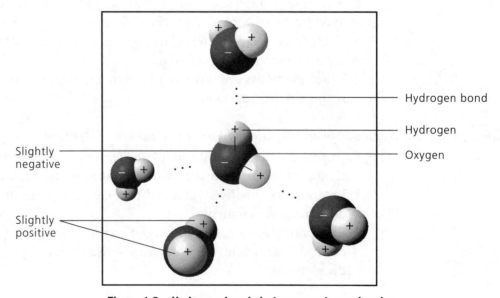

Figure 1.2   Hydrogen bonds between water molecules

- **Ionic bonds** are ones in which two atoms attract valence electrons so unequally that the more electronegative atom steals the electron away from the less electronegative atom.
    - An **ion** is the resulting charged atom or molecule.
    - **Ionic bonds** occur because these ions will be either positively or negatively charged, and will be attracted to each other by these opposite charges.
- **Hydrogen bonds** are relatively weak bonds that form between the partial positively charged hydrogen atom of one molecule and the strongly electronegative oxygen or nitrogen of *another* molecule.
- **Van der Waals interactions** are very weak, transient connections that are the result of asymmetrical distribution of electrons within a molecule. These weak interactions contribute to the three-dimensional shape of molecules.

Concept 2.4 *Chemical reactions make and break chemical bonds*

- A **chemical reaction** shows the **reactants**, which are the starting materials, an arrow to indicate their conversion into the **products**, the ending materials. *Example*: $6 CO_2 + 6 H_2O \longrightarrow C_6H_{12}O_6 + 6 O_2$.
- The chemical reaction above also shows the number of molecules involved. This is the coefficient in front of each molecule. You will note that the number of atoms of each element is the same on each side of the reaction.
- Some chemical reactions are reversible, which is indicated with a double-headed arrow: $3 H_2 + N_2 \rightleftharpoons 2 NH_3$.
- **Chemical equilibrium** is the point at which the forward and reverse reactions offset one another exactly. Their concentrations have stabilized at a particular ratio, though they are not necessarily equal.

Chapter 3: Water and Life

YOU MUST KNOW

- The importance of hydrogen bonding to the properties of water.
- Four unique properties of water, and how each contributes to life on Earth.
- How to interpret the pH scale.
- How changes in pH can alter biological systems.
- The importance of buffers in biological systems.

Concept 3.1 *Polar covalent bonds in water molecules result in hydrogen bonding*

- The **structure of water** is the key to its special properties. Water is made up of one atom of oxygen and two atoms of hydrogen, bonded to form a molecule.
- Water molecules are **polar**. The oxygen region of the molecule has a partial negative charge, and each hydrogen has a partial positive charge.

■ **Hydrogen bonds** form between water molecules. The slightly negative oxygen atom from one water molecule is attracted to the slightly positive hydrogen end of *another* water molecule.

■ Each water molecule can form a maximum of four hydrogen bonds at a time.

Concept 3.2 *Four emergent properties of water contribute to Earth's suitability for life*

■ **Hydrogen bonds** are the key to each of these properties. This is what makes water so unique.

1. **Cohesion.** Cohesion is the linking of like molecules. Think "water molecule joined to water molecule" and visualize a water strider walking on top of a pond due to the *surface tension* that is the result of this property.

   - **Adhesion** is the clinging of one substance to another. Think "water molecule attached to some other molecule" such as water droplets adhering to a glass windshield.
   - **Transpiration** is the movement of water molecules up the very thin xylem tubes and their evaporation from the stomata in plants. The water molecules cling to each other by *cohesion*, and to the walls of the xylem tubes by *adhesion*.

2. Moderation of temperature is possible because of water's high specific heat.

   - **Specific heat** is the amount of heat required to raise or lower the temperature of a substance by 1°C. Relative to most other materials, the temperature of water changes less when a given amount of heat is lost or absorbed. This high specific heat makes the temperature of Earth's oceans relatively stable and able to support vast quantities of both plant and animal life.

3. Insulation of bodies of water by floating ice.

   - Water is less dense as a solid than in its liquid state, whereas the opposite is true of most other substances. Because ice is less dense than liquid water, ice floats. This keeps large bodies of water from freezing solid and therefore moderates temperature.

4. Water is an important *solvent*. (The substance that something is dissolved in is called the *solvent*, whereas the substance being dissolved is the *solute*. Together they are called the *solution*.)

   - **Hydrophilic** substances are water-soluble. These include ionic compounds, polar molecules (e.g., sugars), and some proteins.
   - **Hydrophobic** substances such as oils are nonpolar and do not dissolve in water.

Concept 3.3 *Acidic and basic conditions affect living organisms*

■ The **pH** scale runs between 0 and 14 and measures the relative acidity and alkalinity of aqueous solutions. (See Figure 1.3.)

ACIDIC | BASIC

0   7.0   14

Figure 1.3   pH scale

- **Acids** have an excess of H^+ ions and a pH below 7.0. $[H^+] > [OH^-]$
- **Bases** have an excess of OH^- ions, and pH above 7.0. $[H^+] < [OH^-]$
- Pure water is neutral, which means it has a pH of 7. $[H^+] = [OH^-]$
- **Buffers** are substances that minimize changes in pH. They accept H^+ from solution when they are in excess and donate H^+ when they are depleted.
- **Carbonic acid (H_2CO_3)** is an important buffer in living systems. It moderates pH changes in blood plasma and the ocean.

Chapter 4: Carbon and the Molecular Diversity of Life

YOU MUST KNOW
• The properties of carbon that make it so important.

Concept 4.1   Organic chemistry is the study of carbon compounds

- The major elements of life are C, H, O, N, S, and P, sometimes recalled with the acronym for a person's name: **P.S. COHN**.
- All *organic compounds* contain carbon, and most also contain hydrogen.
- Once thought to be made only in living cells, artificial synthesis of organic compounds is possible. A classic experiment done by Stanley Miller in 1953 showed that complex organic molecules could arise spontaneously. See the figure in your text, and note the conditions and compounds that might have been part of the early conditions on Earth.

Concept 4.2   Carbon atoms can form diverse molecules by bonding to four other atoms

- Carbon is unparalleled in its ability to form molecules that are large, complex, and diverse. Why?

  - It has 4 valence electrons.
  - It can form up to 4 covalent bonds.
  - These can be single, double, or triple covalent bonds.
  - It can form large molecules.
  - These molecules can be chains, ring-shaped, or branched.

- **Isomers** are molecules that have the same molecular formula but differ in their arrangement of these atoms. These differences can result in molecules that are very different in their biological activities. *Examples*: glucose and fructose (both have the molecular formula of $C_6H_{12}O_6$).

Concept 4.3 A few chemical groups are key to the functioning of biological molecules

■ **Functional groups** attached to the carbon skeleton have diverse properties. The behavior of organic molecules is dependent on the identity of their functional groups.

■ Some common functional groups are listed below:

Functional Group Name/Structure	Organic Molecules with the Functional Group and Items of Note about Functional Group
Hydroxyl, —OH	Alcohols such as ethanol, methanol; helps dissolve molecules such as sugars
Carboxyl, —COOH	Carboxylic acids such as fatty acids and sugars; acidic properties because it tends to ionize; source of H^+ ions
Carbonyl, ⟨CO	Ketones and aldehydes such as sugars
Amino, —NH_2	Amines such as amino acids
Phosphate, PO_3	Organic phosphates, including ATP, DNA, and phospholipids
Sulfhydryl, —SH	This group is found in some amino acids; forms disulfide bridges in proteins
Methyl, —CH_3	Addition of a methyl group affects expression of genes

Chapter 5: The Structure and Function of Large Biological Molecules

> ### YOU MUST KNOW
>
> - The role of **dehydration reactions** in the formation of organic compounds and **hydrolysis** in the digestion of organic compounds.
> - How to recognize the four biologically important organic compounds (carbohydrates, lipids, proteins, and nucleic acids) by their structural formulas.
> - The cellular functions of the four groups of organic compounds.
> - The four structural levels of proteins and how changes at any level can affect the activity of the protein.
> - How proteins reach their final shape (**conformation**), the **denaturing** impact that heat and pH can have on protein structure, and how these changes may affect the organism.

Concept 5.1 Macromolecules are polymers, built from monomers

■ **Polymers** are long chain molecules made of repeating subunits called **monomers**. *Examples*: Starch is a polymer composed of glucose monomers. Proteins are polymers composed of amino acid monomers. (See Figure 1.4.)

Figure 1.4 Synthesis and breakdown of polymers

- **Dehydration reactions** create polymers from monomers. Two monomers are joined by removing one molecule of water. *Example*: $C_6H_{12}O_6 + C_6H_{12}O_6 \rightarrow H_{22}O_{11} + H_2O$.
- **Hydrolysis** occurs when water is added to split large molecules. This occurs in the reverse of the above reaction.

Concept 5.2 Carbohydrates serve as fuel and building material

- **Carbohydrates** include both simple sugars (glucose, fructose, galactose, etc.) and polymers such as starch made from these and other subunits. All carbohydrates exist in a ratio of 1 carbon: 2 hydrogen: 1 oxygen or CH_2O.
- **Monosaccharides** are the monomers of carbohydrates. *Examples*: glucose ($C_6H_{12}O_6$) and ribose ($C_5H_{10}O_5$). Notice the 1:2:1 ratio discussed above.
- **Polysaccharides** are polymers of monosaccharides. *Examples*: starch, cellulose, and glycogen.
- Two functions of polysaccharides are **energy storage** and **structural support**.

    1. **Energy-storage polysaccharides**

        - **Starch** is a storage polysaccharide found in plants (e.g., potatoes).
        - **Glycogen** is a storage polysaccharide found in animals, vertebrate muscle cells, and liver cells.

    2. **Structural support polysaccharides**

        - **Cellulose** is a major component of plant cell walls.
        - **Chitin** is found in the exoskeleton of arthropods, such as lobsters and insects and the cell walls of fungi. It gives cockroaches their "crunch."

Concept 5.3 Lipids are a diverse group of hydrophobic molecules

- Lipids are all **hydrophobic**. They aren't polymers, as they are assembled from a variety of components. *Examples*: **waxes, oils, fats,** and **steroids**.
- **Fats** (also called triglycerides) are made up of a **glycerol** molecule and three **fatty acid** molecules.
- **Fatty acids** include hydrocarbon chains of variable lengths. These chains are nonpolar and therefore hydrophobic.

    - **Saturated fatty acids**

        - have no double bonds between carbons
        - tend to pack solidly at room temperature
        - are linked to cardiovascular disease
        - are commonly produced by animals
        - *Examples*: butter and lard

    - **Unsaturated fatty acids**

        - have some C=C (carbon double bonds); this results in kinks
        - tend to be liquid at room temperature

- are commonly produced by plants
- *Examples*: corn oil and olive oil

▌ **Functions of lipids**

- *Energy storage.* Fats store twice as many calories/gram as carbohydrates!
- *Protection* of vital organs and *insulation*. In humans and other mammals, fat is stored in **adipose cells.**

▌ **Phospholipids** make up cell membranes. They

- have a glycerol backbone (head), which is hydrophilic.
- have two fatty acid tails, which are hydrophobic.
- are arranged in a bilayer in forming the cell membrane, with the hydrophilic heads pointing toward the watery cytosol or extracellular environment, and hydrophobic tails sandwiched in between (see Figure 1.5).

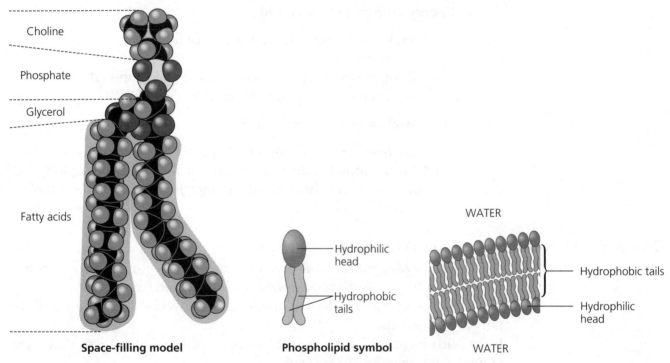

Figure 1.5 The structure of a phospholipid

▌ **Steroids** are made up of four rings that are fused together.

- **Cholesterol** is a steroid. It is a common component of cell membranes.
- **Estrogen** and **testosterone** are steroid hormones.

Concept 5.4 Proteins include a diversity of structures, resulting in a wide range of functions

▌ **Proteins** are polymers made up of amino acid monomers.
▌ **Amino acids** contain a central carbon bonded to a carboxyl group, an amino group, a hydrogen atom, and an R group (variable group or side chain). (See Figure 1.6.)

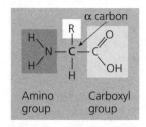

Figure 1.6   The structure of an amino acid

■ **Peptide bonds** link amino acids. They are formed by dehydration synthesis.
■ **There are four levels of protein structure** (see Figure 1.7):

■ **Primary structure** is the unique sequence in which amino acids are joined.
■ **Secondary structure** refers to one of two three-dimensional shapes that are the result of hydrogen bonding.

■ **Alpha (α) helix** is a coiled shape, much like a slinky.
■ **Beta (β) pleated sheet** is an accordion shape.

■ **Tertiary structure** results in a complex globular shape, due to interactions between R groups, such as hydrophobic interactions, van der Waals interactions, hydrogen bonds, and disulfide bridges.

■ Globular proteins such as enzymes are held in position by these R group interactions.

■ **Quaternary structure** refers to the association of two or more polypeptide chains into one large protein. Hemoglobin is a globular protein with quaternary structure, as it is composed of four chains.

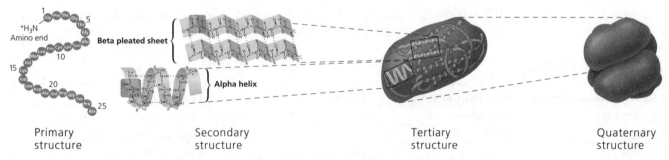

| Primary structure | Secondary structure | Tertiary structure | Quaternary structure |

Figure 1.7   Levels of protein structure

■ **Protein shape is crucial to protein function.** When a protein does not fold properly, its function is changed. This can be the result of a single amino acid substitution, such as that seen in the abnormal hemoglobin typical of sickle-cell disease.
■ **Chaperonins** are protein molecules that assist in the proper folding of proteins within cells. They provide an isolating environment in which a polypeptide chain may attain final conformation.
■ A protein is **denatured** when it loses its shape and ability to function due to **heat**, a **change in pH**, or some other disturbance.

Concept 5.5 Nucleic acids store, transmit, and help express hereditary information

■ **DNA** (deoxyribonucleic acid) and **RNA** (ribonucleic acid) are the two nucleic acids. Their monomers are nucleotides.
■ **Nucleotides** are made up of three parts (see Figure 1.8):

■ **Nitrogenous base** (adenine, thymine, cytosine, guanine, and uracil)
■ **Pentose** (5-carbon) sugar (deoxyribose in DNA or ribose in RNA)
■ **Phosphate group**

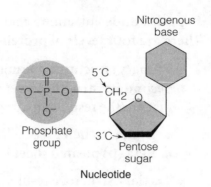

Figure 1.8　The components of nucleic acids

■ **DNA** is the molecule of heredity.

- ■ It is double-stranded helix.
- ■ Its nucleotides are adenine, thymine, cytosine, and guanine.
- ■ Adenine nucleotides will hydrogen bond to thymine nucleotides, and cytosine to guanine.

■ **RNA** is single-stranded. Its nucleotides are adenine, uracil, cytosine, and guanine. Note that it does not have thymine.

SUMMARY TABLE

Macromolecules/Polymers	Monomers/Components	Examples	Functions
Carbohydrates	Monosaccharides	Sugars, starch, glycogen, cellulose	Energy, energy storage; structural
Lipids	Fatty acids and glycerol	Fats, oils	Important energy source; insulation
Proteins	Amino acids	Hemoglobin, pepsin	Enzymes, movement
Nucleic Acids	Nucleotides (sugar, phosphate group, nitrogenous base)	DNA, RNA	Heredity; code for amino acid sequence

Level 1: Knowledge/Comprehension Questions

1. Which list of components characterizes RNA?
   (A) a PO_3 group, deoxyribose, and uracil
   (B) a PO_3 group, ribose, and uracil
   (C) a PO_3 group, ribose, and thymine
   (D) a PO_2 group, deoxyribose, and uracil
   (E) a PO_2 group, deoxyribose, and thymine

2. Which of the following molecules would contain a polar covalent bond?
   (A) Cl_2
   (B) NaCl
   (C) H_2O
   (D) CH_4
   (E) $C_6H_{12}O_6$

3. Which of the following statements regarding carbon is *false*?
   (A) Carbon has a tendency to form covalent bonds.
   (B) Carbon has the ability to bond with up to four other atoms.
   (C) Carbon has the capacity to form single and double bonds.
   (D) Carbon has the ability to bond together to form extensive, branched, or unbranched "carbon skeletons."
   (E) Carbon has the capacity to form polar bonds with hydrogen.

4. Three terms associated with the travel of water from the roots up through the vascular tissues of plants are
   (A) adhesion, cohesion, and translocation.
   (B) adhesion, cohesion, and transcription.
   (C) cohesion, hybridization, and transpiration.
   (D) cohesion, adhesion, and transpiration.
   (E) transpiration, neutralization, and adhesion.

Directions: The group of questions below consists of five lettered choices followed by a list of numbered phrases or sentences. For each numbered phrase or sentence, select the one choice that is most closely related to it. Each choice may be used once, more than once, or not at all.

Questions 5–9
   (A) Lipids
   (B) Peptide bonds
   (C) Alpha helix
   (D) Unsaturated fatty acids
   (E) Cellulose

5. Contain one or more double bonds which "kink" the carbon backbone

6. The major class of biological molecules that are not polymers

7. Linkages between the monomers of proteins

8. A secondary structure of proteins

9. A structural carbohydrate found in plants

10. The process by which protein conformation is lost or broken down is
    (A) dehydration synthesis.
    (B) translation.
    (C) denaturation.
    (D) hydrolysis.
    (E) protein synthesis.

11. An organic compound that is composed of carbon, hydrogen, and oxygen in a 1:2:1 ratio is known as a
    (A) lipid.
    (B) carbohydrate.
    (C) salt.
    (D) nucleic acid.
    (E) protein.

12. If three molecules of a fatty acid that has the formula $C_{16}H_{22}O_2$ are joined to a molecule of glycerol ($C_3H_8O_3$), then the resulting molecule would have the formula
    (A) $C_{48}H_{96}O_6$.
    (B) $C_{48}H_{98}O_8$.
    (C) $C_{51}H_{68}O_6$.
    (D) $C_{51}H_{106}O_8$.
    (E) $C_{51}H_{104}O_9$.

13. Which of the macromolecules below could be structural parts of the cell, enzymes, or involved in cell movement or communication?
    (A) nucleic acids
    (B) proteins
    (C) lipids
    (D) carbohydrates
    (E) minerals

14. Which macromolecule is the main component of all cell membranes?
    (A) DNA
    (B) phospholipids
    (C) carbohydrates
    (D) steroids
    (E) glucose

15. The partial negative charge at one end of a water molecule is attracted to a partial positive charge of another water molecule. What is this type of attraction called?
    (A) a polar covalent bond
    (B) an ionic bond
    (C) a hydration shell
    (D) a hydrogen bond
    (E) a hydrophobic bond

16. Polymers of carbohydrates and proteins are all synthesized from monomers by
    (A) the joining of monosaccharides.
    (B) hydrolysis.
    (C) dehydration reactions.
    (D) ionic bonding of monomers.
    (E) cohesion.

17. If the pH of a solution is decreased from 7 to 6, it means that the
    (A) concentration of H^+ has decreased to 1/10 of what it was at pH 7.
    (B) concentration of H^+ has increased 10 times what it was at pH 7.
    (C) concentration of OH^- has increased 10 times what it was at pH 7.
    (D) concentration of OH^- has increased by 1/7 of what it was.
    (E) solution has become more basic.

18. Which of the following is NOT considered to be an emergent property of water?
    (A) cohesion
    (B) transpiration
    (C) moderation of temperature
    (D) insulation of bodies of water by floating ice
    (E) a versatile solvent

19. Which two functional groups are always found in amino acids?
    (A) amine and sulfhydryl
    (B) carbonyl and carboxyl
    (C) carboxyl and amine
    (D) alcohol and aldehyde
    (E) ketone and amine

20. Hydrolysis is involved in which of the following?
    (A) formation of starch
    (B) hydrogen bond formation between nucleic acids
    (C) peptide bond formation of proteins
    (D) the hydrophilic interactions of lipids
    (E) the digestion of maltose to glucose

Level 2: Application/Analysis/Synthesis Questions

1. The hydrogen bonds shown in this figure are each
   (A) between two hydrogen atoms.
   (B) between two oxygen atoms.
   (C) between an oxygen and a hydrogen atom of the same water molecule.
   (D) between an oxygen and a hydrogen atom of different water molecules.

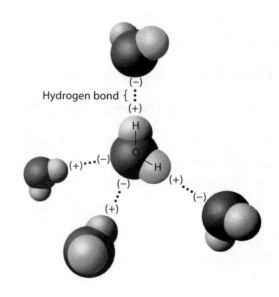

2. These two molecules are structural isomers and can be predicted to have different functions. What is the difference between them?
   (A) the number of carbon atoms
   (B) the number of oxygen atoms
   (C) the number of hydrogen atoms
   (D) the location of a double-bonded oxygen atom

Glucose (an aldose)

Fructose (a ketose)

Questions 3 and 4 refer to the following art.

Carboxyl group · Amino group · Dehydration reaction · Peptide bond · Amino acid · Amino acid · H_2O · Dipeptide

3. How are these two amino acids attached together?
   (A) amino group to amino group
   (B) amino group to carboxylic acid group
   (C) carboxylic acid group to carboxylic acid group
   (D) carbon atom to carbon atom

4. If the dipeptide above were to be digested, how would it be reduced to amino acids?
   (A) by a dehydration reaction
   (B) by reduction in digestive fluid pH
   (C) through removal of functional groups
   (D) through a hydrolysis reaction

After reading the following paragraph, answer questions 5 and 6.

You're the manager of a factory that produces enzyme-washed blue jeans (the enzymes lighten the color of the denim, giving a faded appearance). When the most recent batch of fabric came out of the enzyme wash, however, the color wasn't light enough to meet your standards. Your quality control laboratory wants to do some tests to determine why the wash enzymes didn't perform as expected.

5. Which hypothesis is most likely to be productive for their initial investigation?
   (A) The nucleotide chain of the enzymes may be incorrectly formed.
   (B) The dye in the fabric may have hydrolyzed the fatty acids in the enzymes.
   (C) The polysaccharides in the enzymes may have separated in the wash water.
   (D) The three-dimensional structure of the proteins may have been altered.

6. Based on your understanding of enzyme structure, which of the following would you recommend that they also investigate?
   (A) the temperature of the liquid in the washing vat
   (B) the pH of the liquid in the washing vat
   (C) the manufacturer of the fabric
   (D) both A and B

7. The molecular formula for glucose is $C_6H_{12}O_6$. What would be the molecular formula for a polymer made by lining ten glucose molecules together by dehydration reactions?

(A) $C_{60}H_{120}O_{60}$

(B) $C_6H_{12}O_6$

(C) $C_{60}H_{102}O_{51}$

(D) $C_{60}H_{100}O_{50}$

8. Which of the following pairs of base sequences could form a short stretch of a normal double helix of DNA?

(A) 5′-purine-pyrimidine-purine-pyrimidine-3′ with 3′-purine-pyrimidine-purine-pyrimidine-5′

(B) 5′-AGCT-3′ with 5′-TCGA-3′

(C) 5′-GCGC-3′ with 5′-TATA-3′

(D) 5′-ATGC-3′ with 5′-GCAT-3′

Free-Response Question

1. *The selectively permeable plasma membrane is composed of phospholipids and protein, which allow for its unique functions.*

   (a) **Describe** the structure and properties of phospholipids and *explain* the important roles of phospholipids in the plasma membrane.

   (b) **Explain** why proteins are an important component of the cell membrane, based on their structure and properties.

The Cell

Chapter 6: A Tour of the Cell

YOU MUST KNOW

- Three differences between prokaryotic and eukaryotic cells.
- The structure and function of organelles common to plant and animal cells.
- The structure and function of organelles found only in plant cells or only in animal cells.

Concept 6.2 *Eukaryotic cells have internal membranes that compartmentalize their functions*

▊ The table below organizes the major characteristics of prokaryotic and eukaryotic cells.

Characteristics	Prokaryotic Cells	Eukaryotic Cells
Plasma membrane	yes	yes
Cytosol with organelles	yes	yes
Ribosomes	yes	yes
Nucleus	no	yes
Size	1 µm–10 µm	10 µm–100 µm
Internal membranes	no	yes

▊ Prokaryotic cells are found in the domains Bacteria and Archaea. Eukaryotic cells belong to the domain Eukarya and include animals, fungi, plants, and protists.

▊ Three key details to remember about prokaryotes include

- Chromosomes are grouped together in a region called the nucleoid, but there is no nuclear membrane and therefore no true nucleus.
- No membrane-bounded organelles are found in the cytosol. (Ribosomes are found, but they are not membrane bound.)
- From the table above, notice how much smaller prokaryotes are than eukaryotes.

- Three corresponding details about eukaryotic cells:

  - A membrane-enclosed nucleus contains the cell's chromosomes.
  - Many membrane-bounded organelles are found in the cytoplasm.
  - On average, eukaryotes are much larger than prokaryotes.

- Use Figures 2.1 and 2.2 to locate each component of a plant or animal cell as they are reviewed.
- The **plasma membrane** forms the boundary for a cell. It is selectively permeable and permits the passage of materials into and out of the cell.
- The plasma membrane is made up of *phospholipids*, *proteins*, and associated *carbohydrates*.
- The *surface area to volume ratio* becomes less favorable as a cell increases in size. The total volume grows proportionately more than the surface area. Since a cell acquires resources through the plasma membrane, cell size is limited.

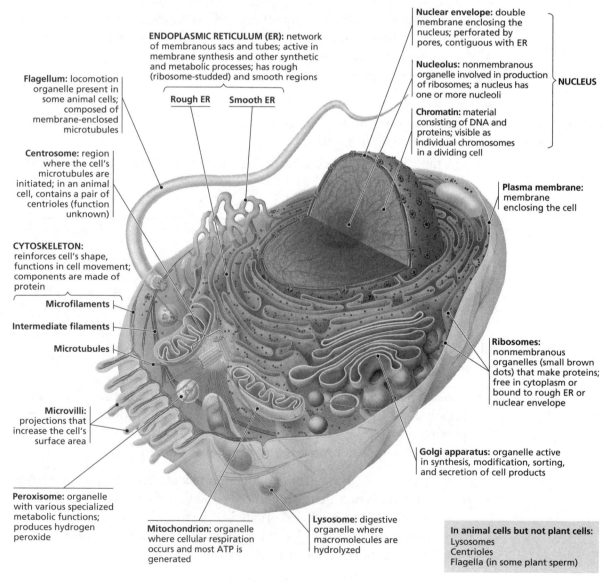

Figure 2.1   Animal cell structure

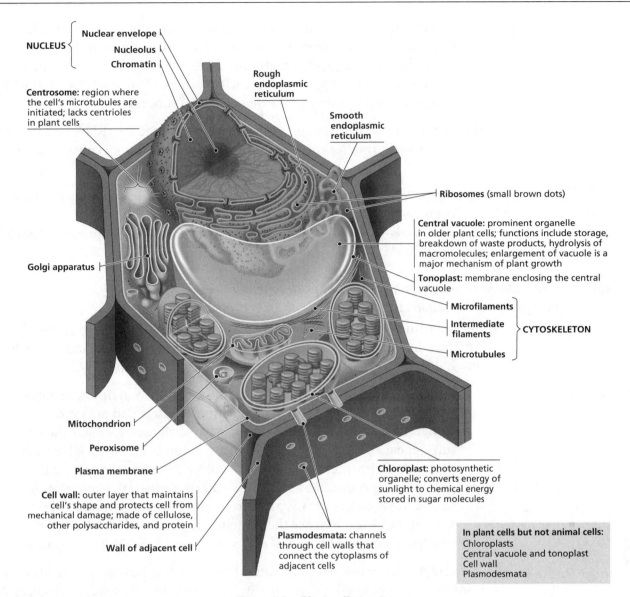

NUCLEUS
Nuclear envelope
Nucleolus
Chromatin

Rough endoplasmic reticulum

Smooth endoplasmic reticulum

Centrosome: region where the cell's microtubules are initiated; lacks centrioles in plant cells

Ribosomes (small brown dots)

Central vacuole: prominent organelle in older plant cells; functions include storage, breakdown of waste products, hydrolysis of macromolecules; enlargement of vacuole is a major mechanism of plant growth

Golgi apparatus

Tonoplast: membrane enclosing the central vacuole

Microfilaments

Intermediate filaments

Microtubules

CYTOSKELETON

Mitochondrion

Peroxisome

Plasma membrane

Chloroplast: photosynthetic organelle; converts energy of sunlight to chemical energy stored in sugar molecules

Cell wall: outer layer that maintains cell's shape and protects cell from mechanical damage; made of cellulose, other polysaccharides, and protein

Wall of adjacent cell

Plasmodesmata: channels through cell walls that connect the cytoplasms of adjacent cells

In plant cells but not animal cells:
Chloroplasts
Central vacuole and tonoplast
Cell wall
Plasmodesmata

Figure 2.2   Plant cell structure

Concept 6.3 *The eukaryotic cell's genetic instructions are housed in the nucleus and carried out by the ribosomes*

▌ The **nucleus** has the following key characteristics:

- The nucleus contains most of the cell's DNA. It is in the nucleus where DNA is used as the template to make messenger RNA (mRNA), which contains the code to produce a protein. Because the nucleus contains the genetic information, it is referred to as the control center of the cell.
- The nucleus is the most noticeable organelle in the cell because of its large relative size. The nucleus is surrounded by a double membrane, the **nuclear envelope**. Note that the nuclear envelope is continuous with the rough endoplasmic reticulum. The nuclear envelope contains **nuclear pores** that control what may enter or leave the nucleus.

- **Chromatin** is the complex of DNA and protein housed in the nucleus that makes up the chromosomes. As a cell gets ready for cell division, the diffuse threads of chromatin condense into visible chromosomes.
- The **nucleolus** is a region of the nucleus where ribosomal RNA (rRNA) complexes with proteins to form ribosomal subunits.

▌ **Ribosomes** are protein factories. They are composed of rRNA and protein, and are sites of protein synthesis in the cell. Each ribosome consists of a large and small subunit.

- *Free ribosomes* are found floating in the cytosol and generally produce proteins that are used within the cell.
- *Bound ribosomes* are attached to the endoplasmic reticulum, and make proteins destined for export from the cell.

Concept 6.4 *The endomembrane system regulates protein traffic and performs metabolic functions in the cell*

▌ **Endoplasmic reticulum (ER)** makes up more than half the total membrane structure in many cells. The ER is a network of membranes and sacs whose internal area is called the *cisternal space*. There are two types of ER:

- **Smooth ER** has three primary functions: synthesis of lipids, metabolism of carbohydrates, and detoxification of drugs and poisons.
- **Rough ER** is so called because its associated ribosomes make the structure appear rough under the microscope. Ribosomes associated with ER synthesize proteins that are generally secreted by the cell. As the proteins are produced by the ER-bound ribosomes, the polypeptide chains travel across the ER membrane and into the cisternal space. Within the cisternal space the proteins are packaged into *transport vesicles,* which bud off the ER and move toward the Golgi apparatus.

▌ The **Golgi apparatus** operates something like the postal system—proteins from the transport vesicles are modified, stored, and shipped. As Figures 2.1 and 2.2 show, the Golgi apparatus consists of flattened sacs of membranes, again called cisternae, arranged in stacks. Golgi stacks have polarity—the *cis* face receives vesicles, whereas the *trans* face ships vesicles.

▌ **Lysosomes** are membrane-bound sacs of hydrolytic enzymes that can digest large molecules, including proteins, polysaccharides, fats, and nucleic acids. They have digestive enzymes that break down macromolecules to organic monomers that are released into the cytosol and thus recycled by the cell. The digestive or hydrolytic enzymes work best in the acidic environment found in lysosomes. If a lysosome breaks open or leaks, the enzymes are not very active in the neutral pH of the cell. This is a good example of the importance of cell compartmentalization.

▌ **Vacuoles** are membrane-bound vesicles. *Food vacuoles* such as those formed by phagocytosis of protists are one example, as are the *contractile vacuoles* that maintain water balance in *Paramecia* and other protists.

- **Central vacuoles** in plant cells may concentrate and contain compounds not found in the cytosol. A large central vacuole is one of the striking differences between plant and animal cells. In plants, a vacuole can make up as much as 80% of the cell.

Concept 6.5 Mitochondria and chloroplasts change energy from one form to another

- **Mitochondria** are the sites of cellular respiration, the metabolic process that uses oxygen to generate ATP by extracting energy from sugars, fats, and other fuels. Study Figure 2.3 to learn the structure.

  - Mitochondria consist of an *outer* and *inner membrane*. The inner membrane is highly folded. These *cristae* (folds) increase the surface area, enhancing the productivity of cellular respiration.
  - The inner compartment, the *mitochondrial matrix*, is fluid-filled.

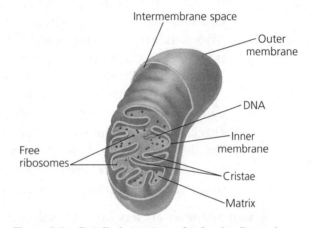

Figure 2.3   Detailed structure of animal cell membrane

- **Chloroplasts**, found in plants and algae, are the sites of photosynthesis.
- The *endosymbiont theory* proposes that both mitochondria and chloroplasts share a similar origin. This theory states that these organelles descended from prokaryotic cells once engulfed by ancestors of eukaryotic cells. There are several lines of evidence for this:

  - Both organelles have a double-membrane structure.
  - Both organelles have their own ribosomes and DNA.
  - Both reproduce independently within the cell.

- **Peroxisomes** are single-membrane-bound compartments in the cell responsible for various metabolic functions that involve the transfer of hydrogen from compounds to oxygen, producing hydrogen peroxide (H_2O_2). Peroxisomes break down fatty acids to be sent to the mitochondria for fuel and detoxify alcohol by transferring hydrogen from the poison to oxygen.
- This is an excellent example of how the cell's compartmental structure is crucial to its functions: The enzymes that produce hydrogen peroxide and those that dispose of this toxic compound are separate from other cellular components that could be damaged.

Concept 6.6 The cytoskeleton is a network of fibers that organizes structures and activities in the cell

▍ The **cytoskeleton** is a network of protein fibers that runs throughout the cytoplasm, where it is responsible for support, motility, and regulating some biochemical activities. Three types of fibers make up the cytoskeleton:

- **Microtubules**, made of the protein tubulin, are the largest of the cytoskeleton fibers. Microtubules shape and support the cell and also serve as tracks along which organelles equipped with *motor molecules* can move. They also separate chromosomes during mitosis and meiosis (forming the spindle) and are the structural components of cilia and flagella (found primarily in animal cells).
- **Microfilaments** are composed of the protein actin. Much smaller than microtubules, microfilaments function in smaller-scale support. When coupled with the motor molecule *myosin*, microfilaments can be involved with movement. *Examples*: amoeboid movement, cytoplasmic streaming, and contraction of muscle cells.
- **Intermediate filaments** are slightly larger than microfilaments and smaller than microtubules. Intermediate fibers are more permanent fixtures in the cell, where they are important in maintaining the shape of the cell and fixing the position of certain organelles.

▍ **Centrosomes** are a region located near the nucleus, from which microtubules grow (the area is also called the microtubule-organizing center). Centrosomes contain centrioles in animal cells.

▍ **Centrioles** are located within the centrosomes of animal cells, where they replicate before cell division.

▍ A specialized arrangement of microtubules is responsible for the beating of flagella and cilia.

- **Flagella** are usually long and few in number. Many unicellular eukaryotic organisms are propelled through the water by flagella, as are the sperm of animals, algae, and some plants.
- **Cilia** are usually much shorter and more numerous than flagella. Cilia can also be used in locomotion or, when held in place as part of a tissue layer, they can move fluid over the surface of the tissue. For example, the lining of the trachea moves mucus-trapped debris out of the lungs in this manner.

▍ Though different in length, number per cell, and beating pattern, cilia and flagella share a common ultrastructure. Nearly all eukaryotic cilia and flagella have nine pairs of microtubules surrounding a central core of two microtubules. This arrangement is referred to as the "9 + 2" pattern.

▍ **Extracellular matrix (ECM)** of animal cells is situated just external to the plasma membrane; it is composed of glycoproteins secreted by the cell (most prominent of which is collagen). The ECM greatly strengthens tissues and serves as a conduit for transmitting external stimuli into the cell, which can turn genes on and modify biochemical activity.

▌ Animal cells have three types of intercellular junctions:

- ▪ **Tight junctions** are sections of animal cell membrane where two neighboring cells are fused, making the membranes watertight.
- ▪ **Desmosomes** fasten adjacent animal cells together, functioning like rivets to fasten cells into strong sheets.
- ▪ **Gap junctions** provide channels between adjacent animal cells through which ions, sugars, and other small molecules can pass.

Concept 6.7 *Extracellular components and connections between cells help coordinate cellular activities*

▌ The **cell wall** of a plant protects the plant and helps maintain its shape. The primary component of cell walls is the carbohydrate *cellulose*.

▌ **Plasmodesmata** are channels that perforate adjacent plant cell walls and allow the passage of some molecules from cell to cell.

▌ To help you summarize structures that are found in either plant cells or animal cells, study the following chart.

Cellular Structure	
Plant Cells Only	**Animal Cells Only**
Central vacuoles	Lysosomes
Chloroplasts	Centrioles
Cell wall of cellulose	Flagella, cilia
Plasmodesmata	Extracellular matrix
	Desmosomes, tight and gap junctions

STUDY TIP Know the structure and function of each organelle and whether it is found in a plant cell, animal cell, or both. Be able to predict and justify how a change in a cellular organelle would affect the function of the entire cell or organism.

Chapter 7: Membrane Structure and Function

YOU MUST KNOW

- Why membranes are selectively permeable.
- The role of phospholipids, proteins, and carbohydrates in membranes.
- How water will move if a cell is placed in an isotonic, hypertonic, or hypotonic solution and be able to predict the effect of different environments on the organism.
- How electrochemical gradients are formed and function in cells.

Concept 7.1 *Cellular membranes are fluid mosaics of lipids and proteins*

▌ The cell or **plasma membrane** is **selectively permeable**; that is, it allows some substances to cross it more easily than others.

▌ Membranes are predominantly made of phospholipids and proteins held together by weak interactions that cause the membrane to be fluid. The *fluid mosaic model* of the cell membrane describes the membrane as fluid, with proteins embedded in or associated with the phospholipid bilayer. Figure 2.4 shows the current model of an animal cell's plasma membrane. Find each part of the membrane as the three primary organic molecules of the membrane are described:

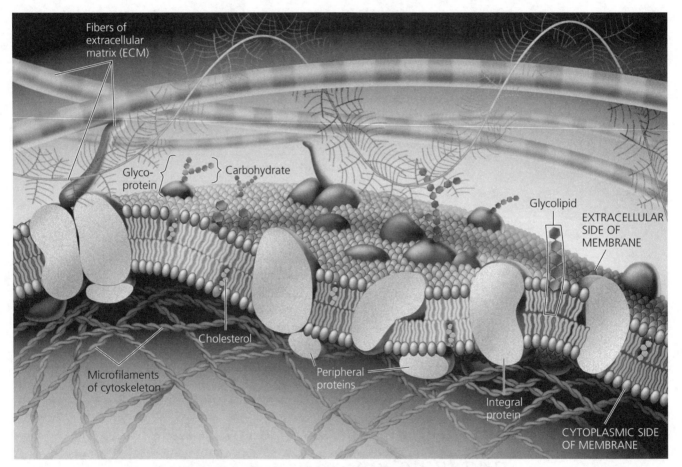

Figure 2.4   Structure of an animal cell's plasma membrane

■ The **phospholipids** in the membrane provide a hydrophobic barrier that separates the cell from its liquid environment. Hydrophilic molecules cannot easily enter the cell, but hydrophobic molecules can enter much more easily; hence, the selectively permeable nature of the membrane.

■ There are both integral proteins and peripheral proteins in the cell membrane. **Integral proteins** are those that are completely embedded in the membrane, some of which are transmembrane proteins that span the membrane completely. **Peripheral proteins** are loosely bound to the membrane's surface.

- **Carbohydrates** on the membrane are crucial in cell-cell recognition (which is necessary for proper immune function) and in developing organisms (for tissue differentiation). Cell-surface carbohydrates vary from species to species and are the reason that blood transfusions must be type-specific.

Concept 7.2 *Membrane structure results in selective permeability*

- Nonpolar molecules—such as hydrocarbons, carbon dioxide, and oxygen—are hydrophobic and can dissolve in the phospholipid bilayer and cross the membrane easily.
- The hydrophobic core of the membrane impedes the passage of ions and polar molecules, which are hydrophilic. However, hydrophilic substances can avoid the lipid bilayer by passing through **transport proteins** that span the membrane (see Figure 2.4).
- Perhaps the most important molecule to move across the membrane is water. Water moves through special transport proteins termed **aquaporins**. Aquaporins greatly accelerate the speed (3 billion water molecules per aquaporin per second!) at which water can cross membranes.

Concept 7.3 *Passive transport is diffusion of a substance across a membrane with no energy investment*

- In **passive diffusion,** a substance travels from where it is more concentrated to where it is less concentrated, diffusing down its **concentration gradient**. Hydrocarbons, carbon dioxide, and oxygen are hydrophobic substances that can pass easily across the cell membrane by passive diffusion. This type of diffusion requires that no work be done, and it relies only on the thermal motion energy intrinsic to the molecule in question. It is called "passive" because the cell expends no energy in moving the substances.
- The diffusion of water across a selectively permeable membrane is **osmosis**. A cell has one of three water relationships with the environment around it.

  - In an **isotonic solution** there will be no net movement of water across the plasma membrane. Water crosses the membrane, but at the same rate in both directions.
  - In a **hypertonic solution** the cell will lose water to its surroundings. The *hyper-* prefix refers to more solutes in the water around the cell; hence, the movement of water to the higher (hyper-) concentration of solutes. In this case the cell loses water to the environment, will shrivel, and may die.
  - In a **hypotonic solution** water will enter the cell faster than it leaves. The *hypo-* prefix refers to fewer solutes in the water around the cell; hence, the movement of water into the cell where solutes are more heavily concentrated. In this case the cell will swell and may burst.
  - Because plant cells have walls, you can observe changes in firmness as water moves in or leaves. A plant cell that is limp is said to be *flaccid*, one that has lost water and is shriveled is *plasmolyzed*, and cells that have full vacuoles and are very firm are *turgid*.

- **Ions** and **polar molecules** cannot pass easily across the membrane. The process by which ions and hydrophilic substances diffuse across the cell membrane with the help of transport proteins is called **facilitated diffusion**. Transport proteins are specific (like enzymes) for the substances they transport. They work in one of two ways:

  - They provide a hydrophilic channel through which the molecules in question can pass.
  - They bind loosely to the molecules in question and carry them through the membrane.

Concept 7.4 Active transport uses energy to move solutes against their gradients

- In **active transport**, substances are moved against their concentration gradient—that is, from the region where they are *less* concentrated to the region where they are *more* concentrated. This type of transport requires energy, usually in the form of ATP.
- A common example of active transport is the **sodium-potassium pump**. This transmembrane protein pumps sodium out of the cell and potassium into the cell. The sodium-potassium pump is necessary for proper nerve transmission and is a major energy consumer in your body as you read this.
- The inside of the cell is negatively charged compared with outside the cell. The difference in electric charge across a membrane is expressed in voltage and termed the **membrane potential**. Because the inside of the cell is negatively charged, a positively charged ion on the outside, like sodium, is attracted to the negative charges inside the cell. Thus, two forces drive the diffusion of ions across a membrane:

  - A *chemical force*, which is the ion's concentration gradient.
  - A *voltage gradient* across the membrane, which attracts positively charged ions and repels negatively charged ions.

  This combination of forces acting on an ion forms an **electrochemical gradient**.
- A transport protein that generates voltage across the membrane is called an **electrogenic pump**. The sodium-potassium pump and the proton pump are examples of electrogenic pumps.

▌In **cotransport**, an ATP pump that transports a specific solute indirectly drives the active transport of other substances. In this process, the substance that was initially pumped across the membrane—an H⁺ pumped by a proton pump, for example—can do work as it moves back across the membrane by diffusion and brings with it a second compound, like sucrose, against its gradient. This process is analogous to water that has been pumped uphill and performs work as it flows back down. Figure 2.5 will help you visualize the process. Note that this process has generated an electrochemical gradient, a source of potential energy that performs cell work.

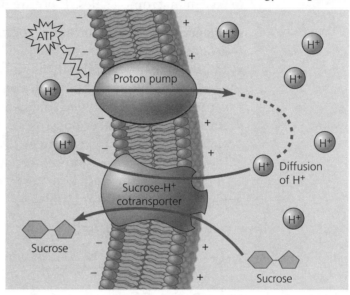

Figure 2.5   Cotransport

To review transport, try your hand at the following questions using Figure 2.6.
1. Which figure represents *active transport*? There are two ways you should be able to tell.
2. Which section shows *simple diffusion*?
3. Which section shows *facilitated diffusion* with a *carrier protein*?
4. Which section shows *facilitated diffusion* with a *channel protein*?

(Answers are at the end of this section.)

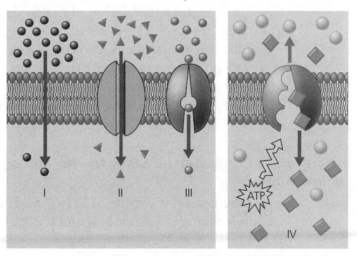

Figure 2.6   Passive and active transport

Concept 7.5 Bulk transport across the plasma membrane occurs by exocytosis and endocytosis

▊ Large molecules are moved across the cell membrane through exocytosis and endocytosis.

  ▪ In **exocytosis**, vesicles from the cell's interior fuse with the cell membrane, expelling their contents.

  ▪ In **endocytosis**, the cell forms new vesicles from the plasma membrane; this is basically the reverse of exocytosis, and this process allows the cell to take in macromolecules. There are three types of endocytosis:

    ■ **Phagocytosis** ("cellular eating") occurs when the cell wraps pseudopodia around a solid particle and brings it into the cell.

    ■ In **pinocytosis** ("cellular drinking"), the cell takes in small droplets of extracellular fluid within small vesicles. Pinocytosis is not specific, because any and all included solutes are taken into the cells.

    ■ **Receptor-mediated endocytosis** is a very specific process. Certain substances (generally referred to as *ligands*) bind to specific receptors on the cell's surface (these receptors are usually clustered in coated pits), and this causes a vesicle to form around the substance and then to pinch off into the cytoplasm.

Answers to questions for Figure 2.6:

1. IV (It shows the use of ATP, and molecules being moved against a concentration gradient.)
2. I
3. III
4. II

Chapter 8: An Introduction to Metabolism

YOU MUST KNOW
• Examples of endergonic and exergonic reactions.
• The key role of ATP in energy coupling.
• That enzymes work by lowering the energy of activation.
• The catalytic cycle of an enzyme that results in the production of a final product.
• Factors that influence enzyme activity.

Concept 8.1 An organism's metabolism transforms matter and energy, subject to the laws of thermodynamics

▊ **Metabolism** is the totality of an organism's chemical reactions. Metabolism as a whole manages the material and energy resources of the cell.

- A **catabolic pathway** leads to the release of energy by the breakdown of complex molecules to simpler compounds. *Example*: Catabolic pathways occur when your digestive enzymes break down food and release energy.
- **Anabolic pathways** consume energy to build complicated molecules from simpler ones. *Example*: Anabolic pathways occur when your body links together amino acids to form muscle protein in response to physical exercise.

▌ **Energy** is defined as the capacity to do work. Anything that is moving is said to possess **kinetic energy**. An object at rest can possess **potential energy** if it has stored energy as a result of its position or structure. **Chemical energy**, a form of potential energy, is stored in molecules, and the amount of chemical energy a molecule possesses depends on its chemical bonds.

▌ **Thermodynamics** is the study of energy transformations that occur in matter.

- The **first law of thermodynamics** states that the energy of the universe is constant and that energy *can* be transferred and transformed, but it *cannot* be created or destroyed.
- The **second law of thermodynamics** states that every energy transfer or transformation increases the **entropy**, or the amount of disorder or randomness, in the universe.

Concept 8.2 *The free-energy change of a reaction tells us whether or not the reaction occurs spontaneously*

▌ **Free energy** is defined as the part of a system's energy that is able to perform work when the temperature of a system is uniform.

▌ ΔG is a symbol for a change in free energy.

- An **exergonic reaction** is one in which energy is released. Exergonic reactions occur spontaneously (that does not necessarily mean quickly) and release free energy to the system. $\Delta G < 0$.
- An **endergonic reaction** is one that requires energy in order to proceed. Endergonic reactions absorb free energy; that is, they require free energy from the system. $\Delta G > 0$.
- Is the breakdown of glucose in cellular respiration exergonic or endergonic? (ΔG is -686 kcal/mol.)

Concept 8.3 *ATP powers cellular work by coupling exergonic reactions to endergonic reactions*

▌ A key feature in the way cells manage their energy resources to do cell work is **energy coupling**, the use of an exergonic process to drive an endergonic one.

▌ The primary source of energy for cells in energy coupling is **ATP (adenosine triphosphate)**. ATP is made up of the nitrogenous base adenine, bonded to ribose and a chain of three phosphate groups. When a phosphate group is hydrolyzed, energy is released in an exergonic reaction.

▌ Work in the cell is done by the release of a phosphate group from ATP. The exergonic release of the phosphate group is used to do the endergonic work of the cell. When ATP transfers one phosphate group through hydrolysis, it becomes **ADP (adenosine diphosphate)**.

Concept 8.4 Enzymes speed up metabolic reactions by lowering energy barriers

▍ **Catalysts** are substances that can change the rate of a reaction without being altered in the process.

▍ **Enzymes** are macromolecules that are biological catalysts. In this chapter all enzymes considered are proteins; however, in Chapters 17 and 25, RNA enzymes—termed *ribozymes*—are discussed.

▍ The **activation energy** of a reaction is the amount of energy it takes to start a reaction—the amount of energy it takes to break the bonds of the reactant molecules. Enzymes speed up reactions *by lowering the activation energy* of the reaction—but without changing the free-energy change of the reaction. The reactant that the enzyme acts on is called a **substrate**. Figure 2.7 graphically depicts how enzymes function.

▍ The **active site** is the part of the enzyme that binds to the substrate. The enzyme and substrate form a complex called an **enzyme-substrate complex** that is generally held together by weak interactions. The substrate is then converted into **products**, and the products are released from the enzyme. Use Figure 2.8 to locate each step in the catalytic cycle of an enzyme.

> *STUDY TIP* Enzymes are a key topic in biology. It is likely that you will do an investigation with enzymes. Focus on factors that affect enzyme action, and be able to pose a question about enzyme function, design an experiment to test this question, predict results, and analyze data from an enzyme experiment.

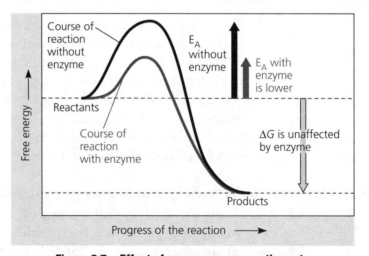

Figure 2.7 Effect of an enzyme on reaction rate

▍ The activity of an enzyme can be affected by several factors:

■ Protein enzymes have complicated three-dimensional shapes that are dramatically affected by changes in **pH** and **temperature**. Changes in the precise shape of an enzyme usually mean the enzyme will not be as effective. Note how the rate of the reaction is altered in the graphs in Figure 2.9 when temperature and pH are not optimal.

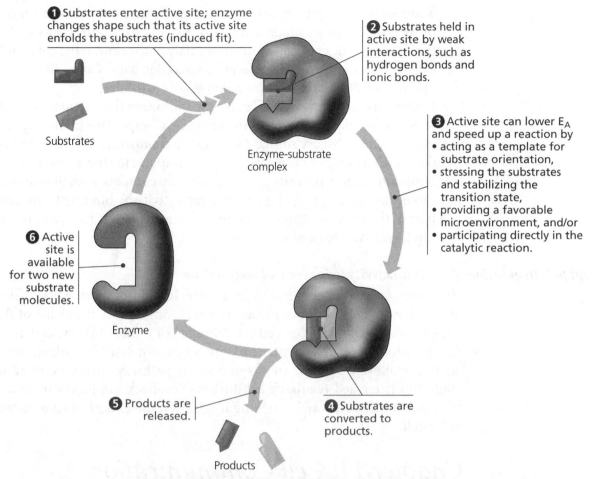

1 Substrates enter active site; enzyme changes shape such that its active site enfolds the substrates (induced fit).

2 Substrates held in active site by weak interactions, such as hydrogen bonds and ionic bonds.

Substrates

Enzyme-substrate complex

3 Active site can lower E_A and speed up a reaction by
• acting as a template for substrate orientation,
• stressing the substrates and stabilizing the transition state,
• providing a favorable microenvironment, and/or
• participating directly in the catalytic reaction.

6 Active site is available for two new substrate molecules.

Enzyme

5 Products are released.

4 Substrates are converted to products.

Products

Figure 2.8   The active site and catalytic cycle of an enzyme

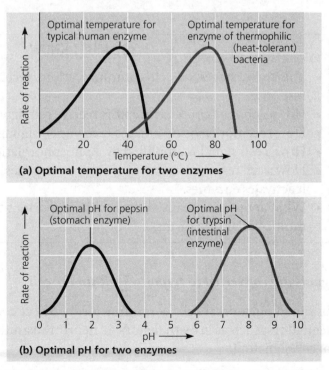

Optimal temperature for typical human enzyme

Optimal temperature for enzyme of thermophilic (heat-tolerant) bacteria

Rate of reaction

0   20   40   60   80   100
Temperature (°C)

(a) Optimal temperature for two enzymes

Optimal pH for pepsin (stomach enzyme)

Optimal pH for trypsin (intestinal enzyme)

Rate of reaction

0   1   2   3   4   5   6   7   8   9   10
pH

(b) Optimal pH for two enzymes

Figure 2.9   Environmental factors affecting enzyme activity

- Many enzymes require nonprotein helpers, termed **cofactors**, to function properly. Cofactors include metal ions like zinc, iron, and copper and function in some crucial way to allow catalysis to occur. If the cofactor is organic, it is more properly referred to as a coenzyme. **Coenzymes** are organic cofactors; vitamins are examples of coenzymes.
- **Competitive inhibitors** are reversible inhibitors that *compete with the substrate for the active site* on the enzyme. Competitive inhibitors are often chemically very similar to the normal substrate molecule and reduce the efficiency of the enzyme as it competes for the active site.
- **Noncompetitive inhibitors** do not directly compete with the substrate molecule; instead, they impede enzyme activity by binding to another part of the enzyme. This causes the enzyme to change its shape, rendering the active site nonfunctional.

Concept 8.5 *Regulation of enzyme activity helps control metabolism*

- Many enzyme regulators bind to an **allosteric** site on the enzyme, which is a specific binding site, but not the active site. Once bound, the shape of the enzyme is changed, and this can either stimulate or inhibit enzyme activity.
- The end product on an enzymatic pathway can switch off its pathway by binding to the allosteric site of an enzyme in the pathway. This type of allosteric inhibition is termed **feedback inhibition**. Feedback inhibition increases the efficiency of the pathway by turning it off when the end product accumulates in the cell.

Chapter 11: Cell Communication

YOU MUST KNOW

- The three stages of cell communication: reception, transduction, and response.
- How G protein-coupled receptors receive cell signals and start transduction.
- How receptor tyrosine kinases receive cell signals and start transduction.
- How a cell signal is amplified by a phosphorylation cascade.
- How a cell response in the nucleus turns on genes, whereas in the cytoplasm it activates enzymes.
- What apoptosis means and why it is important to normal functioning of multicellular organisms.

Concept 11.1 *External signals are converted to responses within the cell*

- In signaling, animal cells communicate by direct contact or by secreting local regulators, such as growth factors or neurotransmitters. There are three stages of cell signaling.

- **Reception**—The target cell's detection of a signal molecule coming from outside the cell.
- **Transduction**—The conversion of the signal to a form that can bring about a specific cellular response.
- **Response**—The specific cellular response to the signal molecule.

Concept 11.2 Reception: A signaling molecule binds to a receptor protein, causing it to change shape

- The binding between a signal molecule (**ligand**) and a **receptor** is highly specific. A conformational change in a receptor is often the initial transduction of the signal. Receptors are found in two places.

  - *Intracellular receptors* are found inside the plasma membrane in the cytoplasm or nucleus. The signal molecule must cross the plasma membrane and therefore must be hydrophobic, like the steroid testosterone, or very small, like nitric oxide (NO).
  - *Plasma membrane receptors* bind to water-soluble ligands.

- A **G protein-coupled receptor** is a membrane receptor that works with the help of a **G protein**. Follow Figure 2.10 to review how these receptors work.

  - Step 1: The ligand or signaling molecule has bound to the G protein-coupled receptor. This causes a conformational change in the receptor so that it may now bind to an inactive G protein, causing a GTP to displace the GDP. This activates the G protein.

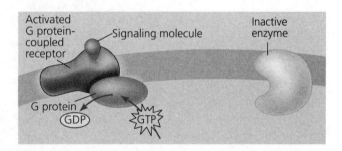

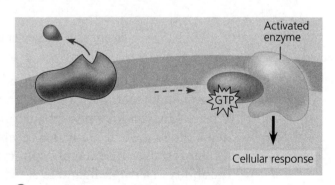

Figure 2.10   G protein-coupled receptor

- Step 2: The G protein binds to a specific enzyme and activates it. When the enzyme is activated, it can trigger the next step in a pathway leading to a cellular response. All the molecular shape changes are temporary. To continue the cellular response, new signal molecules are required.

▌ The **receptor tyrosine kinases** are a second type of membrane protein. Follow Figure 2.11 to review how they function.

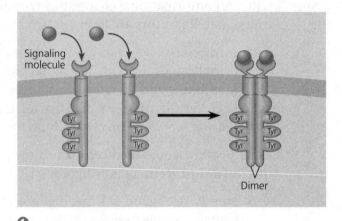

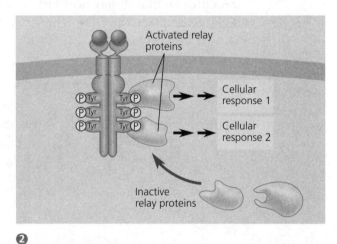

Figure 2.11   Receptor tyrosine kinase

- Step 1 shows the binding of signal molecules to the receptors and the subsequent formation of a dimer. In the dimer configuration each tyrosine kinase adds a phosphate from an ATP molecule.
- Step 2 shows the fully activated receptor protein as it initiates a unique cellular response for each phosphorylated tyrosine.
- The ability of a single ligand to activate multiple cellular responses is a key difference between G protein-coupled receptors and receptor tyrosine kinases.

▌ Specific signal molecules cause **ligand-gated ion channels** in a membrane to open or close, regulating the flow of specific ions.

Concept 11.3 Transduction: Cascades of molecular interactions relay signals from receptors to target molecules in the cell

▌ Signal transduction pathways often involve a **phosphorylation cascade**. Because the pathway is usually a multistep one, the possibility of greatly amplifying the signal exists. At each step enzymes called **protein kinases** phosphorylate and thereby activate many proteins at the next level. This cascade of phosphorylation greatly enhances the signal, allowing for a large cellular response.

▌ **Protein phosphatases** are enzymes that remove phosphate groups and inactive protein kinases. Thus, the signal can be turned on by kinases, and off by phosphatases.

▌ Not all components of signal transduction pathways are proteins. Many signaling pathways involve small, nonprotein water-soluble molecules or ions called **second messengers**. *Calcium ions* and *cyclic AMP* are two common second messengers. The second messengers, once activated, can initiate a phosphorylation cascade resulting in a cellular response.

Concept 11.4 Response: Cell signaling leads to regulation of transcription or cytoplasmic activities

▌ Many signaling pathways ultimately regulate protein synthesis, usually by turning specific genes on or off in the nucleus. Often, the final activated molecule in a signaling pathway functions as a transcription factor.

▌ In the cytoplasm, signaling pathways often regulate the activity of proteins rather than their synthesis. For example, the final step in the signaling pathway may affect the activity of enzymes or cause cytoskeleton rearrangement.

Concept 11.5 Apoptosis integrates multiple cell-signaling pathways

▌ An elaborate example of cell signaling is a program of controlled cell suicide called **apoptosis**. During apoptosis the cell is systematically dismantled and digested. This protects neighboring cells from damage that would occur if a dying cell merely leaked out its digestive and other enzymes.

■ Apoptosis is triggered by signals that activate a cascade of "suicide" proteins in the cells.

■ In vertebrates apoptosis is a normal part of development and is essential for a normal nervous system, for the operation of the immune system, and for normal morphogenesis of hands and feet in humans.

> **STUDY TIP** Cell signaling has emerged as an important topic that explains cell interactions. Devote study time to understanding how the two types of receptors discussed in this chapter function. Also be prepared to explain the role of apoptosis in specific instances, such as formation of fingers and toes, or cancer.

Chapter 12: The Cell Cycle

> **YOU MUST KNOW**
>
> - The structure of the duplicated chromosome.
> - The cell cycle and stages of mitosis.
> - The role of kinases and cyclin in the regulation of the cell cycle.
> - The role of mitosis in the distribution of genetic information.

Concept 12.1 Most cell division results in genetically identical daughter cells

▌ The **cell cycle** is the life of a cell from the time it is first formed from a dividing parent cell until its own division into two cells.

▌ A cell's endowment of DNA, its genetic information, is called its **genome**. Before the cell can divide, the cell's genome must be copied.

 ▪ All eukaryotic organisms have a characteristic number of chromosomes in their cell nuclei. As an example, human **somatic cells** (all body cells except gametes) have 46 chromosomes, which is the diploid chromosome number. *Mitosis* is the process by which somatic cells divide, forming daughter cells that contain the same chromosome number as the parent cell.

 ▪ Human **gametes**—sperm and egg cells—are haploid and have half the number of chromosomes as a diploid cell. Human gametes have 23 chromosomes. A special type of cell division called meiosis (the topic of Chapter 13) results in gametes.

▌ When the chromosomes are replicated, each duplicated chromosome consists of two **sister chromatids** attached by a **centromere**. Figure 2.12 will help you visualize this arrangement.

 ▪ The two sister chromatids have identical DNA sequences.

 ▪ Later, in the process of cell division, the two sister chromatids will separate and move into two new cells. Once the sister chromatids separate, they are considered individual chromosomes.

▌ **Mitosis** is the division of the cell's nucleus. It may be followed by **cytokinesis**, which is the division of the cell's cytoplasm. Where there was one cell, there are now two, each the genetic equivalent of the parent cell.

Concept 12.2 The mitotic phase alternates with interphase in the cell cycle

▌ The primary events of **interphase**, which is 90% of the cell cycle, follow:

 ▪ In G_1 **phase** the cell grows while carrying out cell functions unique to its cell type.

 ▪ In **S phase** the cell continues to carry out its unique functions but does one other important process—it duplicates its chromosomes. This means it faithfully makes a copy of the DNA that makes up the cell's chromosomes.

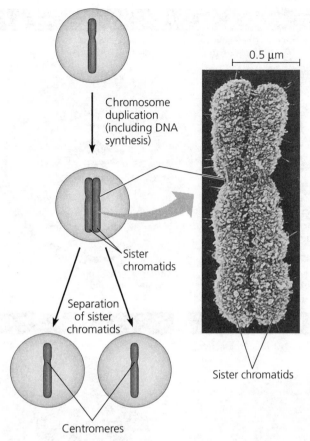

Figure 2.12   Chromosomes and chromatids

- The **G₂ phase** is the gap after the chromosomes have been duplicated and just before mitosis.

WHAT'S IMPORTANT TO KNOW?

The Curriculum Framework says that you do not have to know the specific phases of mitosis. However, many students find it useful to organize their understanding of the process.

Mitosis can be broken down into five phases, not including cytokinesis. At each stage, find the specific references in Figure 2.13. You may be asked to identify stages by diagrams on the AP Biology Exam. To simplify your studying, key features of each phase are given.

- **Prophase:**

  1. The chromatin becomes more tightly coiled into discrete chromosomes.
  2. The nucleoli disappear.
  3. The mitotic spindle (consisting of microtubules extending from the two centrosomes) begins to form in the cytoplasm.

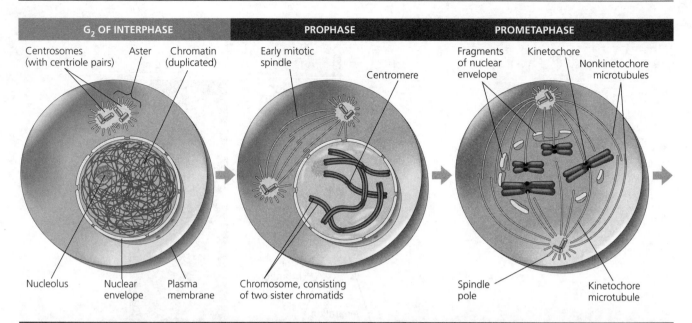

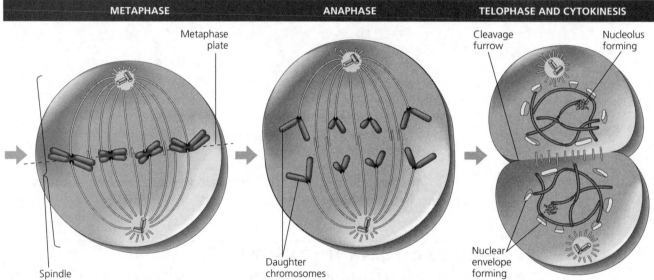

Figure 2.13   The mitotic division of an animal cell

- **Prometaphase:**

    1. The nuclear envelope begins to fragment, allowing the microtubules to attach to the chromosomes.
    2. The two chromatids of each chromosome are held together by protein kinetochores in the centromere region.
    3. The microtubules will attach to the kinetochores.

- **Metaphase:**

    1. The microtubules move the chromosomes to the metaphase plate at the equator of the cell. The microtubule complex is referred to as the *spindle.*
    2. The centrioles have migrated to opposite poles in the cell, riding along on the developing spindle.

- **Anaphase:**

    1. Sister chromatids begin to separate, pulled apart by motor molecules interacting with kinetochore microtubules.
    2. The cell elongates, as the nonkinetochore microtubules ratchet apart, again with the help of motor molecules.
    3. By the end of anaphase, the opposite ends of the cell both contain complete and equal sets of chromosomes.

- **Telophase:**

    1. The nuclear envelopes re-form around the sets of chromosomes located at opposite ends of the cell.
    2. The chromatin fiber of the chromosomes becomes less condensed.
    3. **Cytokinesis** begins, during which the cytoplasm of the cell is divided. In animal cells, a **cleavage furrow** forms that eventually divides the cytoplasm; in plant cells, a **cell plate** forms that divides the cytoplasm.
    4. Prokaryotes replicate their genome by **binary fission** rather than mitosis.

Concept 12.3 *The eukaryotic cell cycle is regulated by a molecular control system*

- The steps of the cycle are controlled by a **cell cycle control system**. This control system moves the cell through its stages by a series of **checkpoints**, during which signals tell the cell either to continue dividing or to stop.
- The major cell cycle checkpoints include the G_1 **phase checkpoint**, G_2 **phase checkpoint**, and **M phase checkpoint**.
- The G_1 **phase checkpoint** seems to be most important. If the cell gets the go-ahead signal at this checkpoint, it usually completes the whole cell cycle and divides. If it does not receive the go-ahead signal, it enters a nondividing phase called the G_0 *phase.*
- **Kinases** are the protein enzymes that control the cell cycle. They exist in the cells at all times but are active only when they are connected to **cyclin** proteins. Thus, they are called **cyclin-dependent kinases (Cdks)**. Specific kinases give the go-ahead signals at the G_1 and G_2 checkpoints.
- As a specific example, cyclin molecules combine with Cdk molecules producing enough molecules of **MPF** to pass the G_2 checkpoint and initiate the events of mitosis.

■ How does the cell stop cell division? During anaphase, MPF switches itself off by starting a process that leads to the destruction of cyclin molecules. Without cyclin molecules Cdk molecules become inactive, bringing mitosis to a close.

■ Normal cell division has two key characteristics:

■ **Density-dependent inhibition**—The phenomenon in which crowded cells stop dividing.

■ **Anchorage dependency**—Normal cells must be attached to a substratum, like the extracellular matrix of a tissue, to divide.

■ Cancer cells exhibit neither density-dependent inhibition nor anchorage dependency. Cancer is covered in more depth in Concept 18.5, but several important points are made in this chapter.

■ **Transformation** is the process that converts a normal cell to a cancer cell.

■ A **tumor** is a mass of abnormal cells within otherwise normal tissue. If the abnormal cells remain at the original site, the lump is called a **benign tumor**. A **malignant tumor** becomes invasive enough to impair the functions of one or more organs. An individual with a malignant tumor is said to have cancer.

■ **Metastasis** occurs when cells separate from a malignant tumor and enter blood or lymph vessels and travel to other parts of the body.

For Additional Review

Compare the process of meiosis with the process of mitosis. In your comparison, include a study of the change in chromosomal number through the cell, the purposes of each process within an organism, and the starting material and product for each. *Note:* The details of meiosis are covered in Chapter 13, Unit 3 of the text.

Level 1: Knowledge/Comprehension Questions

1. Which structure could you observe with a light microscope?
   (A) a ribosome
   (B) a Golgi apparatus
   (C) a nucleus
   (D) an endoplasmic reticulum
   (E) a peroxisome

2. Prokaryotic and eukaryotic cells have all of the following structures in common EXCEPT
   (A) a plasma membrane.
   (B) DNA.
   (C) a nucleoid region.
   (D) ribosomes.
   (E) cytoplasm.

Directions: Questions 3–7 consist of five lettered choices followed by a list of numbered phrases or sentences. For each numbered phrase or sentence, select the one choice that is most closely related to it. Each choice may be used once, more than once, or not at all in each group.

Questions 3–7
   (A) Peroxisomes
   (B) Golgi apparatus
   (C) Lysosomes
   (D) Endoplasmic reticulum
   (E) Mitochondria

3. An organelle that is characterized by extensive, folded membranes and is often associated with ribosomes

4. An organelle with a *cis* and *trans* face, which acts as the packaging and secreting center of the cell

5. The sites of cellular respiration

6. Single-membrane structures in the cell that perform many metabolic functions and produce hydrogen peroxide in the process

7. Large membrane-bound structures that contain hydrolytic enzymes and that are found predominantly in animal cells

8. Which of the following molecules is a typical component of an animal cell membrane?
   (A) starch
   (B) glucose
   (C) nucleic acids
   (D) carbohydrates
   (E) vitamin K

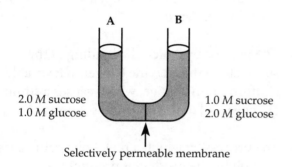

U-Tube Setup

A   B

2.0 *M* sucrose
1.0 *M* glucose

1.0 *M* sucrose
2.0 *M* glucose

Selectively permeable membrane

9. The drawing above shows two solutions of glucose and sucrose in a U-tube containing a semipermeable membrane (which allows the passage of sugars). Which of the following accurately describes what will take place next?
   (A) Glucose will diffuse from side A to side B.
   (B) Sucrose will diffuse from side B to side A.
   (C) No net movement of molecules will occur.
   (D) Glucose will diffuse from side B to side A.
   (E) There will be a net movement of water from side B to side A.

10. Which of the following is an example of passive transport, unaided by proteins, across the cell membrane?
    (A) the stimulation of a muscle cell
    (B) the uptake of glucose by the microvilli of cells lining the stomach
    (C) the movement of insulin across the cell membrane
    (D) the movement of carbon dioxide across the cell membrane
    (E) the selective uptake of hormones across the cell membrane

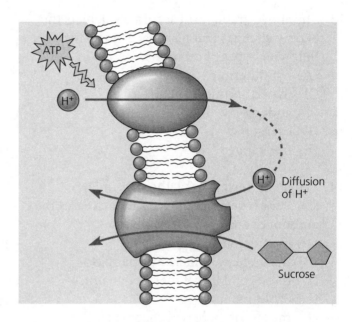

11. The figure above illustrates the process of
    (A) cotransport.
    (B) passive diffusion.
    (C) receptor-mediated endocytosis.
    (D) phagocytosis.
    (E) pinocytosis.

12. Large molecules are moved out of the cell by which of the following processes?
    (A) pinocytosis
    (B) phagocytosis
    (C) receptor-mediated endocytosis
    (D) cytokinesis
    (E) exocytosis

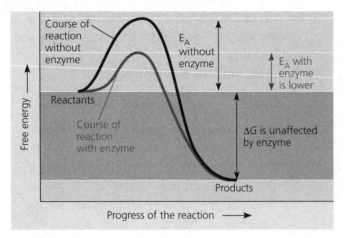

13. The above graph most accurately depicts the energy changes that take place in which of the following types of reactions?
    (A) hypothermic
    (B) hyperthermic
    (C) endergonic
    (D) exergonic
    (E) free range

14. Which of the following theories or laws states that every energy transfer increases the amount of entropy in the universe?
    (A) the free-energy law
    (B) the first law of thermodynamics
    (C) the second law of thermodynamics
    (D) evolutionary theory
    (E) the law of increased chaos

15. Catalysts speed up chemical reactions by
    (A) decreasing the free-energy change of the reaction.
    (B) increasing the free-energy change of the reaction.
    (C) degrading the competitive inhibitors in a reaction.
    (D) lowering the activation energy of the reaction.
    (E) raising the activation energy of the reaction.

Directions: The following group of questions consists of five lettered choices followed by a list of numbered phrases or sentences. For each numbered phrase or sentence, select the one choice that is most closely related to it. Each choice may be used once, more than once, or not at all.

Questions 16–20
    (A) Allosteric interactions
    (B) Feedback inhibition
    (C) Competitive inhibitor
    (D) Noncompetitive inhibitor
    (E) Cooperativity

16. Describes interactions by an enzyme that is capable of either activating or inhibiting an enzymatic reaction

17. A reversible inhibitor that looks similar to the normal substrate and competes for the active site of the enzyme

18. The process by which the binding of the substrate to the enzyme triggers a favorable conformation change, which causes a similar change in all of the proteins' subunits

19. The process by which a metabolic pathway is shut off by the product it produces

20. Binds to the enzyme at a site other than the active site, causing the enzyme to change shape and be unable to bind substrate

21. Which of the following processes could result in the net movement of a substance into a cell, if the substance is more concentrated in the cell than in the surroundings?
    (A) active transport
    (B) facilitated diffusion
    (C) diffusion
    (D) osmosis
    (E) passive transport

22. The activation of receptor tyrosine kinases is characterized by
    (A) channel protein shape change.
    (B) IP$_3$ binding.
    (C) a phosphorylation cascade.
    (D) dimerization and phosphorylation.
    (E) GTP hydrolysis.

23. In cell signaling, how is the flow of specific ions regulated?
    (A) opening and closing of ligand-gated ion channels
    (B) transduction
    (C) cytoskeleton rearrangement
    (D) endocytosis
    (E) phosphorylation cascades

24. What is a G protein?
    (A) a specific type of membrane receptor protein
    (B) a protein on the cytoplasmic side of a membrane that becomes activated by a receptor protein
    (C) a membrane-bound enzyme that converts ATP to cAMP
    (D) a tyrosine kinase relay protein
    (E) a guanine nucleotide that converts between GDP and GTP to activate and inactivate relay proteins

25. Which of the following can activate a protein by transferring a phosphate group to it?
    (A) cAMP
    (B) G protein
    (C) phosphodiesterase
    (D) protein kinase
    (E) protein phosphatase

26. Many signal transduction pathways use second messengers to
    (A) transport a signal through the lipid bilayer portion of the plasma membrane.
    (B) relay a signal from the outside to the inside of the cell.
    (C) relay the message from the inside of the membrane throughout the cytoplasm.
    (D) amplify the message by phosphorylating proteins.
    (E) dampen the message once the signal molecule has left the receptor.

27. Which of the following signal molecules pass through the plasma membrane and bind to intracellular receptors that move into the nucleus and function as transcription factors to regulate gene expression?
    (A) epinephrine
    (B) growth factors
    (C) yeast mating factors α and **a**
    (D) testosterone, a steroid hormone
    (E) neurotransmitter released into synapse between nerve cells

Directions: The group of questions below consists of five lettered choices followed by a list of numbered phrases or sentences. For each numbered phrase or sentence, select the one choice that is most closely related to it. Each choice may be used once, more than once, or not at all in each group.

Questions 28–32
    (A) Telophase
    (B) Interphase
    (C) Cytokinesis
    (D) Prometaphase
    (E) Anaphase

28. Cytokinesis begins during this final stage of mitosis.

29. Division of the cytoplasm of the cell.

30. Sister chromatids begin to separate.

31. The genetic material of the cell replicates to prepare for cell division.

32. Microtubules begin to attach to the centromeres of the sister chromatids.

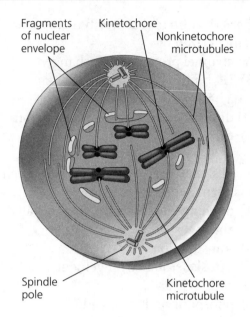

33. What stage of mitosis is represented in the figure to the right?
    (A) prophase
    (B) prometaphase
    (C) metaphase
    (D) anaphase
    (E) telophase

34. After which of the following checkpoints in the cell cycle is the cell first committed to divide?
    (A) G_2 phase checkpoint
    (B) M phase checkpoint
    (C) interphase checkpoint
    (D) G_1 phase checkpoint
    (E) MPF checkpoint

Level 2: Application/Analysis/Synthesis Questions

1. Cells of the pancreas will incorporate radioactively labeled amino acids into proteins. This "tagging" of newly synthesized proteins enables a researcher to track their location. In this case, we are tracking an enzyme secreted by pancreatic cells. What is its most likely pathway?
   (A) ER → lysosomes → vesicles that fuse with plasma membrane
   (B) Golgi → ER → lysosome
   (C) nucleus → ER → Golgi
   (D) ER → Golgi → vesicles that fuse with plasma membrane

2. If the S phase were eliminated from the cell cycle, the daughter cells would
   (A) have half the genetic material found in the parent cell.
   (B) be generally identical to each other.
   (C) be genetically identical to the parent.
   (D) synthesize the missing genetic material on their own.

3. Which figure depicts an animal cell placed in a solution hypotonic to the cell?

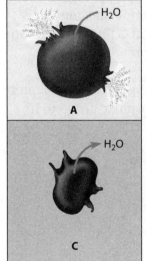
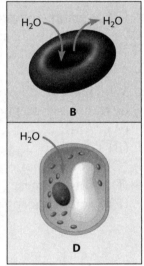

   (A) cell A
   (B) cell B
   (C) cell C
   (D) cell D

4. Which two figures show a cell that is hyper-tonic to its environment?
   (A) cells A and B
   (B) cells A and C
   (C) cells A and D
   (D) cells C and D

After reading the paragraph, answer the question(s) that follow.

Americans spend up to $100 billion annually for bottled water (41 billion gallons). The only beverages with higher sales are carbonated soft drinks. Recent news stories have highlighted the fact that most bottled water comes from municipal water supplies (the same source as your tap water), although it may undergo an extra purification step called reverse osmosis.

Imagine two tanks that are separated by a membrane that is permeable to water, but not to the dissolved minerals present in the water. Tank A contains tap water and Tank B contains the purified water. Under normal conditions, the purified water would cross the membrane to dilute the more concentrated tap water solution. In the reverse osmosis process, pressure is applied to the tap water tank to force the water molecules across the membrane into the pure water tank.

5. After the reverse osmosis system has been op-erating for 30 minutes, the solution in Tank A would
   (A) be hypotonic to Tank B.
   (B) be isotonic to Tank B.
   (C) be hypertonic to Tank B.
   (D) move by passive transport to Tank B.

6. If you shut the system off and pressure was no longer applied to Tank A, you would expect
   (A) the water to flow from Tank A to Tank B.
   (B) the water to reverse flow from Tank B to Tank A.
   (C) the water to flow in equal amounts in both directions.
   (D) the water to flow against the concentration gradient.

7. Earl Sutherland received the Nobel Prize for his discovery of cAMP as a second messenger. Which observation suggested to Sutherland the involvement of a second messenger in epinephrine's effect on liver cells?
   (A) Enzymatic activity was proportional to the amount of calcium added to a cell-free extract.
   (B) Receptor studies indicated that epinephrine was a ligand.
   (C) Glycogen breakdown was observed only when epinephrine was administered to intact cells.
   (D) Glycogen breakdown was observed when epinephrine and glycogen phosphorylase were combined.

8. Protein phosphorylation is commonly in-volved with all of the following EXCEPT
   (A) regulation of transcription by extracellu-lar signaling molecules.
   (B) enzyme activation.
   (C) activation of G protein-coupled receptors.
   (D) activation of receptor tyrosine kinases.

Free-Response Question

1. *Prokaryotic and eukaryotic cells are physiologically different in many ways, but both represent functional, evolutionarily successful cells.*

   (a) It has been theorized that the organelles of eukaryotic cells evolved from prokaryotes living symbiotically within a larger cell. **Compare** and **contrast** the structure of the prokaryotic cell with eukaryotic cell organelles, and make an argument for or against this theory. Be sure to justify your position with factual information.

   (b) **Create** a model to show the path of a protein in a eukaryotic cell from its formation to its secretion from the cell.

Respiration and Photosynthesis

Chapter 9: Cellular Respiration and Fermentation

YOU MUST KNOW

- The summary equation of cellular respiration.
- The difference between fermentation and cellular respiration.
- The role of glycolysis in oxidizing glucose to two molecules of pyruvate.
- The process that brings pyruvate from the cytosol into the mitochondria and introduces it into the citric acid cycle.
- How the process of chemiosmosis utilizes the electrons from NADH and $FADH_2$ to produce ATP.

Oxidation-reduction reactions, fermentation, cellular respiration, and photosynthesis are among the most technically challenging sections of the course. Here, we will focus on the major steps of each of the processes, as well as the results. Questions on the AP Biology Exam are likely to focus on the net results of photosynthesis and respiration—not on the exact reactions that create the products, nor are you expected to know the names of the enzymes involved in the process. As you work through these chapters, compare and contrast the two fundamental cell processes.

Concept 9.1 Catabolic pathways yield energy by oxidizing organic fuels

▮ **Catabolic pathways** occur when molecules are broken down and their energy is released. You should know these two catabolic pathways:

- ▪ **Fermentation** is the partial degradation of sugars that occurs without the use of oxygen.
- ▪ **Cellular respiration** is the most prevalent and efficient catabolic pathway. It is also termed **aerobic respiration** as oxygen is required along with organic fuel.

- Carbohydrates, fats, and proteins can all be broken down to release energy in cellular respiration. However, glucose is the primary nutrient molecule that is used in cellular respiration. The standard way of representing the process of cellular respiration shows glucose being broken down in the following reaction:

$$C_6H_{12}O_6 + 6\,O_2 \rightarrow 6\,CO_2 + 6\,H_2O + \text{Energy (686 kcal/mol of glucose)}$$

- The exergonic release of energy from glucose is used to phosphorylate ADP to ATP. Life processes constantly consume ATP; cellular respiration burns fuels and uses the energy to regenerate ATP.
- The reactions of cellular respiration are of a type termed **oxidation-reduction (redox)** reactions. In redox reactions electrons are transferred from one reactant to another.

  - The loss of one or more electrons from a reactant is called **oxidation**. When a reactant is *oxidized*, it loses electrons and, consequently, energy.
  - The gain of one or more electrons is **reduction**. When a reactant is *reduced*, it gains electrons and, therefore, energy.

- At key steps in cellular respiration, electrons are stripped from glucose. Each electron travels with a proton, thereby forming a hydrogen atom. The hydrogen atoms are not transferred directly to oxygen, as the formula might suggest, but instead are usually passed to an electron carrier, the coenzyme **NAD$^+$** (a derivative of the B vitamin niacin). Within the cell **NAD$^+$** accepts two electrons, plus the stabilizing hydrogen ion, to form **NADH**. Note that NADH has been reduced and therefore has gained energy.
- Figure 3.1 shows the three stages of cellular respiration. Each stage is separately featured in the next three concepts. Use this figure to begin to develop an overall concept of the process of cellular respiration.

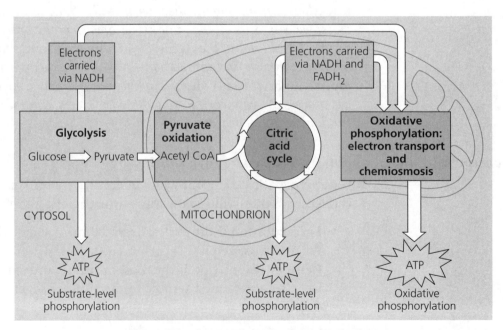

Figure 3.1   An overview of cellular respiration

Concept 9.2 *Glycolysis harvests chemical energy by oxidizing glucose to pyruvate*

▮ In **glycolysis** (which occurs in the cytosol), the degradation of glucose begins as it is broken down into two pyruvate molecules. The six-carbon glucose molecule is split into two three-carbon sugars through a long series of steps.

> ***STUDY TIP*** Use Figure 3.2 as a guide to the important features of glycolysis. It is not necessary to know the enzymes and reactions for each step in glycolysis.

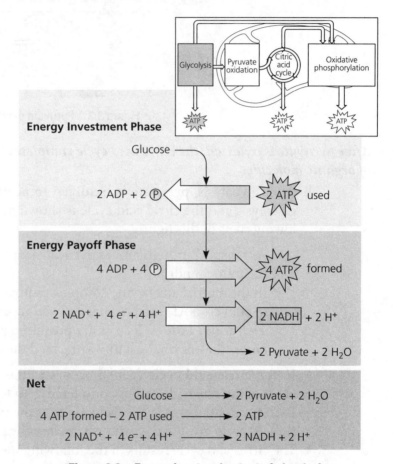

Figure 3.2   Energy input and output of glycolysis

▮ In the course of glycolysis, there is an ATP-consuming phase and an ATP-producing phase. In the ATP-consuming phase, two ATP molecules are consumed, which helps destabilize glucose and make it more reactive. Later in glycolysis, 4 ATP molecules are produced; thus, glycolysis results in a net gain of 2 ATP. Two NADH are also produced, which will be utilized in the electron transport system (see Figure 3.1) to produce ATP.

▮ Notice the *net* energy gain in glycolysis as indicated in Figure 3.2—2 ATP molecules and 2 NADH molecules. Most of the potential energy of the glucose molecule still resides in the two remaining pyruvates, which will now feed into the citric acid cycle, as discussed in the next concept.

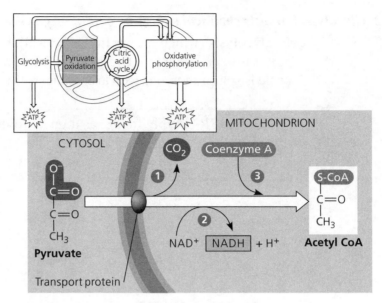

Figure 3.3   Pyruvate oxidation

Concept 9.3 *After pyruvate is oxidized, the citric acid cycle completes the energy-yielding oxidation of organic molecules*

▌ After glycolysis, **pyruvate is oxidized to acetyl CoA**. This junction between glycolysis and the citric acid cycle is shown above in Figure 3.3. Note the following steps in this figure:

  ▪ A transport protein moves pyruvate from the cytosol into the matrix of the mitochondria.
  ▪ In the matrix, an enzyme complex removes a CO_2, strips away electrons to convert NAD^+ to NADH, and adds coenzyme A to form acetyl CoA.
  ▪ Two acetyl CoA molecules are produced per glucose. Acetyl CoA now enters the enzymatic pathway termed the citric acid cycle.

▌ In the **citric acid cycle** (which occurs in the mitochondrial matrix), the job of breaking down glucose is completed with CO_2 released as a waste product. Each turn of the citric acid cycle requires the input of one acetyl CoA. The citric acid cycle must make two turns before the glucose is completely oxidized.

▌ The citric acid cycle results in the following:

  ▪ Each turn of the citric acid cycle produces **2 CO_2, 3 NADH, 1 $FADH_2$, and 1 ATP**.
  ▪ Because each glucose yields *two* pyruvates, the *total* products of the citric acid cycle are usually listed as the result of two cycles:

4 CO_2, 6 NADH, 2 $FADH_2$, and 2 ATP

  ▪ Note that at the end of the citric acid cycle the six original carbons in glucose have been released as CO_2. (You are exhaling this gas as you study.) Only 2 ATP molecules, however, have been produced. Where is all the energy? The energy is held in the electrons in the electron carriers, NADH and $FADH_2$. These electrons will by utilized by the electron transport system, explained in the next concept.

Concept 9.4 During oxidative phosphorylation, chemiosmosis couples electron transport to ATP synthesis

▌ Use Figure 3.4 as a map to understand the process of electron transport.

1. The electron transport chain is embedded in the inner membrane of the mitochondria. Notice that it is composed of three transmembrane proteins that work as hydrogen pumps and two carrier molecules that transport electrons between hydrogen pumps. There are thousands of such electron transport chains in the inner mitochondrial membrane.

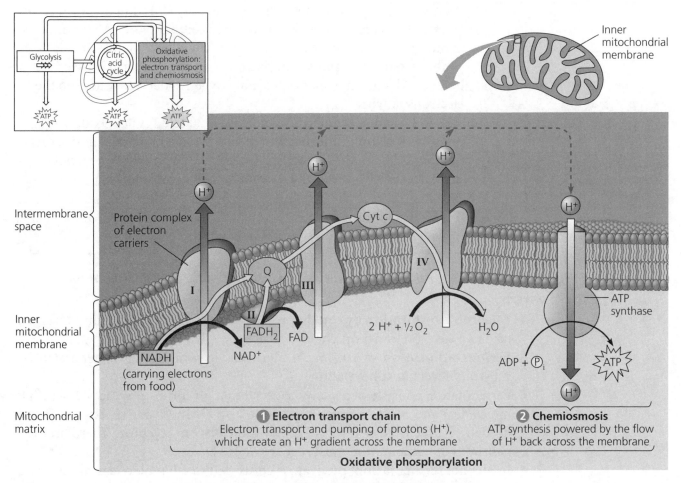

Figure 3.4   ATP production by chemiosmosis

2. The electron transport chain is powered by electrons from the electron carrier molecules NADH and $FADH_2$ ($FADH_2$ is also a B vitamin coenzyme that functions as an electron acceptor in the citric acid cycle). As the electrons flow through the electron chain, the loss of energy by the electrons is used to power the pumping of protons across the inner membrane. At the end of the electron chain, the electrons combine with two hydrogen ions and oxygen to form water. Notice that O_2 is the final electron acceptor, and when it is not available, the transport of electrons comes to a screeching halt! No hydrogen ions are pumped and no ATP is produced.

3. The hydrogen ions flow back down their gradient through a channel in the transmembrane protein known as ATP synthase. **ATP synthase** harnesses the **proton-motive force**—the gradient of hydrogen ions—to phosphorylate ADP, forming ATP. The proton-motive force is in place because the inner membrane of the mitochondria is impermeable to hydrogen ions. Like water behind a dam with its only exit being a spillway, electrons are held behind the inner membrane with their only exit ATP synthase.

4. This process is referred to as chemiosmosis. **Chemiosmosis** is an energy-coupling mechanism that uses energy stored in the form of an H^+ gradient across a membrane to drive cellular work (ATP synthesis in our example). The electron transport chain and chemiosmosis compose **oxidative phosphorylation**. This specific term is used because ADP is phosphorylated and oxygen is necessary to keep the electrons flowing.

5. The ATP yield per molecule of glucose is 30 to 32 ATP. Oxidative phosphorylation produces 26 to 28 of the total. (You may notice this number is less than you might have learned in an earlier text, but more accurately reflects the current biochemical analysis.)

> **STUDY TIP** Sketch this process and explain it verbally. This is a fundamental biological process that you should understand.

Concept 9.5 Fermentation and anaerobic respiration enable cells to produce ATP without the use of oxygen

- **Anaerobic respiration** by certain prokaryotes generates ATP without oxygen using an electron transport chain.
- **Fermentation** is an expansion of glycolysis in which ATP is generated by substrate-level phosphorylation.
- Fermentation consists of glycolysis (recall that glycolysis produces 2 net ATP molecules) and reactions that regenerate NAD^+. In glycolysis oxygen is not needed to accept electrons; NAD^+ is the electron acceptor. Therefore, the pathways of fermentation must regenerate NAD^+.
- In **alcohol fermentation**, pyruvate is converted to ethanol, releasing CO_2 and oxidizing NADH in the process to create more NAD^+.
- In **lactic acid fermentation**, pyruvate is reduced by NADH (NAD^+ is formed in the process), and lactate is formed as a waste product.
- **Facultative anaerobes** can make ATP by aerobic respiration if oxygen is present, but can switch to fermentation under anaerobic conditions. **Obligate anaerobes** cannot survive in the presence of oxygen.

Concept 9.6 Glycolysis and the citric acid cycle connect to many other metabolic pathways

- In addition to glucose and other sugars, proteins and fats are often used to generate ATP through cellular respiration.
- **Phosphofructokinase** is an allosteric enzyme that acts as a regulator of respiration.

Chapter 10: Photosynthesis

```
YOU MUST KNOW

• The summary equation of photosynthesis including the source and fate
  of the reactants and products.
• How leaf and chloroplast anatomy relates to photosynthesis.
• How photosystems convert solar energy to chemical energy.
• How linear electron flow in the light reactions results in the formation
  of ATP, NADPH, and $O_2$.
• How chemiosmosis generates ATP in the light reactions.
• How the Calvin cycle uses the energy molecules of the light reactions
  to produce G3P.
```

Concept 10.1 Photosynthesis converts light energy to the chemical energy of food

■ Before you look at the molecular details of photosynthesis, it is important to think of photosynthesis in an ecological context.

- Life on Earth is solar powered by autotrophs. **Autotrophs** are "self-feeders"; they sustain themselves without eating anything derived from other organisms. Autotrophs are the ultimate source of organic compounds and are therefore known as *producers*.
- **Heterotrophs** live on compounds produced by other organisms and are thus known as *consumers*. Animals immediately come to mind as heterotrophs, but also remember that decomposers like fungi and many prokaryotes are heterotrophs. Heterotrophs are dependent on the process of photosynthesis for both food and oxygen.

■ **Chloroplasts** are the specific sites of photosynthesis in plant cells.

- Chloroplasts are organelles that are mostly located in the cells that make up the **mesophyll** tissue found in the interior of the leaf.
- Use Figure 3.5 to become familiar with the structure of chloroplasts. An envelope of two membranes encloses the **stroma**, which is a dense fluid-filled area. Within the stroma is a vast network of interconnected membranous sacs called **thylakoids**. The thylakoids segregate the stroma from another compartment, the **thylakoid space**.
- **Chlorophyll** is located in the thylakoid membranes and is the light-absorbing pigment that drives photosynthesis and gives plants their green color.

■ The exterior of the lower epidermis of a leaf contains many tiny pores called **stomata**, through which carbon dioxide enters and oxygen and water vapor exit the leaf.

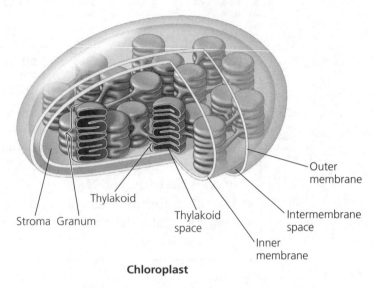

Chloroplast

Figure 3.5   Chloroplast structure

STUDY TIP Practice drawing a chloroplast in the margin of this page. Label its parts and know what major events occur in each region.

▌ The overall reaction of photosynthesis looks like this:

$$6\,CO_2 + 6\,H_2O + \textbf{Light energy} \rightarrow C_6H_{12}O_6 + 6\,O_2$$

- ▪ Notice that the overall chemical change during photosynthesis is the reverse of the one that occurs during cellular respiration.
- ▪ All the oxygen you breathe was formed in the process of photosynthesis when a water molecule was split! Water is split for its electrons, which are transferred along with hydrogen ions from water to carbon dioxide, reducing it to sugar. This process requires energy (an endergonic process), which is provided by the sun.

▌ Photosynthesis is a chemical process that requires two stages to complete.

▌ The **light reactions** occur in the thylakoid membranes where solar energy is converted to chemical energy. The net products of the light reactions are **NADPH** (which stores electrons), **ATP**, and **oxygen**. Here are the primary events.

1. Light is absorbed by chlorophyll and drives the transfer of electrons from water to $NADP^+$, forming **NADPH**.
2. Water is split during these reactions, and O_2 is released.
3. **ATP** is generated, using chemiosmosis to power the addition of a phosphate group to ADP, a process called **photophosphorylation**.

▌ The **Calvin cycle** occurs in the stroma, where CO_2 from the air is incorporated into organic molecules in **carbon fixation**. The Calvin cycle uses the fixed carbon plus NADPH and ATP from the light reactions in the formation of new sugars.

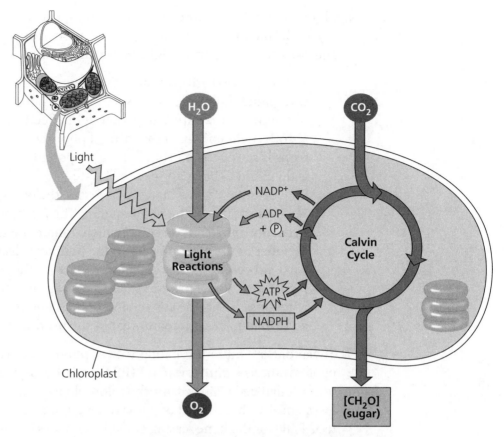

Figure 3.6 Overview of photosynthesis

STUDY TIP Use Figure 3.6 above to help in understanding where reactions occur and the overall purpose of the two stages of photosynthesis. If you understand the big picture, the details will be easier to comprehend.

Concept 10.2 *The light reactions convert solar energy to the chemical energy of ATP and NADPH*

▌ Not surprisingly, light is an important concept in photosynthesis.

- Light is electromagnetic energy, and it behaves as though it is made up of discrete particles, called **photons**—each of which has a fixed quantity of energy.
- Substances that absorb light are called **pigments**, and different pigments absorb light of different wavelengths. Chlorophyll is a pigment that absorbs violet-blue and red light while transmitting and reflecting green light. This is why we see summer leaves as green.
- A graph plotting a pigment's light absorption versus wavelength is called an **absorption spectrum**. The absorption spectrum of chlorophyll provides clues to the effectiveness of different wavelengths for driving photosynthesis. This is confirmed by an action spectrum.
- An **action spectrum** for photosynthesis graphs the effectiveness of different wavelengths of light in driving the process of photosynthesis. Note examples of both of these graphs in your text, Figure 10.10.

▌Photons of light are absorbed by certain groups of pigment molecules in the thylakoid membrane of chloroplasts. These groups are called **photosystems** and consist of two parts: a light-harvesting complex and a reaction center.

- The **light-harvesting complex** is made up of many chlorophyll and carotenoid molecules (accessory pigments in the thylakoid membrane); this allows the complex to gather light effectively. When chlorophyll absorbs light energy in the form of photons, one of the molecule's electrons is raised to an orbital of higher potential energy. The chlorophyll is then said to be in an "excited" state.
- Like a human "wave" at a sports arena, the energy is transferred to the **reaction center** of the photosystem. The reaction center consists of two chlorophyll *a* molecules, which donate the electrons to the second member of the reaction center, the **primary electron acceptor**. The solar-powered transfer of an electron from the reaction-center chlorophyll *a* pair to the primary electron acceptor is the first step of the light reactions. This is the conversion of light energy to chemical energy, and what makes photoautotrophs the producers of the natural world.

▌Thylakoid membranes contain two photosystems that are important to photosynthesis—**photosystem I (PS I)** and **photosystem II (PS II)**. PS I is sometimes designated P700 because the chlorophyll *a* in the reaction center of this photosystem absorbs red light of this wavelength the best; PS II is sometimes referred to as P680 for the same reason. Don't let switches in designation be confusing.

▌Following are the major steps of the light reactions of photosynthesis. The key to the light reactions is a flow of electrons through the photosystems in the thylakoid membrane, a process called **linear (noncyclic) electron flow**. Find each step in Figure 3.7 as you read the following summary:

1. Photosystem II absorbs light energy, allowing the P680 reaction center of two chlorophyll *a* molecules to donate an electron to the primary electron acceptor. The reaction-center chlorophyll is oxidized and now requires an electron.
2. An enzyme splits a water molecule into two hydrogen ions, two electrons, and an oxygen atom. The electrons are supplied to the P680 chlorophyll *a* molecules. The oxygen combines with another oxygen molecule, forming the O_2 that will be released into the atmosphere.
3. The original excited electron passes from the primary electron acceptor of photosystem II to photosystem I through an electron transport chain (similar to the electron chain in cellular respiration).
4. The energy from the transfer of electrons down the electron transport chain is used to pump protons, creating a gradient that is used in chemiosmosis to phosphorylate ADP to ATP. ATP will be used as energy in the formation of carbohydrates in the Calvin cycle.
5. Meanwhile, light energy has also activated PS I, resulting in the donation of an electron to its primary electron acceptor. The electrons just donated by PS I are replaced by the electrons from PS II. (Keep in mind that the ultimate source of electrons is water.)

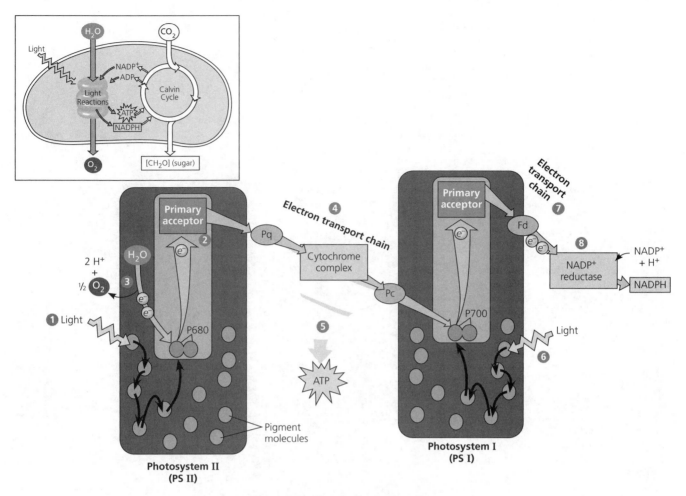

Figure 3.7   Noncyclic electron flow

6. The primary electron acceptor of photosystem I passes the excited electrons along to another electron transport chain, which transmits them to NADP⁺, which is reduced to NADPH—the second of the two important light-reaction products. The high-energy electrons of NADPH are now available for use in the Calvin cycle.

▌ An alternative to linear electron flow is **cyclic electron flow**. Cyclic electron flow uses PS I, but not PS II. Cyclic electron flow uses a short circuit of linear electron flow by cycling the excited electrons back to their original starting point in PS I. Cyclic electron flow produces ATP by chemiosmosis, but no NADPH is produced and no oxygen is released.

▌ Chloroplasts and mitochondria generate ATP by the same basic mechanism: chemiosmosis. Examining Figure 3.8 on the next page will quickly demonstrate the same basic chemiosmotic plan as cellular respiration. Use Figure 3.8 to illustrate the following:

  ▌ An electron transport chain uses the flow of electrons to pump protons across the thylakoid membrane.

  ▌ A proton-motive force is created within the thylakoid space that can be utilized by ATP synthase to phosphorylate ADP to ATP. Notice that the

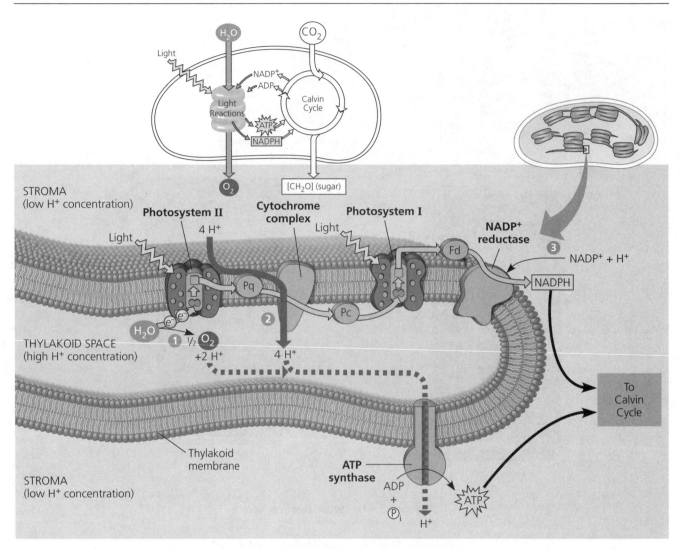

Figure 3.8 Light reactions and chemiosmosis

proton-motive force is generated in three places: (1) hydrogen ions from water; (2) hydrogen ions pumped across the membrane by the cytochrome complex; (3) the removal of a hydrogen ion from the stroma when $NADP^+$ is reduced to NADPH.

▌ Although similar, chemiosmosis in cellular respiration and photosynthesis are not identical. In addition to some spatial differences, the key conceptual difference is that mitochondria use chemiosmosis to transfer chemical energy from food molecules to ATP, whereas chloroplasts transform light energy into chemical energy in ATP. This is the essence of the difference between a consumer and a producer.

TIP FROM THE READERS

Create and use a model of electron transport across either the inner membrane of the mitochondria or a thylakoid membrane to explain the production of ATP. Be able to predict how changes in the model would affect the output of ATP.

Concept 10.3 *The Calvin cycle uses the chemical energy of ATP and NADPH to convert CO_2 to sugar*

❚ In the course of the **Calvin cycle**, carbon enters in the form of CO_2 and leaves in the form of a sugar. The cycle spends ATP as an energy source and consumes NADPH as reducing power for adding high-energy electrons to make the sugar. Use Figure 3.9 to chart each step summarized in the outline below. You must note that in order to net one molecule of G3P, the cycle must go through three rotations and fix three molecules of CO_2.

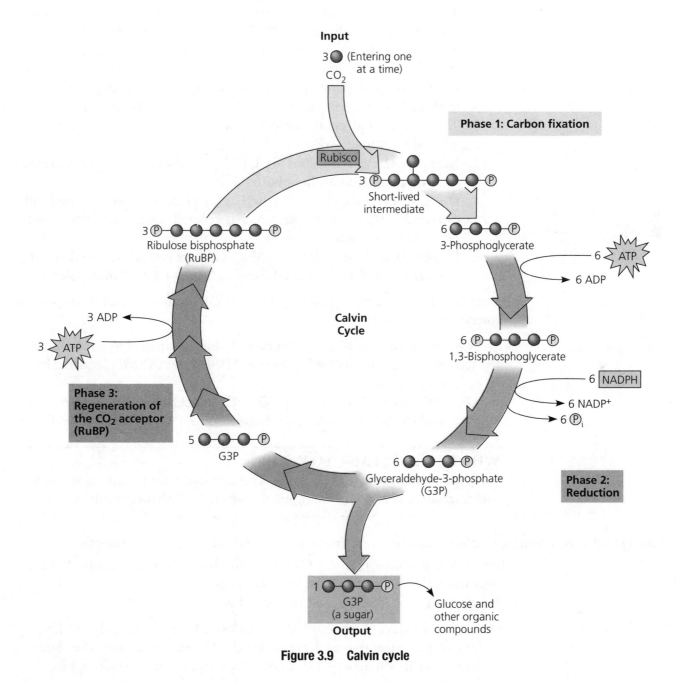

Figure 3.9   Calvin cycle

▌ These are the major steps of the Calvin cycle. It is not important that you memorize the intermediate organic molecules, but it is important to understand the conceptual scheme of reducing CO_2 to a sugar.

1. Three CO_2 molecules are attached to three molecules of the five-carbon sugar **ribulose bisphosphate (RuBP)**. These reactions are catalyzed by the enzyme **rubisco** (probably the most common protein in the biosphere) and produce an unstable product that immediately splits into two three-carbon compounds called 3-phosphoglycerate. At this point carbon has been fixed—the incorporation of CO_2 into an organic compound.
2. The 3-phosphoglycerate molecules are phosphorylated to become 1,3-bisphosphoglycerate.
3. Next, 6 NADPH reduce the six 1,3-bisphosphoglycerate molecules to six **glyceraldehyde 3-phosphate (G3P)**.
4. One G3P leaves the cycle to be used by the plant cell. Two G3P molecules can combine to form glucose, which is generally listed as the final product of photosynthesis.
5. Finally, RuBP is regenerated as the 5 G3P are reworked into three of the starting molecules, with the expenditure of 3 ATP molecules.

▌ In the Calvin cycle, the formation of one net G3P requires the following energy molecules:

■ *Nine molecules of ATP* are consumed (to be replenished by the light reactions) along with *six molecules of NADPH* (also to be replenished by the light reactions).
■ One of the six G3P molecules produced in the Calvin cycle is a net gain, and will be used for biosynthesis or the energy needs of the cell.

> **WHAT'S IMPORTANT TO KNOW?**
> Concept 10.4 is not required for the AP exam, but your teacher may use the information here as a nice example of adaptations to arid conditions.

Concept 10.4 *Alternative mechanisms of carbon fixation have evolved in hot, arid climates*

▌ The overall problem is that CO_2 enters the leaf through stomata, the same pores through which water exits the leaf in transpiration.
▌ The specific problem for C_3 plants is as follows:

■ On hot dry days C_3 plants produce less sugar because the declining levels of CO_2 in the leaf starves the Calvin cycle. This occurs because the plant must keep its stomata closed to conserve water—thus, no CO_2 uptake.
■ Additionally, the enzyme rubisco can bind O_2 in place of CO_2. This causes the oxidation or breakdown of RuBP, resulting in a loss of energy and carbon for the plant—a metabolic process called

photorespiration. Photorespiration can drain away as much as 50% of the carbon fixed by the Calvin cycle!

▌ How can hot, arid regions have any plants? They have metabolic adaptations (as well as structural ones covered in Topic 8) that reduce photorespiration. The two most important of these adaptations are C_4 and CAM plants.

▌ C_4 plants have two kinds of photosynthetic cells: bundle-sheath cells and mesophyll cells. The **bundle-sheath cells** are grouped around the leaf's veins, and the mesophyll cells are dispersed elsewhere around the leaf.

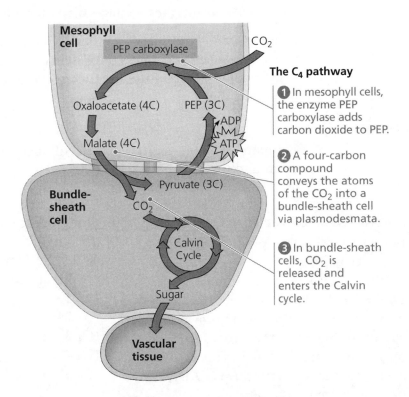

Figure 3.10 C_4 photosynthesis

▌ Use Figure 3.10 above as a guide for the following steps to C_4 photosynthesis:

1. CO_2 is added to **phosphoenolpyruvate** (PEP) to form the four-carbon compound **oxaloacetate** (hence the designation C_4 plant). This reaction is catalyzed by **PEP carboxylase**, which does not combine with O_2 and does not participate in photorespiration. At this point carbon is fixed.

2. The mesophyll cells export the oxaloacetate to the bundle-sheath cells, which break down the oxaloacetate, releasing CO_2. The C_4 pathway acts as a CO_2 pump.

3. The CO_2 in the bundle-sheath cells is then converted into carbohydrates through the normal Calvin cycle.

   • In C_4 plants, the mesophyll cells pump CO_2 into the bundle-sheath cells, keeping the CO_2 concentration high enough so that rubisco will be more likely to bind to CO_2 rather than O_2. This reduces photorespiration. In C_4 plants the two stages of photosynthesis are separated structurally, providing a process that minimizes photorespiration and enhances sugar production.

▌ **CAM photosynthesis** is another adaptation to hot, dry climates.

  ▪ These plants keep their stomata closed during the day to prevent excessive water loss. Of course, this also prevents gas exchange. At night, the stomata open and CO_2 is fixed in organic acids and stored in vacuoles. In the morning when the stomata close, the plant cells release the stored CO_2 from the acids and proceed with photosynthesis.
  ▪ In CAM plants the two stages of photosynthesis are separated temporally.

▌ In both C_4 and CAM photosynthesis, CO_2 is first transformed into an organic intermediate before it enters the Calvin cycle. All of the processes—C_3, C_4, and CAM photosynthesis—use the Calvin cycle; they just have different methods for getting there.

For Additional Review

Compare and contrast the process of chemiosmosis in both the mitochondrion and the chloroplast. Note how the H^+ gradient is established and the orientation of the ATP synthase molecules.

Level 1: Knowledge/Comprehension Questions

1. The purpose of cellular respiration in a eukaryotic cell is to
   (A) synthesize carbohydrates from CO_2.
   (B) synthesize fats and proteins from CO_2.
   (C) break down carbohydrates to provide energy for the cell in the form of ATP.
   (D) break down carbohydrates to provide energy for the cell in the form of ADP.
   (E) provide oxygen to the cell.

$$2 \, K + Br_2 \rightarrow 2 \, K^+ + 2 \, Br^-$$

2. In the course of the above reaction, potassium is
   (A) neutralized.
   (B) oxidized.
   (C) reduced.
   (D) sublimated.
   (E) recycled.

3. The net result of glycolysis is
   (A) 4 ATP and 4 NADH.
   (B) 4 ATP and 2 NADH.
   (C) 2 ATP and 4 NADH.
   (D) 2 ATP and 2 NADH.
   (E) 4 ATP and 8 NADH.

4. One glucose molecule provides enough carbons for two trips through the citric acid cycle. How many molecules of ATP are directly produced in this process?
   (A) 1
   (B) 2
   (C) 3
   (D) 4
   (E) 5

5. The process that produces the greatest amount of ATP during respiration is
   (A) glycolysis.
   (B) fermentation.
   (C) the citric acid cycle.
   (D) oxidative phosphorylation.
   (E) lactic acid formation.

Directions: The group of questions that follow, 6 through 10, consists of five lettered choices followed by a list of numbered phrases or sentences. For each numbered phrase or sentence, select the one choice that is most closely related to it. Each choice may be used once, more than once, or not at all in each group.

Questions 6–10

    (A) Chemiosmosis
    (B) Electron transport chain
    (C) The citric acid cycle
    (D) Glycolysis
    (E) Fermentation

6. The process by which glucose is split into pyruvate

7. The process by which a hydrogen ion gradient is used to produce ATP

8. A process that makes a small amount of ATP and can produce lactic acid as a by-product

9. A series of membrane-embedded electron carriers that ultimately create the hydrogen ion gradient to drive the synthesis of ATP

10. The process by which the chemical breakdown of glucose is completed and CO_2 is produced

11. Groups of photosynthetic pigment molecules situated in the thylakoid membrane are called
    (A) photosystems.
    (B) carotenoids.
    (C) chlorophyll.
    (D) grana.
    (E) CAM plants.

12. The main products of the light reactions of photosynthesis are
    (A) NADPH and $FADH_2$.
    (B) NADPH and ATP.
    (C) ATP and $FADH_2$.
    (D) ATP and CO_2.
    (E) ATP and H_2O.

13. The process in photosynthesis that bears the most resemblance to chemiosmosis and oxidative phosphorylation in cell respiration is called
    (A) cyclic electron flow.
    (B) linear electron flow.
    (C) ATP synthase coupling.
    (D) substrate-level phosphorylation.
    (E) glycolysis.

14. The major product of the Calvin cycle is
    (A) rubisco.
    (B) oxaloacetate.
    (C) ribulose bisphosphate.
    (D) pyruvate.
    (E) glyceraldehyde 3-phosphate.

15. All of the following statements are *false* EXCEPT
    (A) C_3 plants grow better in hot, arid conditions than do C_4 plants.
    (B) C_4 plants grow better in cold, moist conditions than do C_3 plants.
    (C) C_3 plants grow better in hot, arid conditions than do CAM plants.
    (D) CAM plants grow better in cold, moist conditions than do C_3 plants.
    (E) CAM plants and C_4 plants both grow better in hot, arid conditions than do C_3 plants.

16. All of the following statements about photosynthesis are *true* EXCEPT
    (A) the light reactions convert solar energy to chemical energy in the form of ATP and NADPH.
    (B) the Calvin cycle uses ATP and NADPH to convert CO_2 to sugar.
    (C) photosystem I contains P700 chlorophyll *a* molecules at the reaction center; photosystem II contains P680 molecules.
    (D) in chemiosmosis, electron transport chains pump protons (H^+) across a membrane from a region of high H^+ concentration to a region of low H^+ concentration.
    (E) the steps of the Calvin cycle use ATP and NADPH to convert CO_2 to sugar.

17. Which of the following is mismatched with its location?
    (A) light reactions—grana
    (B) electron transport chain—thylakoid membrane
    (C) Calvin cycle—stroma
    (D) ATP synthesis—double membrane surrounding chloroplast
    (E) splitting of water—thylakoid space

18. The chlorophyll known as P680 has its electron "holes" filled by electrons from
    (A) photosystem I.
    (B) photosystem II.
    (C) water.
    (D) NADPH.
    (E) CO_2.

19. CAM plants avoid photorespiration by
    (A) fixing CO_2 into organic acids during the night; these acids then release CO_2 during the day.
    (B) performing the Calvin cycle at night.
    (C) fixing CO_2 into four-carbon compounds in the mesophyll, which release CO_2 in the bundle-sheath cells.
    (D) using PEP carboxylate to fix CO_2 to ribulose bisphosphate (RuBP).
    (E) keeping their stomata open during the day.

20. How many "turns" of the Calvin cycle are required to produce one molecule of glucose?
    (A) 1
    (B) 2
    (C) 3
    (D) 6
    (E) 12

21. What are the final electron acceptors for the electron transport chains in the light reactions of photosynthesis and in cellular respiration?
    (A) O_2 in both
    (B) CO_2 in both
    (C) H_2O in the light reactions and O_2 in respiration
    (D) P700 and $NADP^+$ in the light reactions and NAD^+ or FAD in respiration
    (E) $NADP^+$ in the light reactions and O_2 in respiration

Level 2: Application/Analysis/Synthesis Questions

After reading the paragraph, answer the question(s) that follow.

You're conducting an experiment to determine the effect of different wavelengths of light on the absorption of carbon dioxide as an indicator of the rate of photosynthesis in aquatic ecosystems. If the rate of photosynthesis increases, the amount of carbon dioxide in the environment will decrease and vice versa. You've added an indicator to each solution. When the carbon dioxide concentration decreases, the color of the indicator solution also changes.

Small aquatic plants are placed into three containers of water mixed with carbon dioxide and indicator solution. Container A is placed under normal sunlight, B under green light, and C under red light. The containers are observed for a 24-hour period.

1. Based on your knowledge of the process of photosynthesis, the plant in the container placed under red light would probably
   (A) absorb no CO_2.
   (B) absorb the same amount of CO_2 as the plants under both the green light and normal sunlight.
   (C) absorb less CO_2 than the plants under the green light.
   (D) absorb more CO_2 than the plants under the green light.

2. Carbon dioxide absorption is an appropriate indicator of photosynthesis because
   (A) CO_2 is needed to produce sugars in the Calvin cycle.
   (B) CO_2 is needed to complete the light reactions.
   (C) plants produce oxygen gas by splitting CO_2.
   (D) the energy in CO_2 is used to produce ATP and NADPH.

3. In this drawing of a chloroplast, which structure represents the location of the Calvin cycle?

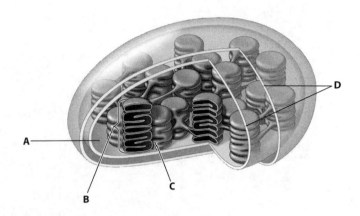

(A) structure A
(B) structure B
(C) structure C
(D) structure D

4. In the figure below, which step of the citric acid cycle requires both NAD^+ and ADP as reactants?
   (A) step 1
   (B) step 2
   (C) step 3
   (D) step 4

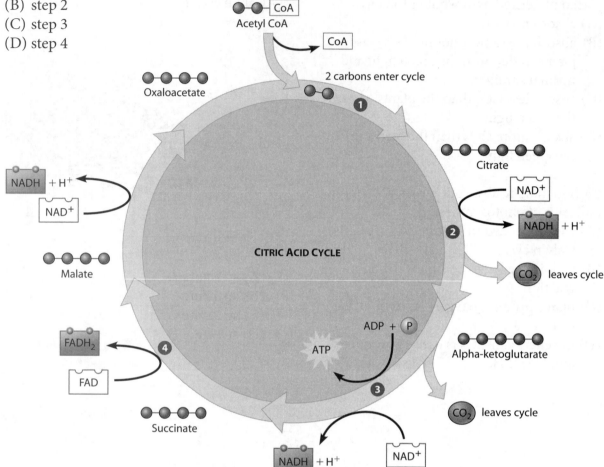

After reading the paragraph, answer the question(s) that follow.

As a scientist employed by the FDA, you've been asked to sit on a panel to evaluate a pharmaceutical company's application for approval of a new weight loss drug called Fat Away. The company has submitted a report summarizing the results of their animal and human testing. In the report, it was noted that Fat Away works by affecting the electron transport chain. It decreases the synthesis of ATP by making the mitochondrial membrane permeable to H^+, which allows H^+ to leak from the intermembrane space to the mitochondrial matrix. This effect leads to weight loss.

5. The method of weight loss described for Fat Away shows that the drug is acting as a metabolic
   (A) feedback inhibitor.
   (B) oxygen carrier.
   (C) redox promoter.
   (D) uncoupler.

6. Fat Away prevents ATP from being made by
   (A) destroying the H^+ gradient that allows ATP synthase to work.
   (B) preventing glycolysis from occurring.
   (C) preventing the conversion of NADH to NAD^+.
   (D) slowing down the citric acid cycle.

Free-Response Question

1. *Cellular respiration and photosynthesis are basic cellular processes. Below are key events in cellular respiration and/or photosynthesis.*

   (a) **Explain** how a photosystem converts light energy to chemical energy.

   (b) **Explain** the specific role of glycolysis in cellular respiration.

   (c) **Describe** the function of water in both cellular respiration and photosynthesis.

Mendelian Genetics

Chapter 13: Meiosis and Sexual Life Cycles

YOU MUST KNOW

- The differences between asexual and sexual reproduction.
- The role of meiosis and fertilization in sexually reproducing organisms.
- The importance of homologous chromosomes to meiosis.
- How the chromosome number is reduced from diploid to haploid through the stages of meiosis.
- Three important differences between mitosis and meiosis.
- The importance of crossing over, independent assortment, and random fertilization to increasing genetic variability.

Concept 13.1 Offspring acquire genes from parents by inheriting chromosomes

▌ **Genes** are segments of DNA that code for the basic units of heredity and are transmitted from one generation to the next. In animals and plants, reproductive cells that transmit genes from one generation to the next are called **gametes**.

▌ A **locus** (plural, *loci*) is the location of a gene on a chromosome. See Figure 4.1a.

  ▪ In **asexual reproduction** a single parent is the sole parent and passes copies of all its genes to its offspring. In asexual reproduction the new offspring arise by mitosis and have virtually exact copies of the parent's genome. An individual that reproduces asexually gives rise to a **clone**, a group of genetically identical individuals.

  ▪ In **sexual reproduction**, two individuals (parents) contribute genes to the offspring. This form of reproduction results in greater genetic variation in the offspring than asexual reproduction.

Concept 13.2 Fertilization and meiosis alternate in sexual life cycles

▌ A **life cycle** is the generation-to-generation sequence of stages in the reproductive history of an organism, from conception to production of its own offspring.

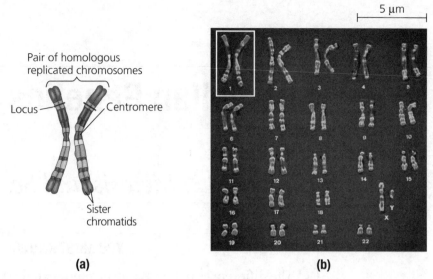

Figure 4.1 Metaphase chromosome (a) and human karyotype (b)

■ **Somatic cells** are any cells in the body that are not gametes. Each somatic cell in humans has 46 chromosomes. Liver cells and neurons are somatic cells.

■ The **karyotype** of an organism refers to a picture of its complete set of chromosomes, arranged in pairs of homologous chromosomes from the largest pair to the smallest pair. Figure 4.1b is a karyotype made from a human somatic cell. Notice that the 46 chromosomes are paired into 23 homologous chromosomes.

■ In **homologous chromosomes** both chromosomes of each pair carry genes that control the same inherited characteristics. If a gene for eye color is found at a specific locus on one chromosome, its homologs will have the same gene at the same locus.

  ▪ Homologous chromosomes are similar in length and centromere position, and they have the same staining pattern.
  ▪ One homologous chromosome from each pair is inherited from each parent; in other words, half of the set of 46 chromosomes in your somatic cells was inherited from your mother, and the other half was inherited from your father.

■ Exceptions to the rule that all chromosomes are part of a homologous pair may be found with the sex chromosomes—in humans, it is the **X** and **Y**. Human females have a homologous pair of chromosomes, XX, but males have one X chromosome and one Y chromosome. Nonsex chromosomes are called **autosomes**.

■ *What sex did the somatic cell come from that was used to make the karyotype in Figure 4.1?* (Check your answer in Part V.)

■ **Gametes**—meaning sperm and ova (eggs)—are haploid cells. Haploid cells contain half the number of chromosomes of somatic cells. In humans, gametes contain 22 autosomes plus a single sex chromosome (either X or Y), giving them a haploid number of 23. The haploid number of chromosomes is symbolized by *n*.

▌ **Meiosis** and **fertilization** are the key events in sexually reproducing life cycles. The human life cycle in Figure 4.2 is typical of a sexually reproducing animal.

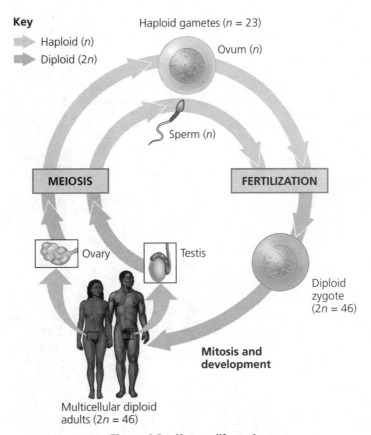

Figure 4.2   Human life cycle

▌ During **fertilization** (the combination of a sperm cell and an egg cell), one haploid gamete from the father fuses with one haploid gamete from the mother. The result is a fertilized egg called a **zygote**. It is **diploid** (has two sets of chromosomes) and may be symbolized by 2*n*.

▌ **Meiosis** is the type of cell division that reduces the numbers of sets of chromosomes from two to one. Fertilization restores the diploid number as the gametes are combined. Fertilization and meiosis alternate in the life cycles of sexually reproducing organisms.

Concept 13.3 *Meiosis reduces the number of chromosome sets from diploid to haploid*

▌ Meiosis and mitosis look similar—both are preceded by the replication of the cell's DNA, for instance, but in meiosis this replication is followed by *two* stages of cell division, meiosis I and meiosis II.

▌ The final result of meiosis is *four* **daughter cells**, each of which has *half* as many chromosomes as the parent cell.

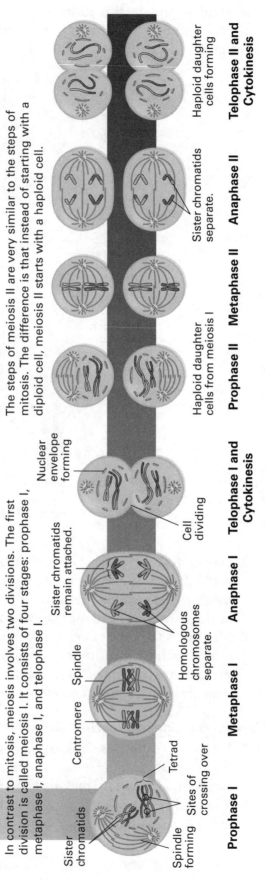

Centrosomes

Centrioles

Chromatin

Nuclear
envelope

Interphase

Just as in mitosis, the cell duplicates its
DNA. Each chromosome then consists
of two identical sister chromatids that
can be seen more clearly in prophase.

Meiosis I:

In contrast to mitosis, meiosis involves two divisions. The first
division is called meiosis I. It consists of four stages: prophase I,
metaphase I, anaphase I, and telophase I.

Sister chromatids remain attached.

Centromere

Spindle

Homologous
chromosomes
separate.

Nuclear
envelope
forming

Cell
dividing

Sister
chromatids

Spindle
forming

Sites of
crossing over

Tetrad

Prophase I   **Metaphase I**   **Anaphase I**   **Telophase I and
Cytokinesis**

Meiosis II:

The steps of meiosis II are very similar to the steps of
mitosis. The difference is that instead of starting with a
diploid cell, meiosis II starts with a haploid cell.

Haploid daughter
cells from meiosis I

Sister chromatids
separate.

Haploid daughter
cells forming

Prophase II   **Metaphase II**   **Anaphase II**   **Telophase II and
Cytokinesis**

Figure 4.3   Meiosis in an animal cell

Carefully follow the stages in Figure 4.3 as they are explained:

▌ **Interphase:** Each of the chromosomes makes a copy of itself; that is, each chromosome replicates its DNA, roughly doubling the amount of DNA in the cell. The centrosome also divides during this phase.

> **STUDY TIP** Understanding prophase I is critical to understanding meiosis. Study the unique events of prophase I carefully!

▌ **Meiosis I:** The first cellular division in meiosis is referred to as meiosis I.

■ **Prophase I:** The chromosomes condense, resulting in two sister chromatids attached at their centromeres.

■ **Synapsis** occurs—that is, the joining of homologous chromosomes along their length. This newly formed structure is called a *tetrad* and precisely aligns the homologous chromosomes gene by gene. This perfect alignment is necessary for the next step—crossing over.

■ In **crossing over** the DNA from one homolog is cut and exchanged with an exact portion of DNA from the other homolog. Essentially, a small part of the DNA from one parent is exchanged with the DNA from the other parent. *The result of crossing over is to increase genetic variation.* Where crossing over has occurred (two to three times per homologous pair), crisscrossed regions termed **chiasmata** form, which hold the homologs together until anaphase I.

■ After crossing over, the centrioles move away from each other, the nuclear envelope disintegrates, and spindle microtubules attach to the kinetochores forming on the chromosomes that begin to move to the metaphase plate of the cell.

> **ORGANIZE YOUR THOUGHTS**
>
> In Prophase I:
>
> 1. Synapsis occurs, forming tetrads.
> 2. Crossing over occurs between *homologous chromosomes* in the tetrads.
> 3. Crossing over increases genetic variation.
> 4. Areas of crossing over form chiasmata.
> 5. The nuclear envelope disintegrates, allowing the spindle to attach to the homologs.

■ **Metaphase I:** The homologous pairs of chromosomes are lined up at the metaphase plate, and microtubules from each pole attach to each member of the homologous pairs in preparation for pulling them to opposite ends of the cell. How many homologous pairs are found in the cell in Figure 4.3? (There are four chromosomes and two homologous pairs.)

- **Anaphase I:** The spindle apparatus helps to move the chromosomes toward opposite ends of the cell; sister chromatids stay connected and move together toward the poles.
- **Telophase I** and **cytokinesis:** The homologous chromosomes move until they reach the opposite poles. Each pole, then, contains a haploid set of chromosomes, with each chromosome still consisting of two sister chromatids.

    - Cytokinesis is the division of the cytoplasm and occurs during telophase. A **cleavage furrow** occurs in animal cells, and **cell plates** (the forming new cell wall) occur in plant cells. Both result in the formation of two haploid cells.

- **Meiosis II:** The second cellular division in meiosis is referred to as meiosis II.

    - **Prophase II:** A spindle apparatus forms, and sister chromatids move toward the metaphase plate.
    - **Metaphase II:** The chromosomes are lined up on the metaphase plate, and the kinetochores of each sister chromatid prepare to move to the opposite poles.
    - **Anaphase II:** The centromeres of the sister chromatids separate, and individual chromosomes move to opposite ends of the cell.
    - **Telophase II** and **cytokinesis:** The chromatids have moved all the way to opposite ends of the cell, nuclei reappear, and cytokinesis occurs. Each of the four daughter cells has the haploid number of chromosomes and is genetically different from the other daughter cells and from the parent cell.

> *STUDY TIP* Be prepared to cite the following three examples when asked to explain differences between mitosis and meiosis.

- Three events occur during meiosis I that do not occur during mitosis.

    1. Synapsis and crossing over normally do not occur during mitosis.
    2. At metaphase I, paired homologous chromosomes (tetrads) are positioned on the metaphase plate, rather than individual replicated chromosomes, as in mitosis.
    3. At anaphase I, duplicated chromosomes of each homologous pair separate but the sister chromatids of each duplicated chromosome stay attached. In mitosis, the chromatids separate.

> *TIP FROM THE READERS*
> When is chromosome number reduced? When does the cell go from diploid to haploid? Be sure you know and understand this! Refer again to Figure 4.3 and you should see that chromosome number is reduced during **meiosis I**. Each chromosome consists of two sister chromatids but the homologous pairs have separated.

Concept 13.4 Genetic variation produced in sexual life cycles contributes to evolution

> **STUDY TIP** There are three important processes that contribute to variation. They are given below. Be able to list and explain them.

▌ **Crossing over:** During prophase I the exchange of genetic material on homologous chromosomes between nonsister chromatids occurs. Use Figure 4.3 to help make this unique feature of meiosis clear. Notice that all four chromatids that make up the tetrad are different due to crossing over. In metaphase II when sister chromatids separate, each chromatid is unique, thus increasing variation.

▌ **Independent assortment of chromosomes:** In metaphase I, when the homologous chromosomes are lined up on the metaphase plate, they can pair up in any combination, with any of the homologous pairs facing either pole. This means that there is a 50% chance that a particular daughter cell will get a maternal chromosome or a paternal chromosome from the homologous pair.

▌ **Random fertilization:** Because each egg and sperm is different, as a result of independent assortment and crossing over, each combination of egg and sperm is unique.

Chapter 14: Mendel and the Gene Idea

> **YOU MUST KNOW**
>
> * Terms associated with genetics problems: P, F_1, F_2, dominant, recessive, homozygous, heterozygous, phenotype, and genotype.
> * How to derive the proper gametes when working a genetics problem.
> * The difference between an allele and a gene.
> * How to read a pedigree.
> * How to use data sets to determine Mendelian patterns of inheritance.

Concept 14.1 Mendel used the scientific approach to identify two laws of inheritance

▌ True-breeding parents in a genetic cross are called the **P (parental) generation**; their offspring are called the **F_1 (first filial) generation**. If the F_1 population is crossed, their offspring are called the **F_2 (second filial) generation**.

▌ The following are four related concepts that make up Mendel's model explaining the 3:1 inheritance pattern that he observed among F_2 offspring.

    1. **Alternative versions of genes cause variations in inherited characteristics among offspring.** For example, consider flower color in peas.

The gene for flower color in pea plants comes in two versions: white and purple. These alternative versions of the gene, called **alleles**, are the result of slightly different DNA sequences.

2. **For each character, every organism inherits one allele from each parent.**
3. **If the two alleles are different, then the *dominant allele* will be fully expressed in the offspring, whereas the *recessive allele* will have no noticeable effect on the offspring.**
4. **The two alleles for each character separate during gamete production.** If the parent has two of the same alleles, then the offspring will all get that version of the gene, but if the parent has two different alleles for a gene, each offspring has a 50% chance of getting one of the two alleles. This is Mendel's **law of segregation.**

> **STUDY TIP** Use Figure 4.4 to find each of the four basic concepts of Mendel's model.

▌ **Law of independent assortment** was Mendel's second law. It states that each pair of alleles will segregate (separate) independently during gamete formation.

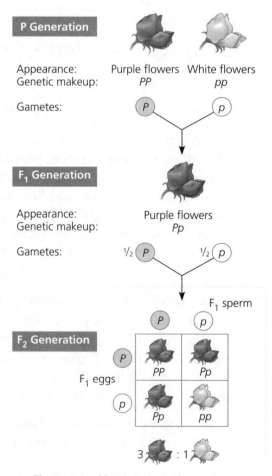

Figure 4.4   Mendel's law of segregation

- **Homozygous** organisms have two of the same alleles for a particular trait. If the dominant allele for a trait is designated as *R* (dominant traits are generally capitalized), and the recessive allele is designated *r* (recessive traits are generally not capitalized), then an individual could be homozygous for the dominant trait (*RR)* or homozygous for the recessive trait (*rr*).
- A **heterozygous** organism has two different alleles for a trait (*Rr*).
- **Phenotype** refers to an organism's expressed physical traits, and **genotype** refers to an organism's genetic makeup. For example, the phenotype of a seed might be round, and its genotype could be *RR* or *Rr*.
- A **testcross** is done to determine if an individual showing a dominant trait is homozygous or heterozygous. A homozygous dominant parent will yield all dominant phenotypes in the offspring (*RR* × *rr*), whereas a heterozygous parent will give a ratio of one dominant trait to one recessive trait.
- A **monohybrid cross** is a cross involving the study of only one character (e.g., flower color), whereas a **dihybrid cross** is a cross intended to study two characters (e.g., flower color and seed shape).
- The diagram below shows the results of a dihybrid cross. In this case, in the parental generation two homozygous plants are crossed: one homozygous dominant for light gray and round seeds (*YYRR*) and one homozygous recessive for dark gray and wrinkled seeds (*ppyy*). The only gamete type the first parent can produce is *YR*, and the only gamete the second parent can produce is *yr*. The F_1 generation, therefore, is composed of individuals with genotype *YyRr*. Crossing *YyRr* with a second *YyRr* gives an F_2 generation that completes the cross and looks like Figure 4.5.

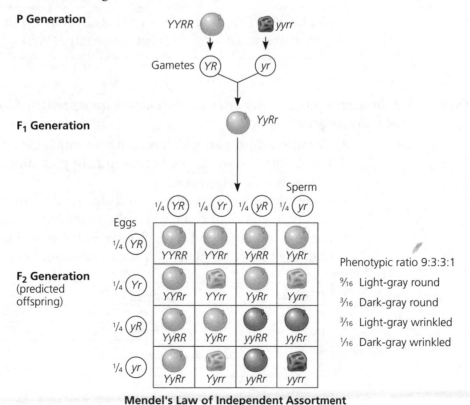

Mendel's Law of Independent Assortment

Figure 4.5   Mendel's law of independent assortment

Concept 14.2 The laws of probability govern Mendelian inheritance

❚ Understanding how to predict genetic crosses involves familiarity with the basic laws of probability. There are two laws that you will use directly in solving genetics problems:

- ▪ **The rule of multiplication:** When calculating the probability that two or more independent events will occur together in a specific combination, multiply the probabilities of each of the two events. Thus, the probability of a coin landing face up two times in two flips is $^1/_2 \times ^1/_2 = ^1/_4$. If you cross two organisms with the genotypes *AABbCc* and *AaBbCc*, the probability of an offspring having the genotype *AaBbcc* is $^1/_2 \times ^1/_2 \times ^1/_4 = ^1/_{16}$.
- ▪ **The rule of addition:** When calculating the probability that any of two or more mutually exclusive events will occur, you need to add together their individual probabilities. For example, if you are tossing a die, what is the probability that it will land on either the side with 4 spots or the side with 5 spots? ($^1/_6 + ^1/_6 = ^1/_3$)

Concept 14.3 Inheritance patterns are often more complex than predicted by simple Mendelian genetics

❚ **Complete dominance** is dominance in which the heterozygote and the homozygote for the dominant allele are indistinguishable. A *Yy* yellow seed is just as yellow as a *YY* yellow seed.

❚ **Codominance** occurs when two alleles are dominant and affect the phenotype in two different but equal ways. The traditional example for this type of dominance is human blood types. Notice in the chart below that in type AB blood both alleles are completely expressed and so are codominant.

Phenotype	Genotype	Antibody Expressed
A	I^AI^A or I^Ai	Anti-B
B	I^BI^B or I^Bi	Anti-A
AB	I^AI^B	None
O	ii	Anti-A, Anti-B

- **Incomplete dominance** is a type of dominance in which the F_1 hybrids have an appearance that is in between that of the two parents. For example, if two plants, one with white flowers and one with red flowers, were crossed and all of the offspring had pink flowers, you could conclude that the trait for flower color exhibits incomplete dominance. Breeding two of the hybrids with incomplete dominance gives a flower ratio of 1 red: 2 pink: 1 white.
- **Multiple alleles** occur when a gene has more than two alleles. Again, a good example of this is seen in human blood types. The chart shows the three alleles in human blood types: I^A, I^B, and i. Notice how the three alleles combine to form different blood types.
- **Pleiotropy** is the property of a gene that causes it to have multiple phenotypic effects. For example, sickle-cell disease has multiple symptoms all due to a single defective gene.
- In **epistasis**, a gene at one locus alters the effects of a gene at another locus. For example, an individual may have genes for heavy skin pigmentation, but if a separate gene that produces the pigment is defective, the genes for pigment deposition will not be expressed. This would lead to a condition known as albinism.
- In **polygenic inheritance**, two or more genes have an additive effect on a single character in the phenotype (such as height or skin color in humans). When several genes are involved, the phenotype usually is described by a bell-shaped curve, with fewer individuals at each extreme and most individuals clustered in the middle.

Concept 14.4 Many human traits follow Mendelian patterns of inheritance

- A **pedigree** is a diagram that shows the relationship between parents and offspring across two or more generations. (See Figure 4.6.) In a typical pedigree, circles represent females, and squares represent males. White open circles or squares indicate that the individual did not or does not express a particular trait, whereas the shaded ones indicate that the individual expresses or expressed that trait. Through the patterns they reveal, pedigrees can help determine the genome of individuals that comprise them; pedigrees can also help predict the genome of future offspring.

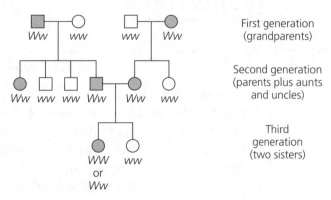

Figure 4.6   Human pedigree analysis

- **Recessively inherited disorders** require two copies of the defective gene for the disorder to be expressed. Examples include the following:

  - **Cystic fibrosis** is caused by a mutation in an allele that codes for a cell membrane protein that functions in the transport of chloride ions into and out of cells. The resulting high extracellular levels of chloride cause mucus to be thicker and stickier, leading to organ malfunction and recurrent bacterial infections.
  - **Tay-Sachs** disease is caused by an allele that codes for a dysfunctional enzyme, which is unable to break down certain lipids in the brain. As these lipids accumulate in the brain cells, the child suffers from blindness, seizures, and degeneration of brain function, leading to death.
  - **Sickle-cell disease** is caused by an allele that codes for a mutant hemoglobin molecule that forms long rods when the oxygen levels in the blood are low. These long rods cause the red blood cell to sickle, clogging small blood vessels and leading to pain, organ damage, and even paralysis.

- **Lethal dominant alleles** require only one copy of the allele in order for the disorder to be expressed. Usually, only late-acting lethal alleles are passed on.
- **Huntington's disease** is caused by a lethal dominant allele. It is a degenerative disease of the nervous system, which usually doesn't affect the individual until he or she is over 40 years old.
- Genetic testing may be used on a fetus to detect certain genetic disorders. Two common tests are amniocentesis and chorionic villus sampling (CVS).

  - **Amniocentesis** occurs when the physician removes amniotic fluid from around the fetus. The amniotic fluid can be utilized to detect some genetic disorders, and the cells in the fluid can be cultured for a karyotype.
  - **Chorionic villus sampling** involves using a narrow tube inserted through the cervix to suction out a tiny sample of the placenta that contains only fetal cells. A karyotype can immediately be developed from these cells.

Chapter 15: The Chromosomal Basis of Inheritance

YOU MUST KNOW

- How the chromosome theory of inheritance connects the physical movement of chromosomes in meiosis to Mendel's laws of inheritance.
- The unique pattern of inheritance in sex-linked genes.
- How alteration of chromosome number or structurally altered chromosomes (deletions, duplications, etc.) can cause genetic disorders.
- How genomic imprinting and inheritance of mitochondrial DNA are exceptions to standard Mendelian inheritance.

Concept 15.1 Mendelian inheritance has its physical basis in the behavior of chromosomes

▌ The **chromosome theory of inheritance** states that genes have specific locations (called *loci*) on chromosomes and that it is chromosomes that segregate and assort independently. It is important to connect this physical movement of chromosomes in meiosis to Mendel's laws of inheritance.

    ■ A **sex-linked gene** is one located on a sex chromosome (X or Y in humans). After the chromosome theory of inheritance was formed, *Thomas Hunt Morgan* discovered the existence of sex-linked genes.

Concept 15.2 Sex-linked genes exhibit unique patterns of inheritance

▌ In humans, there are two types of sex chromosomes, X and Y. Normal females have two X chromosomes, whereas normal males have one X and one Y chromosome.

▌ Sex-linked genes may be either X-linked or Y-linked. Figure 4.7 shows the unique pattern of inheritance in X-linked genes. In addition to tracking the gene from one generation to the next, it is also necessary to track the sex of the offspring. Figure 4.7 is based on work with *Drosophila* (fruit flies) performed by Thomas Hunt Morgan, but the pattern is the same in humans.

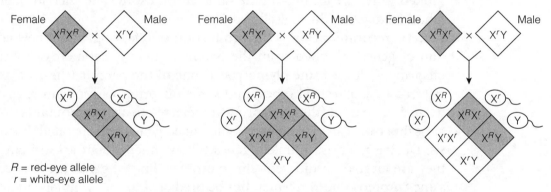

Figure 4.7   Patterns of inheritance with sex-linked traits

▌ Each egg or ovum contains an X chromosome; there are two types of sperm: those with an X chromosome and those with a Y chromosome. In fertilization, there is a 50% chance that a sperm carrying an X or Y will reach and penetrate the egg first. Thus, gender is determined by chance and by the male sperm cell in humans.

▌ Fathers pass X-linked genes on to their daughters but not to their sons; fathers pass the Y chromosome to their sons.

▌ Females will express an X-linked trait exactly like any other trait, but males, with only one X chromosome, will express the allele on the X chromosome they inherited from their mother. The terms *homozygous* and *heterozygous* do not apply to a male pattern of sex-linked genes.

▌ The vast majority of genes on the X chromosome are not related to sex.

- Several X-linked disorders have medical significance:

  - **Duchenne muscular dystrophy** is an X-linked disorder characterized by a progressive weakening of the muscles and loss of coordination. Affected individuals rarely live past their early 20s.
  - **Hemophilia** is an X-linked disorder characterized by having blood with an inability to clot normally, caused by the absence of proteins required for blood clotting.

- **X-inactivation** regulates gene dosage in females. Although female mammals inherit two X chromosomes, one of the X chromosomes (randomly chosen) in each cell of the body becomes inactivated during embryonic development by **methylation**. As a result, males and females have the same effective dose of genes with loci on the X chromosome.

- The inactive chromosome condenses into a **Barr body**, which lies along the inside of the nuclear envelope. Still, females are not affected as heterozygote carriers of problematic alleles, because half of their sex chromosomes are normal and produce the necessary proteins.

Concept 15.3 *Linked genes tend to be inherited together because they are located near each other on the same chromosome*

- **Linked genes** are located on the same chromosome and therefore tend to be inherited together during cell division.

- **Genetic recombination** is the production of offspring with a new combination of genes inherited from the parents. Many genetic crosses yield some offspring with the same phenotype as one of the parents (these offspring are referred to as **parental types**) and some offspring with phenotypes different from either parent (these offspring are referred to as **recombinants**).

- **Crossing over** can explain why some linked genes get separated during meiosis. During meiosis, unlinked genes follow independent assortment because they are located on different chromosomes. Linked genes are located on the same chromosome and would not be predicted to follow independent assortment. However, sometimes genetic crosses give results that seem to indicate some independent assortment has occurred, even when genes are on the same chromosome. These results are not due to independent assortment but can be explained by crossing over. Research further indicates that the farther apart two genes are on a chromosome, the higher the probability that crossing over will occur between them. The likelihood of crossing over between different genes on the same chromosome is expressed as a percent.

- A **linkage map** is a genetic map that is based on the percentage of crossover events.

- A **map unit** is equal to a 1% recombination frequency. Map units are used to express relative distances along the chromosome.

Concept 15.4 *Alterations of chromosome number or structure cause some genetic disorders*

▎ **Nondisjunction** occurs when the members of a pair of homologous chromosomes do not separate properly during meiosis I, or sister chromatids don't separate properly during meiosis II.

▎ As a result of nondisjunction, one gamete receives two copies of the chromosome, while the other gamete receives none. In the next step, if the faulty gametes engage in fertilization, the offspring will have an incorrect chromosome number. This is known as **aneuploidy**.

▎ Fertilized eggs that have received three copies of the chromosome in question are said to be **trisomic**; those that have received just one copy of a chromosome are said to be **monosomic** for the chromosome.

▎ **Polyploidy** is the condition of having more than two complete sets of chromosomes, forming a 3*n* or 4*n* individual. Rare in animals, this condition is fairly frequent in plants.

▎ Portions of a chromosome may also be lost or rearranged, resulting in the following mutations:

  ▪ A **deletion** occurs when a chromosomal fragment is lost, resulting in a chromosome with missing genes.

  ▪ A **duplication** occurs when a chromosomal segment is repeated.

  ▪ An **inversion** occurs when a chromosomal fragment breaks off and reattaches to its original position—but backward, so that the part of the fragment that was originally at the attachment point is now at the end of the chromosome.

  ▪ A **translocation** occurs when the deleted chromosome fragment joins a *nonhomologous* chromosome.

▎ Human disorders caused by chromosomal alterations include the following:

  ▪ **Down syndrome:** An aneuploid condition that is the result of having an extra chromosome 21 (trisomy 21). Down syndrome includes characteristic facial features, short stature, heart defects, and developmental delays.

  ▪ **Klinefelter syndrome:** An aneuploid condition in which a male possesses the sex chromosomes XXY (an extra X). Klinefelter males have male sex organs but are sterile.

  ▪ **Turner syndrome:** A monosomic condition in which the female has just one sex chromosome, an X. Turner syndrome females are sterile because the reproductive organs do not mature. Turner syndrome is the only known viable monosomy in humans.

Concept 15.5 *Some inheritance patterns are exceptions to standard Mendelian inheritance*

▎ In mammals, the phenotypic effect of a gene may depend on which allele is inherited from each parent, a phenomenon called **genomic imprinting**.

▎ Genes that are present in mitochondria and plastids are inherited only from the mother because the zygote's cytoplasm comes from the egg.

1. A couple has six children, all daughters. If the woman has a seventh child, what is the probability that the seventh child will be a daughter?
   (A) $^6/_7$
   (B) $^1/_7$
   (C) $^1/_{36}$
   (D) $^1/_{49}$
   (E) $^1/_2$

2. If alleles R and S are on two different chromosomes, and the probability of gamete R segregating into a gamete is $^1/_4$, while the probability of allele S segregating into a gamete is $^1/_2$, what is the probability that both will segregate into the same gamete?
   (A) $^1/_4 \times ^1/_2$
   (B) $^1/_4 \div ^1/_2$
   (C) $^1/_4 + ^1/_2$
   (D) $^1/_4 + ^1/_4$
   (E) $^1/_2$

3. In llamas, coat color is controlled by a gene that exists in two allelic forms. If a homozygous yellow llama is crossed with a homozygous brown llama, the offspring have gray coats. If two of the gray-coated offspring were crossed, what percentage of their offspring would have brown coats?
   (A) 100%
   (B) 75%
   (C) 50%
   (D) 25%
   (E) 0%

4. Which of the following is NOT true of meiosis?
   (A) During metaphase, spindle microtubules first come into contact with chromosomes.
   (B) The chromosome number in the newly formed cells is half that of the parent cell.
   (C) The homologous chromosomes line up along the metaphase plate, or equator of the cell.
   (D) The cytoplasm of the cell and all its organelles are divided approximately in half.
   (E) In anaphase II, the sister chromatids travel to opposite ends of the cell.

5. In rabbits, the trait for short hair (S) is dominant, and the trait for long hair (s) is recessive. The trait for green eyes (G) is dominant, and the trait for blue eyes (g) is recessive. A cross between two rabbits produces a litter of six short-haired rabbits with green eyes, and two short-haired rabbits with blue eyes. What is the most likely genotype of the parent rabbits in this cross?
   (A) *ssgg* $\times$ *ssgg*
   (B) *SSGG* $\times$ *SSGG*
   (C) *SsGg* $\times$ *SsGg*
   (D) *SsGg* $\times$ *SSGg*
   (E) *ssGG* $\times$ *ssGG*

6. In humans, hemophilia is an X-linked recessive trait. If a man and a woman have a son who is affected with hemophilia, which of the following is definitely *true*?
   (A) The mother carries an allele for hemophilia.
   (B) The father carries an allele for hemophilia.
   (C) The father is afflicted with hemophilia.
   (D) Both parents carry an allele for hemophilia.
   (E) The boy's paternal grandfather has hemophilia.

7. Which of the following explains a significantly low rate of crossing over between two genes?
   (A) They are located far apart on the same chromosome.
   (B) They are located on separate but homologous chromosomes.
   (C) The genes code for proteins that have similar functions.
   (D) The genes code for proteins that have very different functions.
   (E) The genes are located very close together on the same chromosome.

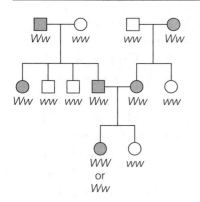

First generation (grandparents)

Second generation (parents plus aunts and uncles)

Third generation (two sisters)

8. In the pedigree above, circles represent females and squares represent males; those who express a particular trait are shaded, whereas those who do not are not shaded. Which pattern of inheritance best describes the pedigree for this trait?
   (A) X-linked recessive
   (B) X-linked dominant
   (C) autosomal recessive
   (D) autosomal dominant
   (E) codominant

Questions 9–10 refer to an individual with type O blood, whose mother has type A blood.

9. The father must have which of the following blood types?
   (A) A, B, or O
   (B) AB or A
   (C) AB or B
   (D) AB only
   (E) O only

10. If the type O individual were to mate with a person with type AB blood, which of the following is the best calculation of the ratio of the offspring?
    (A) $3\, I^A i{:}1\, I^B i$
    (B) $2\, I^A i{:}1\, I^B i$
    (C) $I^A i{:}I^B i$
    (D) $1\, I^A i{:}2\, I^A I^B{:}1\, I^B i$
    (E) $9\, I^A I^B{:}3\, I^A i{:}3\, I^B i{:}1\, O$

11. Two yellow mice with the genotype Yy are mated. After many offspring, $^2/_3$ are yellow and $^1/_3$ are not yellow (a 2:1 ratio). Mendelian genetics dictates that this cross should produce offspring that were $^1/_4\, YY$ (yellow), $^1/_2\, Yy$ (yellow), and $^1/_4\, yy$ (not yellow). What is the most likely conclusion from this experiment?
    (A) The mice did not bear enough offspring for the ratio calculation to be specific.
    (B) Y is lethal in the homozygous form and caused death early in development.
    (C) Nondisjunction occurred.
    (D) A mutation masked the effects of the Y allele.
    (E) A mutation masked the effects of the y allele.

12. All of the following contribute to genetic recombination EXCEPT
    (A) random fertilization.
    (B) independent assortment.
    (C) crossing over.
    (D) gene linkage.
    (E) random gene mutation.

13. In cucumbers, warty (W) is dominant over dull (w), and green (G) is dominant over orange (g). A cucumber plant that is homozygous for warty and green is crossed with one that is homozygous for dull and orange. The F_1 generation is then crossed. If a total of 144 offspring is produced in the F_2 generation, which of the following is the closest to the number of dull green cucumbers expected?
    (A) 3
    (B) 10
    (C) 28
    (D) 80
    (E) 111

14. The restoration of the diploid chromosome number after halving in meiosis is due to
    (A) synapsis.
    (B) fertilization.
    (C) mitosis.
    (D) DNA replication.
    (E) chiasmata.

15. During the first meiotic division (meiosis I),
    (A) homologous chromosomes separate.
    (B) the chromosome number becomes haploid.
    (C) crossing over between nonsister chromatids occurs.
    (D) paternal and maternal chromosomes assort randomly.
    (E) all of the above occur.

16. A cell with a diploid number of 6 could produce gametes with how many different combinations of maternal and paternal chromosomes?
    (A) 6
    (B) 8
    (C) 12
    (D) 64
    (E) 128

17. The DNA content of a diploid cell is measured in the G_1 phase. After meiosis I, the DNA content of one of the two cells produced would be
    (A) equal to that of the G_1 cell.
    (B) twice that of the G_1 cell.
    (C) one-half that of the G_1 cell.
    (D) one-fourth that of the G_1 cell.
    (E) impossible to estimate due to independent assortment of homologous chromosomes.

18. A synaptonemal complex would be found during
    (A) prophase I of meiosis.
    (B) fertilization or syngamy of gametes.
    (C) metaphase II of meiosis.
    (D) prophase of mitosis.
    (E) anaphase I of meiosis.

19. Meiosis II is similar to mitosis because
    (A) sister chromatids separate.
    (B) homologous chromosomes separate.
    (C) DNA replication precedes the division.
    (D) they both take the same amount of time.
    (E) haploid cells are produced.

20. Which of the following is NOT true of homologous chromosomes?
    (A) They behave independently in mitosis.
    (B) They synapse during the S phase of meiosis.
    (C) They travel together to the metaphase plate in prometaphase of meiosis I.
    (D) Each parent contributes one set of homologous chromosomes to an offspring.
    (E) Crossing over between nonsister chromatids of homologous chromosomes is indicated by the presence of chiasmata.

21. Given the following recombination frequencies, what is the correct order of the genes on the chromosome? A-B, 8 map units; A-C, 28 map units; A-D, 25 map units; B-C, 20 map units; B-D 33 map units
    (A) A-B-C-D
    (B) D-C-A-B
    (C) A-D-C-B
    (D) B-A-C-D
    (E) D-A-B-C

22. X-linked conditions are more common in men than in women because
    (A) men acquire two copies of the defective gene during fertilization.
    (B) men need to inherit only one copy of the recessive allele for the condition to be fully expressed.
    (C) women simply do not develop the disease regardless of their genetic composition.
    (D) the sex chromosomes are more active in men than in women.
    (E) the genes associated with the X-linked conditions are linked to the X chromosome, which determines maleness.

Level 2: Application/Analysis/Synthesis Questions

1. A black guinea pig crossed with an albino guinea pig produces 12 black offspring. When the albino is crossed with a second black one, 7 blacks and 5 albinos are obtained. What is the best explanation for this genetic situation? Write genotypes for the parents, gametes, and offspring.

2. In pea plants, pod color may be green (G) or yellow (g), while the pod shape may be inflated (I) or constricted (i). Two pea plants heterozygous for the characters of pod color and pod shape are crossed. Draw a Punnett square to determine the phenotypic ratios of the offspring.

3. In some plants, a true-breeding, red-flowered strain gives all pink flowers when crossed with a white-flowered strain: $C^R C^R$ (red) $\times$ $C^W C^W$ (white) $\rightarrow$ $C^R C^W$ (pink). The placement of the flower can be determined by the dominant allele for an axial flower (A), while a terminal flower is determined by the recessive allele (a). What will be the phenotypic ratio of the F_1 generation resulting from the following cross: axial-red (both genes are homozygous) $\times$ terminal white? What will the phenotypic ratio be in the F_2?

4. If these four cells resulted from cell division of a single cell with diploid chromosome number $2n = 4$, what best describes what just occurred?

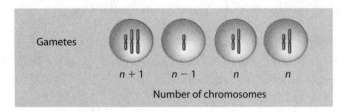

Gametes

$n + 1$ $n - 1$ n n

Number of chromosomes

(A) normal meiosis
(B) translocation
(C) inversion
(D) nondisjunction

After reading the paragraph, answer the question(s) that follow.

A woman has been trying to conceive for several years, unsuccessfully. At a fertility clinic, they discover that she has blocked fallopian tubes. Using modern technologies, some of her eggs are removed, fertilized with her husband's sperm, and implanted into her uterus. The procedure is successful, but the couple discovers that their new son is color-blind and has type O blood. The woman claims that the child can't be theirs since she has type A blood and her husband has type B. Also, neither parent is color-blind, although one grandparent (the woman's father) is also color-blind.

5. As a genetic counselor, you would explain to the parents that
(A) the eggs must have been accidentally switched, since the baby's blood type has to match one of his parents.
(B) each parent could have contributed one recessive allele, resulting in type O blood.
(C) the eggs must have been accidentally switched, since a type A parent and a type B parent can have any type children except O.
(D) it is possible for the baby to have type O blood, since type O is inherited through a dominant allele.

6. In regard to the baby's color blindness, a sex-linked recessive trait, you explain that
(A) color blindness often appears randomly, even if neither parent is color-blind.
(B) the baby's father must have a recessive allele for color blindness.
(C) since color blindness is sex-linked, a son can inherit color blindness if his mother has the recessive color blindness allele.
(D) the eggs must have been accidentally switched, since males inherit sex-linked traits only from their fathers.

7. Independent orientation of chromosomes at metaphase I results in an increase in the number of
(A) gametes.
(B) homologous chromosomes.
(C) possible combinations of characteristics.
(D) sex chromosomes.

Molecular Genetics

Chapter 16: The Molecular Basis of Inheritance

> **YOU MUST KNOW**
>
> - The structure of DNA.
> - The knowledge about DNA gained from the work of Watson, Crick, Wilkins, and Franklin; Avery, MacLeod, and McCarty; and Hershey and Chase.
> - The major steps of replication.
> - The difference between replication, transcription, and translation.
> - The general differences between the bacterial chromosomes and eukaryotic chromosomes.
> - How DNA packaging can affect gene expression.

Concept 16.1 DNA is the genetic material

▌ Once chromosomes were known to carry genes, the next question became which of the two organic compounds that make chromosomes, DNA or protein, was the genetic material?

▪ In 1952 **Alfred Hershey and Martha Chase** answered this question utilizing *bacteriophages*—viruses that infect bacteria. Bacteriophages were excellent organisms for this study, in part because they are made of only two organic compounds, DNA and protein. Hershey and Chase used a radioactive isotope of phosphorus to tag the DNA in one culture of bacteriophages and radioactive sulfur to tag the protein in a second culture. Their results clearly showed that only the DNA entered bacteria infected by the virus; the radioactive protein never entered the cell. This research convinced scientists that DNA must be the genetic material.

▌ The next big question centered on the structure of DNA. Would the structure of DNA give any clues as to how it functioned as the genetic material?

▪ **James Watson and Francis Crick** were the first to solve the puzzle of the structure of DNA. Critical to their success was the work of Rosalind Franklin and Maurice Wilkins, both working in the field of X-ray crystallography.

- **X-ray crystallography** is a process used to visualize molecules three-dimensionally. X-rays are diffracted as they pass through the molecule, and they bounce back to produce patterns that can be interpreted through mathematical equations. Through this technique, a rough blueprint of the molecule was formed.
- Watson and Crick's model determined four major features of DNA. Find each major point by following Figure 5.1 as the model is explained.

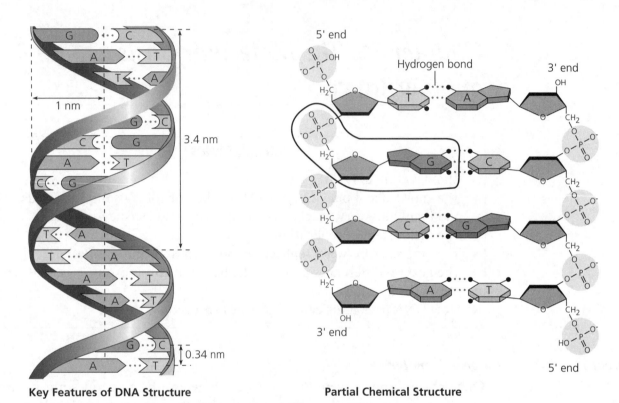

Key Features of DNA Structure   Partial Chemical Structure

Figure 5.1   Structure of DNA

- DNA is a **double helix**, which can be described as a twisted ladder with rigid rungs. The side, or backbone, is made up of sugar-phosphate components, whereas the rungs are made up of pairs of nitrogenous bases.
- Notice that a single nucleotide is circled in Figure 5.1. It is composed of a sugar (deoxyribose) attached to a phosphate and a nitrogen base.
- The nitrogenous bases of DNA are adenine (A), thymine (T), guanine (G), and cytosine (C). In DNA, adenine pairs only with thymine, and guanine pairs only with cytosine.
- Notice that the chain on the right side of the model runs in one direction, while the left side of the chain runs in the opposite, upside-down direction. The strands are termed **antiparallel**. The left side runs 5′ to 3′ while the opposite strand runs 3′ to 5′. (Recall that the carbons are numbered, and you will see that the number 5 carbon and number 3 carbon and the resultant nucleotides are flipped relative to each other.) Nucleic acid strands are always antiparallel, whether they are DNA/DNA or DNA/RNA or RNA/RNA interactions.

Concept 16.2 Many proteins work together in DNA replication and repair

▌ **Replication** is the making of DNA from an existing DNA strand. DNA replication is *semiconservative*. This means that at the end of replication, each of the daughter molecules has one old strand, derived from the parent strand of DNA, and one strand that is newly synthesized. Study Figure 5.2 to see the pattern of semiconservative replication.

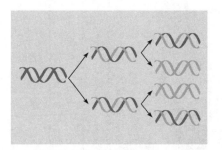

Figure 5.2   Semiconservative replication

▌ The replication of DNA includes six major points:

1. The replication of DNA begins at sites called the *origins of replication*.
2. Initiation proteins bind to the origin of replication and separate the two strands, forming a *replication bubble*. DNA replication then proceeds in both directions along the DNA strand until the molecule is copied.
3. A group of enzymes called **DNA polymerases** catalyzes the elongation of new DNA at the replication fork.
4. DNA polymerase adds nucleotides to the growing chain one by one, working in a 5′ to 3′ direction, matching adenine with thymine and guanine with cytosine.
5. Recall that the strands of DNA are antiparallel. This means that DNA replication occurs continuously along the 5′ to 3′ strand, which is called the **leading strand**. The strand that runs 3′ to 5′ is copied in series of segments and termed the **lagging strand**. Read steps 1–3 in Figure 5.3 to visualize this process.
6. The lagging strand is synthesized in separate pieces called **Okazaki fragments**, which are then sealed together by **DNA ligase** (step 4, Figure 5.3), forming a continuous DNA strand.

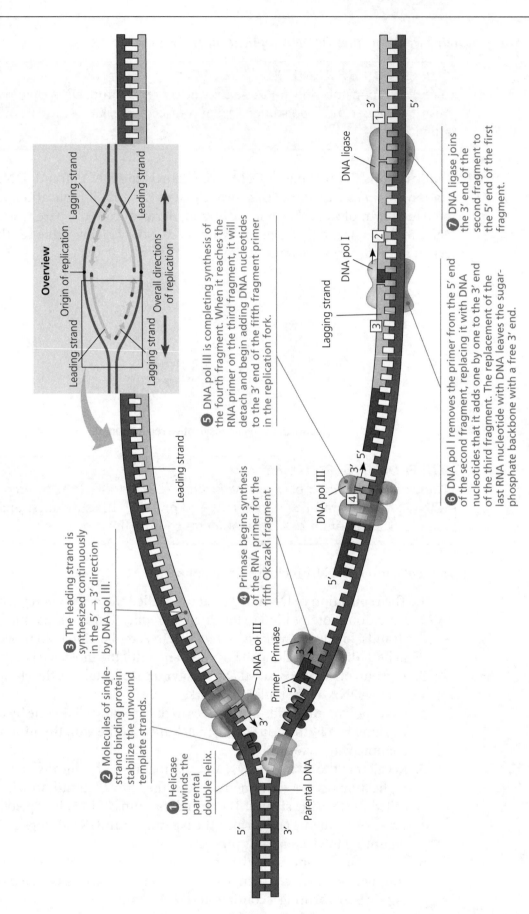

Overview

Origin of replication

Lagging strand   Lagging strand

Leading strand   Leading strand

Leading strand

Lagging strand

Overall directions of replication

1 Helicase unwinds the parental double helix.

Parental DNA

2 Molecules of single-strand binding protein stabilize the unwound template strands.

3 The leading strand is synthesized continuously in the 5' → 3' direction by DNA pol III.

Leading strand

DNA pol III

Primer

Primase

3'

5'

3'

4 Primase begins synthesis of the RNA primer for the fifth Okazaki fragment.

DNA pol III

4

3' 5'

5'

5 DNA pol III is completing synthesis of the fourth fragment. When it reaches the RNA primer on the third fragment, it will detach and begin adding DNA nucleotides to the 3' end of the fifth fragment primer in the replication fork.

Lagging strand

DNA pol I

2

3

6 DNA pol I removes the primer from the 5' end of the second fragment, replacing it with DNA nucleotides that it adds one by one to the 3' end of the third fragment. The replacement of the last RNA nucleotide with DNA leaves the sugar-phosphate backbone with a free 3' end.

DNA ligase

1

3'

5'

7 DNA ligase joins the 3' end of the second fragment to the 5' end of the first fragment.

Figure 5.3 DNA replication

5'

3'

- There are several different factors contributing to the accuracy of DNA replication:
  - The specificity of base pairing (A = T, G = C)
  - **Mismatch repair**, in which special repair enzymes fix incorrectly paired nucleotides
  - **Nucleotide excision repair**, in which incorrectly placed nucleotides are excised or removed by enzymes termed **nucleases**, and the gap left over is filled in with the correct nucleotides
- The fact that DNA polymerase can add nucleotides only to the 3′ end of a molecule means that it would have no way to complete the 5′ end of the DNA molecule at the end of the chromosome. Every time the chromosome is replicated for mitosis, a small portion of the tip of the chromosome is removed. To avoid losing the terminal genes, the linear ends of eukaryotic chromosomes are "capped" with **telomeres**, short, repetitive nucleotide sequences that do not contain genes.

Concept 16.3 *A chromosome consists of a DNA molecule packed together with proteins*

- A bacterial chromosome is one double-stranded, circular DNA molecule associated with a small amount of protein.
- Eukaryotic chromosomes are linear DNA molecules associated with large amounts of protein.
- In eukaryotic cells, DNA and proteins are packed together as **chromatin**. Eukaryotic DNA shows four levels of packaging. Visualize each level of packaging by studying Figure 5.4 (see next page) as you read the information provided there.
- As DNA becomes more highly packaged, it becomes less accessible to transcription enzymes. This reduces gene expression. In interphase cells, most chromatin is in the highly extended form (**euchromatin**) and is available for transcription, but some remains more condensed (**heterochromatin**). Heterochromatin is largely inaccessible to transcription enzymes and, thus, generally is not transcribed. Barr bodies are heterochromatin.

Chapter 17: From Gene to Protein

YOU MUST KNOW

- The key terms *gene expression*, *transcription*, and *translation*.
- The major events of transcription.
- How eukaryotic cells modify RNA after transcription.
- The steps to translation.
- How mutations can change the amino acid sequence of a protein.

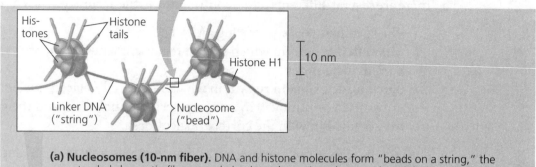

(a) Nucleosomes (10-nm fiber). DNA and histone molecules form "beads on a string," the extended chromatin fiber seen during interphase. A nucleosome has eight histone molecules with the amino end (tail) of each projecting outward. A different type of histone, H1, can bind to DNA next to a nucleosome, where it helps to further compact the 10-nm fiber.

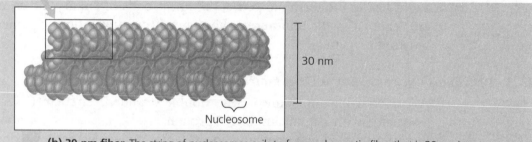

(b) 30-nm fiber. The string of nucleosomes coils to form a chromatin fiber that is 30 nm in diameter (tails not shown). This form is also seen during interphase.

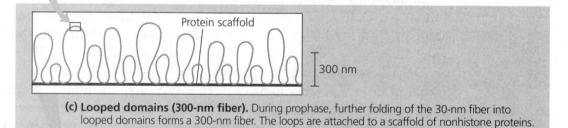

(c) Looped domains (300-nm fiber). During prophase, further folding of the 30-nm fiber into looped domains forms a 300-nm fiber. The loops are attached to a scaffold of nonhistone proteins.

(d) Metaphase chromosome. The chromatin folds further, resulting in the maximally compacted chromosome seen at metaphase. Each metaphase chromosome consists of two chromatids.

Figure 5.4   Levels of chromatin packing

Concept 17.1 Genes specify proteins via transcription and translation

- **Gene expression** is the process by which DNA directs the synthesis of proteins (or, in some cases, RNAs).
- The **one gene–one polypeptide hypothesis** states that each gene codes for a polypeptide, which can be—or can constitute a part of—a protein.
- **Transcription** is the synthesis of RNA using DNA as a template. It takes place in the nucleus of eukaryotic cells.
- **Messenger RNA**, or **mRNA**, is produced during transcription. It carries the genetic message of DNA to the protein-making machinery of the cell in the cytoplasm, the *ribosome*.
- In eukaryotes, transcription results in pre-mRNA, which undergoes **RNA processing** to yield the final mRNA.
- In prokaryotes, transcription results directly in mRNA, which is not processed. Transcription and translation can occur simultaneously.
- **Translation** is the production of a polypeptide chain using the mRNA transcript and occurs at the ribosomes.
- The instructions for building a polypeptide chain are written as a series of three-nucleotide groups; this is called a *triplet code*.
- During transcription, only one strand of the DNA is transcribed, and it is called the **template strand**. The mRNA that is produced is said to be *complementary* to the original DNA strand. The mRNA base triplets are called **codons**. They are written in the 5′ to 3′ direction.
- The genetic code is *redundant*, meaning that more than one codon codes for each of the 20 amino acids. The codons are read based on a consistent reading frame—the groups of three must be read in the correct groupings in order for translation to be successful.

Concept 17.2 Transcription is the DNA-directed synthesis of RNA: a closer look

- **RNA polymerase** is an enzyme that separates the two DNA strands and connects the RNA nucleotides as they base-pair along the DNA template strand.
- The RNA polymerases can add RNA nucleotides only to the 3′ end of the strand, so RNA elongates in the 5′ to 3′ direction. As RNA nucleotides are added, remember that uracil replaces thymine when base pairing to adenine.
- The DNA sequence at which RNA polymerase attaches is called the **promoter**, whereas the DNA sequence that signals the end of transcription is called the **terminator**.
- A **transcription unit** is the entire stretch of DNA that is transcribed into an RNA molecule. A transcription unit may code for a polypeptide or an RNA, like transfer RNA or ribosomal RNA.

There are three main stages of transcription:

1. **Initiation:** In bacteria, RNA polymerase recognizes and binds to the promoter. In eukaryotes, RNA polymerase II, the specific RNA polymerase that transcribes mRNA, cannot bind to the promoter without supporting help from proteins known as transcription factors. **Transcription factors** assist the binding of RNA polymerase to the promoter and, thus, the initiation of transcription. The whole complex of RNA polymerase II and transcription factors is called a **transcription initiation complex** (see Figure 5.5).

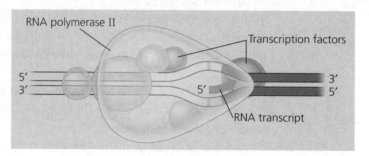

Figure 5.5   Transcription initiation complex

2. **Elongation:** RNA polymerase moves along the DNA, continuing to untwist the double helix. RNA nucleotides are continually added to the 3′ end of the growing chain. As the complex moves down the DNA strand, the double helix re-forms, with the new RNA molecule straggling away from the DNA template. Find these key steps in Figure 5.6.

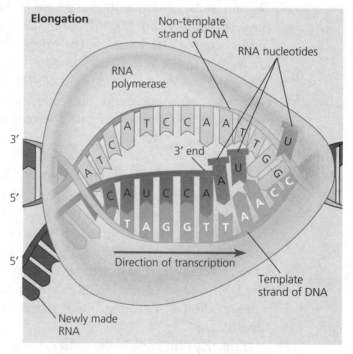

Figure 5.6   Elongation stage of transcription

3. **Termination:** After RNA polymerase transcribes a terminator sequence in the DNA, the RNA transcript is released, and the polymerase detaches.

Concept 17.3 Eukaryotic cells modify RNA after transcription

▌In eukaryotes, there are a couple of key post-transcriptional modifications to RNA—the addition of a **5′ cap** and the addition of a **poly-A tail**.

▌The 5′ cap and the poly-A tail facilitate the export of mRNA from the nucleus, help protect the mRNA from degradation by enzymes, and facilitate the attachment of the mRNA to the ribosome.

▌**RNA splicing** also takes place in eukaryotic cells. In RNA splicing, large portions of the newly synthesized RNA strand are removed. The sections of the mRNA that are spliced out are called **introns**, and the sections that remain—and subsequently spliced together by a *spliceosome*—are called **exons**. Use Figure 5.7 to help you visualize exons and introns.

▌One amazing thing about how spliceosomes work is the role of a special kind of RNA, termed **small nuclear RNA** (snRNA). snRNA plays a major role in catalyzing the excision of the introns and joining of exons. When RNA serves a catalytic role, the molecule is termed a **ribozyme**. For many years it was thought that only proteins could be catalytic, but the discovery of ribozymes totally changed that idea!

▌Another rethinking that has taken place came with the realization that we have only about 20,000 genes to make approximately 100,000 polypeptides.

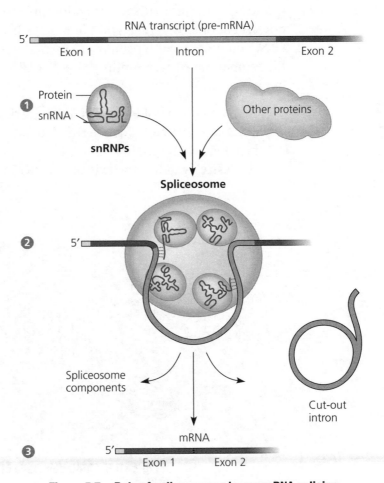

Figure 5.7   Role of spliceosomes in pre-mRNA splicing

One gene can often make more than one polypeptide. An intron removed in the production of one polypeptide can be an exon in a second polypeptide made from the same gene! Alternative RNA splicing allows for different combinations of exons, resulting in more than one polypeptide per gene.

Concept 17.4 Translation is the RNA-directed synthesis of a polypeptide: a closer look

- In addition to mRNA, two additional types of RNA play important roles in translation: transfer RNA (tRNA) and ribosomal RNA (rRNA).
- **tRNA** functions in transferring amino acids from a pool of amino acids in the cell's cytoplasm to a ribosome. The ribosome accepts the amino acid from tRNA and incorporates the amino acid into a growing polypeptide chain.
- Each type of tRNA is specific for a particular amino acid; at one end it loosely binds the amino acid, and at the other end it has a nucleotide triplet called an **anticodon**, which allows it to pair specifically with a complementary codon on the mRNA.
- A **codon** is an mRNA triplet. Since there are four different nucleotides (A, T, C, and G), taking them three at a time results in 64 different codons.
- The mRNA is read codon by codon, and one amino acid is added to the chain for each codon read.
- The rules for base pairing between the third base of a codon and the corresponding base of a tRNA anticodon are not as strict as those for DNA and mRNA codons. This relaxation of base-pairing rules is called **wobble**.
- **rRNA** complexes with proteins to form the two subunits that form ribosomes. Ribosomes have three binding sites for tRNA (locate each tRNA binding site in Figure 5.8).

  - A **P site**, which holds the tRNA that carries the growing polypeptide chain.
  - An **A site**, which holds the tRNA that carries the amino acid that will be added to the chain next.
  - An **E site**, which is the exit site for tRNA.

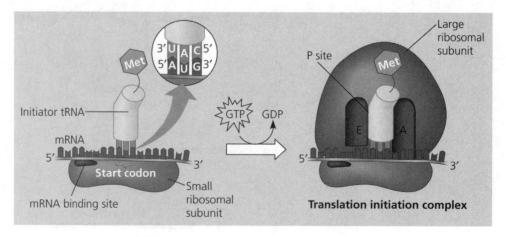

Figure 5.8   Initiation stage of translation

Translation, like transcription, can be divided into three stages:

1. **Initiation:** Organize initiation into these three steps. Use Figure 5.8 to find each step.

    A. A small ribosomal subunit binds to mRNA in such a way that the first codon of the mRNA strand, which is always AUG, is placed in the proper position.
    B. tRNA with anticodon UAC, which carries the amino acid methionine, hydrogen bonds to the first codon (initiation factors are proteins that assist in holding all this together).
    C. Large subunit of ribosome attaches, allowing the tRNA with methionine to attach to the P site. Notice that the A site is now available to the tRNA that will bring the second amino acid.

2. **Elongation:** Elongation also has three steps. Use Figure 5.9 to follow each of the three steps of elongation.

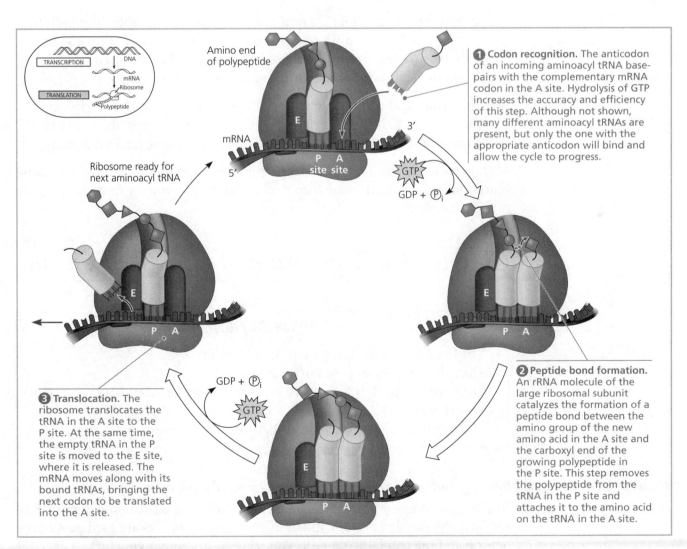

Figure 5.9   Protein synthesis

3. **Termination:** A stop codon in the mRNA is reached and translation stops. A protein called release factor binds to the stop codon, and the polypeptide is freed from the ribosome.

▌ Polypeptides then fold to assume their specific conformation, and they are sometimes modified further to render them functional. The destination of a protein is often determined by the sequence of about 20 amino acids at the leading end of the polypeptide chain. The **signal peptide**, the sequence of the leading 20 or so amino acids, serves as a sort of cellular zip code, directing proteins to their final destination.

Concept 17.5 Mutations of one or a few nucleotides can affect protein structure and function

▌ Mutations are alterations in the genetic material of the cell; **point mutations** are alterations of just one base pair of a gene. They come in two basic types:

  ▪ A **nucleotide-pair substitution** is the replacement of one nucleotide and its partner with another pair of nucleotides.

    ▪ **Missense mutations** are those substitutions that enable the codon to still code for an amino acid, although it might not be the correct one.
    ▪ **Nonsense mutations** are those substitutions that change a regular amino acid codon into a stop codon, ceasing translation.

  ▪ **Insertions** and **deletions** refer to the additions and losses of nucleotide pairs in a gene. If they interfere with the codon groupings, they can cause a **frameshift mutation**, which causes the mRNA to be read incorrectly.

▌ **Mutagens** are substances or forces that interact with DNA in ways that cause mutations. X-rays and other forms of radiation are known mutagens, as are certain chemicals.

Chapter 18: Regulation of Gene Expression

YOU MUST KNOW

- The functions of the three parts of an operon.
- The role of repressor genes in operons.
- The impact of DNA methylation and histone acetylation on gene expression.
- The role of gene regulation in embryonic development and cancer.

Concept 18.1 Bacteria often respond to environmental change by regulating transcription

▌ In bacteria, genes are often clustered into units called *operons*. Figure 5.10 shows a repressible operon with an inactive repressor. Locate each part of the operon and the regulatory gene as you read the accompanying text.

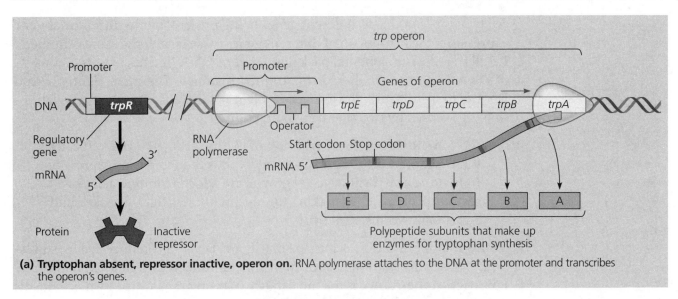

(a) **Tryptophan absent, repressor inactive, operon on.** RNA polymerase attaches to the DNA at the promoter and transcribes the operon's genes.

Figure 5.10   A repressible operon

- An **operon** consists of three parts:

    - An **operator** that controls the access of RNA polymerase to the genes. The operator is found within the promoter site or between the promoter and the protein coding genes of the operon.
    - The **promoter**, which is where RNA polymerase attaches.
    - The **genes of the operon**. This is the entire stretch of DNA required for all the enzymes produced by the operon.

- Located some distance from the operon is a regulatory gene. **Regulatory genes** produce repressor proteins that may bind to the operator site. When a regulatory protein occupies the operator site, RNA polymerase is blocked from the genes of the operon. In this situation the operon is off.

- A **repressible operon** is normally on but can be inhibited. This type of operon is normally anabolic, building an essential organic molecule. The repressor protein produced by the regulatory gene is inactive. If the organic molecule being produced by the operon is provided to the cell, the molecule can act as a **corepressor** and bind to the repressor protein, activating it. The activated repressor protein binds to the operator site, shutting down the operon. This is the type of operon shown in Figure 5.10. The *lac* operon is inducible.

- An **inducible operon** is normally off but can be activated. This type of operon is normally catabolic, breaking down food molecules for energy. The repressor protein produced by the regulatory gene is active. To turn an inducible operon on, a specific small molecule, called an **inducer**, binds to and inactivates the repressor protein. With the repressor out of the operator site, RNA polymerase can access the genes of the operon.

Concept 18.2 Eukaryotic gene expression is regulated at many stages

- The expression of eukaryotic genes can be turned off and on at any point along the pathway from gene to functional protein. Further, the differences between cell types are not due to different genes being present, but to **differential gene expression**, the expression of different genes by cells with the same genome.

▌ Recall that the fundamental packaging unit of DNA, the nucleosome, consists of DNA bound to small proteins termed histones. The more tightly bound DNA is to its histones, the less accessible it is for transcription. This relationship is governed by two chemical interactions:

- ■ **DNA methylation** is the addition of methyl groups to DNA. It causes the DNA to be more tightly packaged, thus reducing gene expression.
- ■ In **histone acetylation**, acetyl groups are added to amino acids of histone proteins, thus making the chromatin less tightly packed and encouraging transcription.

▌ Notice that methylation occurs primarily on DNA and reduces gene expression, whereas acetylation occurs on histones and increases gene expression.

▌ **Epigenetic inheritance** is the inheritance of traits transmitted by mechanisms not directly involving the nucleotide sequence. The DNA sequence is not changed, just its expression.

▌ *Transcription initiation* is another important control point in gene expression. At this stage, DNA control elements that bind transcription factors are involved in regulation.

▌ The **transcription initiation complex** greatly enhances gene expression. Study Figure 5.11 (see next page) and note its essential elements. DNA sequences far from the gene, termed **enhancer regions**, are bound to the promoter region by proteins termed **activators**.

▌ The control of gene expression may also occur after transcription and just after translation, when proteins are processed.

▌ Coordinately controlled genes, such as the genes coding for the enzymes of a metabolic pathway, are expressed together. This is possible even though the genes in a given pathway may be scattered on different chromosomes. All of the genes that code for the enzymes of the pathway share the same control elements. In general, eukaryotes do not have operons.

Concept 18.3 Noncoding RNAs play multiple roles in controlling gene expression

▌ Small molecules of single-stranded RNA can complex with proteins and influence gene expression. Two types of RNA, *micro RNAs (miRNA)* and *small interfering RNAs (siRNAs)*, can bind to mRNA and degrade the mRNA or bind to mRNA and block its translation.

Concept 18.4 A program of differential gene expression leads to the different cell types in a multicellular organism

▌ The zygote undergoes transformation through three interrelated processes:

1. **Cell division** is the series of mitotic divisions that increases the number of cells.
2. **Cell differentiation** is the process by which cells become specialized in structure and function.
3. **Morphogenesis** gives an organism its shape.

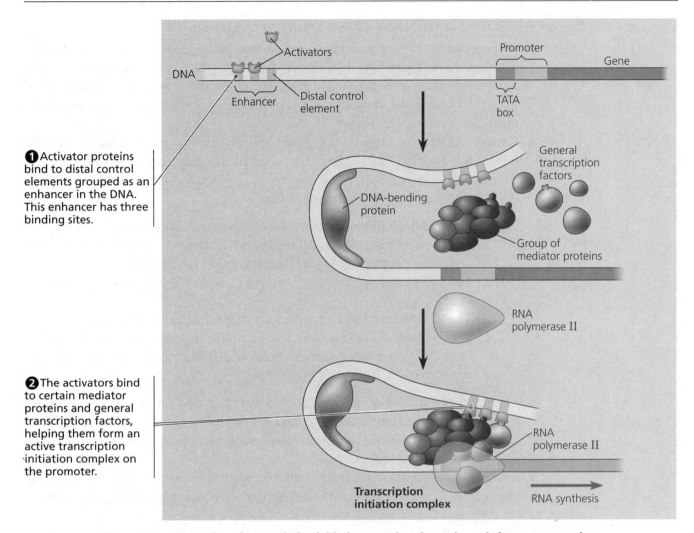

① Activator proteins bind to distal control elements grouped as an enhancer in the DNA. This enhancer has three binding sites.

② The activators bind to certain mediator proteins and general transcription factors, helping them form an active transcription initiation complex on the promoter.

Figure 5.11   **Formation of transcription initiation complex plays a key role in gene expression.**

❚ What controls differentiation and morphogenesis?

1. **Cytoplasmic determinants** are maternal substances in the egg that influence the course of early development. These are distributed unevenly in the early cells of the embryo and result in different effects.
2. **Cell-cell signals** result from molecules, such as growth factors, produced by one cell influencing neighboring cells, a process called **induction**, which causes cells to differentiate.

❚ **Determination** is the series of events that lead to observable differentiation of a cell. Differentiation is caused by cell-cell signals and is irreversible.

❚ **Pattern formation** sets up the body plan and is a result of cytoplasmic determinants and inductive signals. This is what determines head and tail, left and right, back and front. Uneven distribution of substances called **morphogens** plays a role in establishing these axes.

❚ **Homeotic genes** are master control genes that control pattern formation.

Concept 18.5 Cancer results from genetic changes that affect cell cycle control

▮ **Oncogenes** are cancer-causing genes; **proto-oncogenes** are genes that code for proteins that are responsible for normal cell growth. Proto-oncogenes become oncogenes when a mutation occurs that causes an increase in the product of the proto-oncogene, or an increase in the activity of each protein molecule produced by the gene.

▮ **Cancer** can also be caused by a mutation in a gene whose products normally inhibit cell division. These genes are called **tumor-suppressor genes**.

▮ An important tumor-suppressor gene is the *p53 gene*. The product of this gene is a protein that suppresses cancer in four ways:

1. The p53 protein can activate the *p21* gene, whose product halts the cell cycle by binding to cyclin-dependent kinases. This allows time for DNA to be repaired before the resumption of cell division.
2. The p53 protein activates a group of miRNAs, which inhibit the cell cycle.
3. The p53 protein turns on genes directly involved in DNA repair.
4. When DNA damage is too great to repair, the p53 protein activates "suicide" genes whose products cause cell death, a process termed **apoptosis**.

▮ The multistep model of cancer development is based on the idea that cancer results from the accumulation of mutations that occur throughout life. The longer we live, the more mutations that are accumulated and the more likely that cancer might develop.

▮ *Embryonic development* represents what happens when gene regulation proceeds correctly and *cancer* shows what can happen when gene regulation goes awry.

Chapter 19: Viruses

YOU MUST KNOW

- The components of a virus.
- The differences between lytic and lysogenic cycles.

Concept 19.1 A virus consists of a nucleic acid surrounded by a protein coat

▮ Smaller than ribosomes, the tiniest viruses are about 20 nm across.

▮ Genomes can be double- or single-stranded DNA or double- or single-stranded RNA.

▮ The **capsid** is a protein shell that surrounds the genetic material.

▮ Some viruses also have **viral envelopes** that surround the capsid and aid the viruses in infecting their hosts. (See Figure 5.12.)

▮ **Bacteriophages**, or **phages**, are viruses that infect bacterial cells.

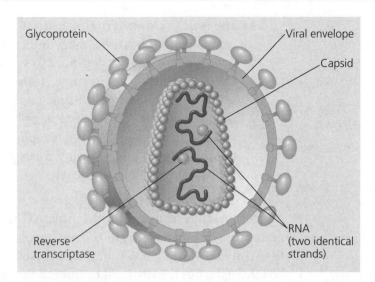

Figure 5.12 Structure of HIV. Note the components of a typical virus: nucleic acid, capsid. This virus also has an envelope, and reverse transcriptase.

Concept 19.2 *Viruses replicate only in host cells*

▎ Viruses have a limited **host range**. This means they can infect only a very limited variety of hosts. *Example*: Human cold virus infects only cells of the upper respiratory tract.

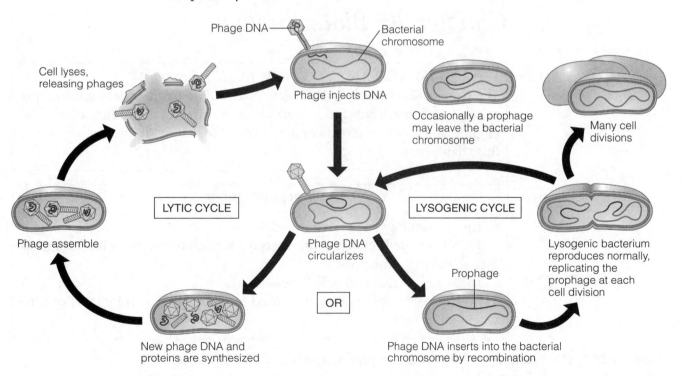

Figure 5.13 Infection of a bacterial cell by a bacteriophage. Note the two cycles.

▎ Viral reproduction occurs only in host cells. Two variations have been studied in bacteriophages. Read the following, then study Figure 5.13 above.

1. The **lytic cycle** ends in the death of the host cell by rupturing it (lysis). In this cycle, a bacteriophage injects its DNA into a host cell and takes over the host cell's machinery to synthesize new copies of the viral DNA as well as protein coats. These self-assemble, and the bacterial cell is lysed, releasing multiple copies of the virus.

2. In the **lysogenic cycle** the bacteriophage's DNA becomes incorporated into the host cell's DNA and is replicated along with the host cell's genome. The viral DNA is known as a **prophage**. Under certain conditions, the prophage will enter the lytic cycle, described on the previous page.

▮ **Retroviruses** are RNA viruses that use the enzyme **reverse transcriptase** to transcribe DNA from an RNA template. The new DNA then permanently integrates into a chromosome in the nucleus of an animal cell. The host transcribes the viral DNA into RNA that may be used to synthesize viral proteins or may be released from the host cell to infect more cells. *Example*: HIV is a retrovirus.

Concept 19.3 Viruses, viroids, and prions are formidable pathogens in animals and plants

▮ **Viroids** are circular RNA molecules several hundred nucleotides in length that infect plants. They cause errors in regulatory systems that control plant growth.

▮ **Prions** are misfolded, infectious proteins that cause the misfolding of normal proteins they contact in various animal species. *Examples* of diseases caused by prions include mad cow disease and, in humans, Creutzfeldt-Jakob disease.

Chapter 20: Biotechnology

> **WHAT'S IMPORTANT TO KNOW?**
> The Curriculum Framework expects that you are familiar with a technique of modern biotechnology, as well as an example of a product of genetic engineering. Your teacher may not cover all of the possibilities described in this chapter.

> **YOU MUST KNOW**
> - The terminology of biotechnology.
> - The steps in gene cloning with special attention to the biotechnology tools that make cloning possible.
> - The key ideas that make PCR possible.
> - How gel electrophoresis can be used to separate DNA fragments or protein molecules.

Concept 20.1 DNA cloning yields multiple copies of a gene or other DNA segment

▮ The key to unlocking the concepts of biotechnology is to understand the terms. Know the following commonly used terms:

▪ **Genetic engineering** is the process of manipulating genes and genomes.

▪ **Biotechnology** is the process of manipulating organisms or their components for the purpose of making useful products.

- **Recombinant DNA** is DNA that has been artificially made, using DNA from different sources—and often different species. An example is the introduction of a human gene into an *E. coli* bacterium.
- **Gene cloning** is the process by which scientists can produce multiple copies of specific segments of DNA that they can then work with in the lab.
- **Restriction enzymes** are used to cut strands of DNA at specific locations (called **restriction sites**). They are derived from bacteria.
- When a DNA molecule is cut by restriction enzymes, the result will always be a set of **restriction fragments**, which will have at least one single-stranded end, called a **sticky end**. Sticky ends can form hydrogen bonds with complementary single-stranded pieces of DNA. These unions can be sealed with the enzyme **DNA ligase**.

▍Follow the steps that may occur to clone a gene in Figure 5.14.

1. *Identify and isolate the gene of interest and a **cloning vector**.* The vector will carry the DNA sequence to be cloned and is often a bacterial plasmid, as shown in Figure 5.14.

In this example, a human gene is inserted into a plasmid from *E. coli*. The plasmid contains the *amp^R* gene, which makes *E. coli* cells resistant to the antibiotic ampicillin.

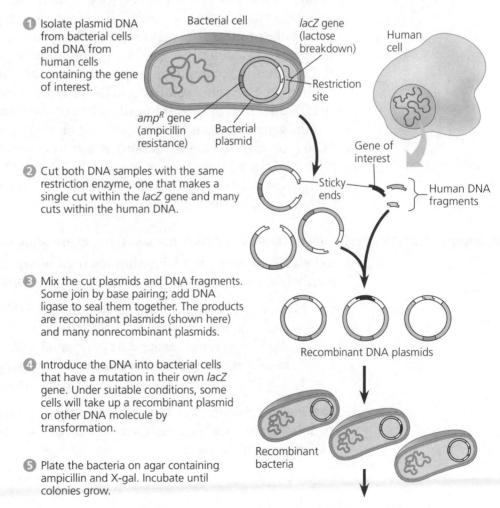

❶ Isolate plasmid DNA from bacterial cells and DNA from human cells containing the gene of interest.

❷ Cut both DNA samples with the same restriction enzyme, one that makes a single cut within the *lacZ* gene and many cuts within the human DNA.

❸ Mix the cut plasmids and DNA fragments. Some join by base pairing; add DNA ligase to seal them together. The products are recombinant plasmids (shown here) and many nonrecombinant plasmids.

❹ Introduce the DNA into bacterial cells that have a mutation in their own *lacZ* gene. Under suitable conditions, some cells will take up a recombinant plasmid or other DNA molecule by transformation.

❺ Plate the bacteria on agar containing ampicillin and X-gal. Incubate until colonies grow.

Bacterial cell

lacZ gene (lactose breakdown)

Human cell

Restriction site

amp^R gene (ampicillin resistance)

Bacterial plasmid

Gene of interest

Sticky ends

Human DNA fragments

Recombinant DNA plasmids

Recombinant bacteria

Figure 5.14   Cloning a human gene in a bacterial plasmid

2. *Cut both the gene of interest and the vector with the same restriction enzyme.* This gives the plasmid and the human gene matching sticky ends.
3. *Join the two pieces of DNA.* Form recombinant plasmids by mixing the plasmids with the DNA fragments. The human DNA fragments can be sealed into the plasmid using DNA ligase.
4. *Get the vector carrying the gene of interest into a host cell.* The plasmids are taken up by the bacterium by *transformation*. The process of transformation is a key part of Investigation 8.
5. *Select for cells that have been transformed.* The bacterial cells carrying the clones must be identified or selected. This can be done by linking the gene of interest to an antibiotic resistance gene or a *reporter gene* such as green fluorescent protein. In AP Investigation 8, we use an ampicillin-resistant plasmid. Any bacterial cells that do not pick up the plasmid by transformation will be killed when grown on agar with the antibiotic ampicillin.

▌ The next problem is finding the gene of interest among the many colonies present after transformation. A process known as **nucleic acid hybridization** can be used to find the gene. If we know at least part of the nucleotide sequence of the gene of interest, we can synthesize a probe complementary to it. For example, if the known sequence is G-G-C-T-A-A, then we would synthesize the complementary probe C-C-G-A-T-T. If we make the probe radioactive or fluorescent, the probe will be easy to track, taking us to the proper gene of interest.

▌ The process just described leads to a genomic library. A **genomic library** is a set of thousands of recombinant plasmid clones, each of which has a piece of the original genome being studied. A **cDNA library** is made up of complementary DNA made from mRNA transcribed by reverse transcriptase. This technique rids the gene of introns but may not contain every gene in the organism.

▌ **PCR** (polymerase chain reaction) is a method used to amplify a particular piece of DNA without the use of cells. PCR is used to amplify DNA when the source is impure or scanty (as it would be at a crime scene). Figure 5.15 shows the basic steps of the PCR procedure.

Concept 20.2 DNA technology allows us to study the sequence, expression, and function of a gene

▌ **Gel electrophoresis** is a lab technique used to separate macromolecules, primarily DNA and proteins. The principles of this separation of DNA include:

1. An *electric current* is applied to the field. DNA is negatively charged and migrates to the positive electrode.
2. *Agarose gel* is used as a matrix to separate molecules by size. The gel allows smaller molecules to move more easily than larger fragments of DNA.
3. The DNA must be stained or tagged for visualization.

Follow Figure 5.16 (on page 140) as the specific steps are shown. Because gel electrophoresis of DNA is a required AP Lab, pay special attention to the concepts explained in the figure.

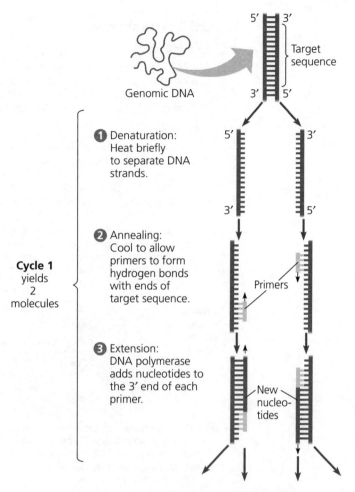

Figure 5.15 Polymerase chain reaction (PCR)

The figure shows the following labels:

5' 3'
Target sequence
Genomic DNA
3' 5'

Cycle 1 yields 2 molecules

1. Denaturation: Heat briefly to separate DNA strands.

5' 3'
3' 5'

2. Annealing: Cool to allow primers to form hydrogen bonds with ends of target sequence.

Primers

3. Extension: DNA polymerase adds nucleotides to the 3' end of each primer.

New nucleotides

▌ Genome-wide studies of gene expression are made possible by the use of **DNA microarray assays**. DNA microarray chips work as follows:

1. Small amounts of single-stranded DNA fragments representing different genes are fixed to a glass slide in a tight grid, termed a *DNA chip*.
2. mRNA molecules from the cells being tested are isolated and converted to cDNA by reverse transcriptase, then tagged with a fluorescent dye.
3. The cDNA bonds to the ssDNA on the chip, indicating which genes are "on" in the cell (actively producing mRNA). This enables researchers, for example, to see differences in gene expression between breast cancer tumors and noncancerous breast tissue.

▌ **Restriction fragment length polymorphisms (RFLPs)** result from small differences in DNA and can be detected by electrophoresis. The difference in banding patterns after electrophoresis allows for diagnosis of disease or is used to answer paternity and identity questions.

Gel Electrophoresis

APPLICATION Gel electrophoresis is used for separating nucleic acids or proteins that differ in size, electrical charge, or other physical properties. DNA molecules are separated by gel electrophoresis in restriction fragment analysis of both cloned genes (see Figure 20.10) and genomic DNA (see Figure 20.11).

TECHNIQUE Gel electrophoresis separates macromolecules on the basis of their rate of movement through a polymeric gel in an electric field: The distance a DNA molecule travels is inversely proportional to its length. A mixture of DNA molecules, usually fragments produced by restriction enzyme digestion (cutting) or PCR amplification, is separated into bands. Each band contains thousands of molecules of the same length.

1 Each sample, a mixture of DNA molecules, is placed in a separate well near one end of a thin slab of gel. The gel is set into a small plastic support and immersed in an aqueous solution in a tray with electrodes at each end.

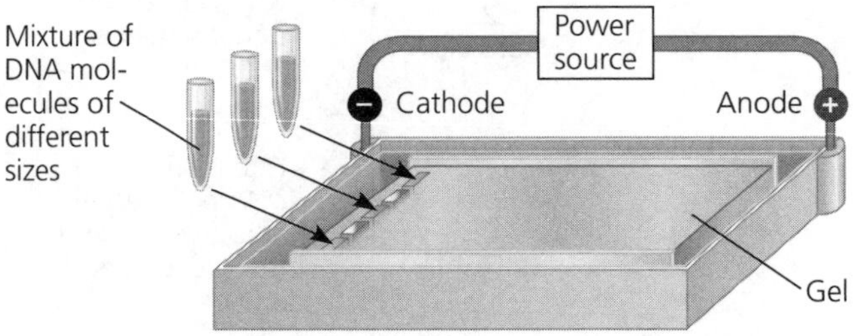

2 When the current is turned on, the negatively charged DNA molecules move toward the positive electrode, with shorter molecules moving faster than longer ones. Bands are shown here in blue, but on an actual gel, the bands would not be visible at this time.

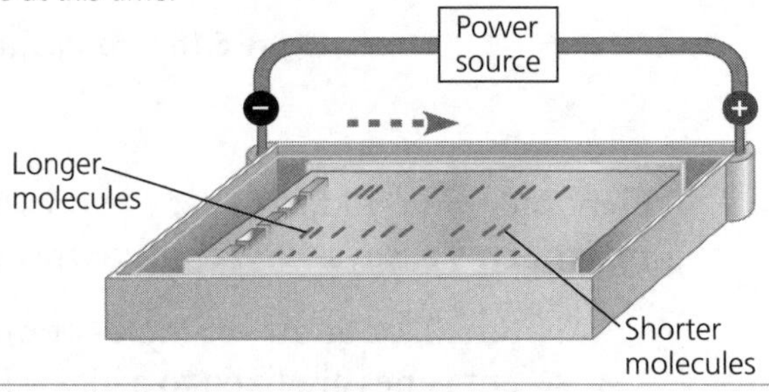

Figure 5.16 Gel electrophoresis

Concept 20.3 *Cloning organisms may lead to production of stem cells for research and other applications*

▌ In animal cloning the nucleus of an egg is removed and replaced with the diploid nucleus of a body cell, a process termed *nuclear transplantation*. The ability of a body cell to successfully form a clone decreases with embryonic development and cell differentiation.

▌ The major goal of most animal cloning is reproduction, but not for humans. In humans, the major goal is the production of **stem cells**. A stem cell can both reproduce itself indefinitely and, under the proper conditions, produce other specialized cells. Stem cells have enormous potential for medical applications.

■ **Embryonic stem cells** are *pluripotent*, which means capable of differentiating into many different cell types. The ultimate aim is to use them for the repair of damaged or diseased organs, such as insulin-producing pancreatic cells for people with diabetes or certain kinds of brain cells for people with Parkinson's disease.

Concept 20.4 *The practical applications of DNA technology affect our lives in many ways*

There are many different uses for DNA technology, some of which are as follows:

■ **Diagnosis of disease:** A number of diseases can be detected by RFLP analysis (e.g., cystic fibrosis, sickle-cell disease) or through amplification of blood samples to test for viruses (e.g., HIV).

■ **Gene therapy:** The alteration of an afflicted individual's genes. Gene therapy holds great potential for treating disorders traceable to a single defective gene, such as cystic fibrosis.

■ **The production of pharmaceuticals:** Gene splicing and cloning can be used to produce large amounts of particular proteins in the lab (e.g., human insulin and growth hormone).

■ **Forensic applications:** DNA samples taken from the blood, skin cells, or hair of alleged criminal suspects can be compared to DNA collected from the crime scene. *Genetic profiles* can be compared and used to identify persons at that crime scene.

■ **Environmental cleanup:** Scientists engineer metabolic capabilities into microorganisms, which are then used to treat environmental problems, such as removing heavy metals from toxic mining sites.

■ **Agricultural applications:** Certain genes that produce desirable traits have been inserted into crop plants to increase their productivity or efficiency. An organism that has acquired by artificial means one or more genes from another species or variety is termed a **genetically modified (GM) organism.** Currently, a debate is in progress over the safety of GM organisms.

Chapter 21: Genomes and Their Evolution

YOU MUST KNOW

- How prokaryotic genomes compare to eukaryotic genomes.
- The activity and role of transposable elements and retrotransposons.
- How evo-devo relates to our understanding of the evolution of genomes.
- The role of homeotic genes and homeoboxes.

Concept 21.2 *Scientists use bioinformatics to analyze genomes and their functions*

■ **Bioinformatics** is the use of computers, software, and mathematical models to process and integrate the incredible volume of data from these sequencing projects. In addition to DNA sequences, protein interactions are analyzed in an approach called *proteomics*.

■ *Systems biology* aims to model the behavior of entire biological systems, and is enhanced by bioinformatics. This has many applications, including medical ones—for example, in the understanding and treatment of cancers.

Concept 21.3 Genomes vary in size, number of genes, and gene density

■ More than 1,200 genomes have now been sequenced. In general, bacteria and archaea have fewer genes than eukaryotes, and the number of genes in eukaryotic genomes is less than was expected. For example, the human genome has only about 20,000 genes.

■ How could so many proteins be made with so few genes? The answer lies in extensive *alternative gene splicing* of RNA transcripts. Recall that this process results in more than one functional protein from a single gene.

Concept 21.4 Multicellular eukaryotes have much noncoding DNA and many multigene families

■ Only a tiny part of the human genome—1.5%—codes for proteins or is transcribed into rRNAs or tRNAs. Much of the rest is **repetitive DNA**, sequences that are present in multiple copies in the genome.

■ **Transposable elements** make up much of the repetitive DNA. These are stretches of DNA that can move from one location to another in the genome with the aid of an enzyme, *transposase*. There are two types:

■ **Transposons** move by means of a DNA intermediate.
■ **Retrotransposons** move by means of a RNA intermediate, and leave a copy at the original site. The process involves *reverse transcriptase*, an enzyme seen before in retroviruses.

■ Transposons can interrupt normal gene function if inserted in the middle of a functional gene, or alter gene expression if inserted into a regulatory element. While these effects may be harmful or lethal, over many generations some may have small beneficial effects. The resulting genetic diversity provides raw material for natural selection.

■ **Multigene families** are collections of two or more identical or very similar genes. A classic example is the *human α-globin and β-globin* gene families. Here, the genes for different human globins are on different chromosomes.

Concept 21.5 Duplication, rearrangement, and mutation of DNA contribute to genome evolution

■ How might genes with novel functions evolve? Duplication events can lead to the evolution of genes with related functions, such as those of the α-globin and β-globin gene families. Mutations and transpositions can occur, and nonfunctional *pseudogenes* may be found in the clusters. Ultimately, new genes with new functions may occur.

Concept 21.6 Comparing genome sequences provides clues to evolution and development

■ **Evo-devo** is a field of biology that compares developmental processes to understand how they may have evolved and how changes can modify existing organismal features or lead to new ones.

■ **Homeotic** genes are master regulatory genes that control placement and spatial organization of body parts by controlling the developmental fate of groups of cells.

■ A **homeobox** is a widely conserved 180-nucleotide sequence found within homeotic genes. When we say that a sequence is *widely conserved*, this means that it is found in many groups (e.g., fungi, animals, and plants) with very few differences. This hints at the relatedness and common evolution of all life-forms.

Level 1: Knowledge/Comprehension Questions

1. Which of the following is NOT a potential control mechanism for regulation of gene expression in eukaryotic organisms?
   (A) the degradation of RNA
   (B) the transport of mRNA from the nucleus
   (C) the lactose operon
   (D) transcription
   (E) gene amplification

2. Which of the following exists as DNA surrounded by a protein coat?
   (A) retrovirus
   (B) virus
   (C) eukaryotic cell
   (D) prokaryotic cell
   (E) ampicillin

3. A goat can produce milk containing the same polymers present in the silk produced by spiders when particular genes from a spider are inserted into the goat's genome. Which of the following reasons describes why this is possible?
   (A) Goats and spiders share a common ancestor and, thus, produce similar protein excretions.
   (B) The opposite is true, too—when genes from a goat are inserted into a spider's genome, the spider produces goats' milk instead of silk.
   (C) The proteins in goats' milk and spiders' silk have the same amino acid sequence.
   (D) The processes of transcription and translation in the cells of spiders and goats are fundamentally similar.
   (E) The processes of transcription and translation in the cells of spiders and goats produce exactly the same proteins anyway.

4. Restriction enzymes are generally used in the laboratory for which of the following reasons?
   (A) restricting the replication of DNA
   (B) restricting the transcription of DNA
   (C) restricting the translation of mRNA
   (D) cutting DNA molecules at specific locations
   (E) cutting DNA into manageable sizes for manipulation

Directions: The group of questions below consists of five lettered choices followed by a list of numbered phrases or sentences. For each numbered phrase or sentence, select the one choice that is most closely related to it. Each choice may be used once, more than once, or not at all.

Questions 5–9
   (A) Transcription
   (B) Translation
   (C) Transposon
   (D) DNA methylation
   (E) Histone acetylation

5. A mobile segment of DNA that travels from one location on a chromosome to another, one element of genetic change

6. The addition of groups to certain bases of DNA after DNA synthesis; this is thought to be an important control mechanism for gene expression

7. The synthesis of polypeptides from the genetic information coded in mRNA

8. The synthesis of RNA from a DNA template

9. The attachment of groups to particular amino acids of specific proteins; this is thought to be an important control mechanism for gene expression

10. The figure below shows which of the following processes?
    (A) the lytic cycle of a phage
    (B) the lysogenic cycle of a phage
    (C) transposition
    (D) retrovirus infection
    (E) mutualism

11. The actions of which of the following enzymes are responsible for ensuring that chromosomes do not decrease in length with every round of replication?
    (A) telomerase
    (B) DNA ligase
    (C) DNA polymerase
    (D) helicase
    (E) primase

12. PCR (polymerase chain reaction) allows target segments of DNA to be produced quickly because it enables lab technicians to do which of the following?
    (A) Isolate gene-source DNA.
    (B) Insert DNA into an appropriate vector.
    (C) Introduce the cloning vector into a host cell.
    (D) Amplify DNA samples.
    (E) Identify clones carrying the gene of interest.

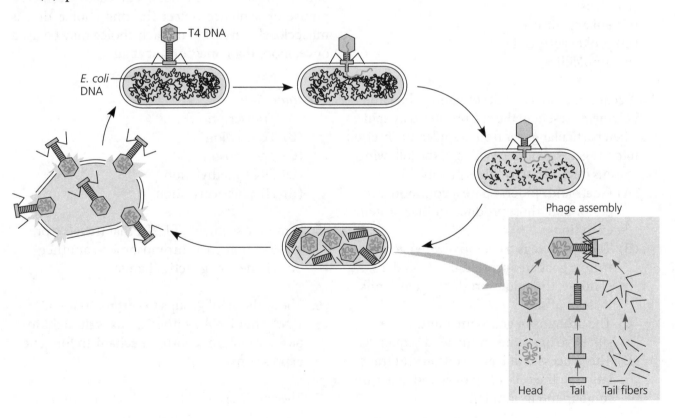

Questions 13–14 refer to an experiment that was performed to separate DNA fragments from three samples radioactively labeled with ^{32}P. The fragments were then separated using gel electrophoresis. The visualized bands are depicted below:

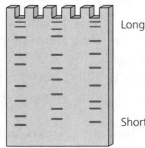

Longer molecules

Shorter molecules

Completed gel

13. When the electric field was applied, the fragments of DNA in each of the three samples migrated to different locations along the gel because
(A) the fragments differed in their levels of radioactivity.
(B) the fragments differed in their charges—some were positively charged, whereas others were negatively charged.
(C) the fragments differed in size.
(D) the fragments differed in polarity.
(E) the fragments differed in solubility.

14. How many sites on DNA were cut by the particular restriction enzyme used in Sample 1 (the leftmost sample)?
(A) 5
(B) 6
(C) 7
(D) 8
(E) 9

15. A bacterium is infected with an experimentally constructed bacteriophage composed of the T2 phage protein coat and T4 phage DNA. The new phages produced would have
(A) T2 protein and T4 DNA.
(B) T2 protein and T2 DNA.
(C) a mixture of the DNA and proteins of both phages.
(D) T4 protein and T4 DNA.
(E) T4 protein and T2 DNA.

16. RNA viruses require their own supply of certain enzymes because
(A) host cells rapidly destroy the viruses.
(B) host cells lack enzymes that can replicate the viral genome.
(C) these enzymes translate viral mRNA into proteins.
(D) these enzymes penetrate host cell membranes.
(E) these enzymes cannot be made in host cells.

17. In genetic engineering, DNA ligase is used for which of the following purposes?
(A) to act as a probe for locating cloned genes
(B) to create breaks in DNA in order to allow foreign DNA fragments to be inserted
(C) to seal up nicks created in newly created recombinant DNA
(D) to ensure that "sticky ends" of like DNA fragments do not re-anneal
(E) in Southern blotting

Directions: The group of questions below consists of five lettered choices followed by a list of numbered phrases or sentences. For each numbered phrase or sentence, select the one choice that is most closely related to it. Each choice may be used once, more than once, or not at all.

Questions 18–22
(A) tRNA
(B) mRNA
(C) Poly-A tail
(D) RNA polymerase
(E) rRNA

18. An example of a post-transcriptional modification

19. Binds to the promoter on DNA to initiate transcription

20. Along with proteins, comprises ribosomes

21. Binds to free amino acids in the cytoplasm

22. Travels out of the nucleus and into the cytoplasm where it serves as a template in translation

23. The expression of eukaryotic genes can be controlled at all the following stages of protein synthesis EXCEPT
    (A) initiation of transcription.
    (B) RNA processing.
    (C) DNA unpacking.
    (D) degradation of protein.
    (E) acetylation of DNA.

24. After eukaryotic transcription takes place, mRNA undergoes several modifications before leaving the nucleus to take part in translation. One of these is the cutting out of nonessential sections of mRNA and the subsequent splicing together of stretches of mRNA necessary for the final functional molecule. Which of the following mRNA sections are spliced together into the finished mRNA molecule?
    (A) introns
    (B) exons
    (C) genes
    (D) coding sequences
    (E) ribozymes

25. In the process of eukaryotic translation, the term *wobble* refers to
    (A) the tendency of the two ribosome subunits to come closer to one another and to separate at different points in translation.
    (B) the tendency of the amino acid loosely attached to the tRNA to move back and forth before finally attaching to the polypeptide chain.
    (C) the fact that the genetic code is redundant.
    (D) the fact that the anticodon and codon bind very loosely.
    (E) the fact that the third nucleotide of a tRNA can form hydrogen bonds with more than one kind of base in the third position of a codon.

26. Which of the following is an example of a missense mutation?
    (A) A nucleotide and its partner are replaced with an "incorrect" pair of nucleotides, which destroys the function of the final protein.
    (B) A nucleotide pair is added into a gene, destroying the reading frame of the genetic message.
    (C) A nucleotide pair is lost from the gene, destroying the reading frame of the genetic message.
    (D) A frameshift mutation occurs, ultimately causing the production of nonfunctional proteins.
    (E) A nucleotide pair substitution occurs, which causes the codon to code for an amino acid that may not be the "correct" one, although translation continues.

27. In analyzing the number of different bases in a DNA sample, which result would be consistent with the base-pairing rules?
    (A) A = G
    (B) A + G = C + T
    (C) A + T = G + T
    (D) A = C
    (E) G = T

28. At the end of DNA replication, each of the daughter molecules has one old strand, derived from the parent strand of DNA, and one strand that is newly synthesized. This explains why DNA replication is described as
    (A) conservative.
    (B) largely conservative.
    (C) nonconservative.
    (D) semiconservative.
    (E) unconservative.

29. The segment of DNA shown below has restriction sites I and II, which create restriction fragments a, b, and c. Which of the following gels produced by electrophoresis would represent the separation and identity of these fragments?

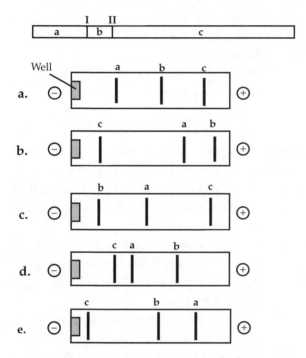

30. Which of the following is a difficulty in getting prokaryotic cells to express eukaryotic genes?
    (A) The signals that control gene expression are different and prokaryotic promoter regions must be added to the vector.
    (B) The genetic code differs because prokaryotes substitute the base uracil for thymine.
    (C) Prokaryotic cells cannot transcribe introns because their genes do not have them.
    (D) The ribosomes of prokaryotes are not large enough to handle long eukaryotic genes.
    (E) The RNA splicing enzymes of bacteria work differently from those of eukaryotes.

31. One of the characteristics of retrotransposons is that
    (A) they code for an enzyme that synthesizes DNA using an RNA template.
    (B) they are found only in animal cells.
    (C) they generally move by a cut-and-paste mechanism.
    (D) they contribute a significant portion of the genetic variability seen within a population of gametes.
    (E) their amplification is dependent on a retrovirus.

32. You have affixed the chromosomes from a cell onto a microscope slide. Which of the following would NOT make a good radioactively labeled probe to help map a particular gene to one of those chromosomes? (Assume DNA of chromosomes and probes is single-stranded.)
    (A) cDNA made from the mRNA transcribed from the gene
    (B) a portion of the amino acid sequence of that protein
    (C) mRNA transcribed from the gene
    (D) a piece of the restriction fragment on which the gene is located
    (E) a sequence of nucleotide bases determined from the genetic code needed to produce a known sequence of amino acids found in the protein product of the gene

33. The human genome appears to have only one-third more genes than the simple nematode, *C. elegans*. Which of the following best explains how the more complex humans can have relatively few genes?
    (A) The unusually long introns in human genes are involved in regulation of gene expression.
    (B) More than one polypeptide can be produced from a gene by alternative splicing.
    (C) Human genes code for many more types of domains.
    (D) The human genome has a high proportion of noncoding DNA.
    (E) The large number of SNPs (single nucleotide polymorphisms) in the human genome provides for a great deal of genetic variability.

34. Multigene families are
    (A) groups of enhancers that control transcription.
    (B) usually clustered at the telomeres.
    (C) equivalent to the operons of prokaryotes.
    (D) sets of genes that are coordinately controlled.
    (E) sets of identical or similar genes that have evolved by gene duplication.

35. Homeotic genes
    (A) encode transcription factors that control the expression of genes responsible for specific anatomical structures.
    (B) are found only in *Drosophila* and other arthropods.
    (C) are the only genes that contain the homeobox domain.
    (D) encode proteins that form anatomical structures in the fly.
    (E) are responsible for patterning during plant development.

Level 2: Application/Analysis/Synthesis Questions
After reading the paragraphs, answer the question(s) that follow.

Exposure to the HIV virus doesn't necessarily mean that a person will develop AIDS. Some people have genetic resistance to infection by HIV. Dr. Stephen O'Brien from the U.S. National Cancer Institute has recently identified a mutant form of a gene, called CCR5, which can protect against HIV infection. The mutation probably originated in Europe among survivors of the bubonic plague. The mutated gene prevents the plague bacteria from attaching to cell membranes and, therefore, from entering and infecting body cells.

Although the HIV virus is very different from the bacterium that causes the plague, both diseases affect the exact same cells and use the same method of infection. The presence of the mutated gene in descendants of plague survivors helps prevent them from contracting AIDS. Pharmaceutical companies are using this information as the basis for a new approach to AIDS prevention. This would be very important in areas of the world where the mutation is scarce or absent, such as Africa.

1. The most likely method by which the mutated CCR5 gene prevents AIDS is by
    (A) covering the cell membrane.
    (B) rupturing the nuclear membrane.
    (C) attacking and destroying the HIV virus particles.
    (D) coding for a protective protein in the cell membrane.

2. Which of the following shows the steps of a viral infection in the proper order?
    (A) Virus locates host cell → enters nucleus → alters host cell DNA → destroys cell membrane.
    (B) Virus locates host cell → alters host cell DNA → host cell produces copies of virus → copies enter host cell nucleus → nucleus leaves cell.
    (C) Virus locates host cell → penetrates cell membrane → enters nucleus → alters host cell DNA → host cell produces copies of virus.
    (D) Virus locates host cell → forms hydrogen bonds → changes DNA to RNA → host cell produces copies of virus.

3. Conjugation, transformation, and transduction are all ways that bacteria
    (A) reduce their DNA content.
    (B) increase the amount of RNA in the cytoplasm.
    (C) increase their genetic diversity.
    (D) alter their oxygen requirements.

After reading the paragraph, answer the question(s) that follow.

Four decades after the end of the Vietnam War, the remains of an Air Force pilot were discovered and returned to the United States. A search of Air Force records identified three families to which the remains might possibly belong. Each family had a surviving twin of a missing service member. The following STR profiles were obtained from the remains of the pilot and the surviving twins from the three families.

	Air Force Pilot	Family 1	Family 2	Family 3
1		—		
2	—	—		—
3	—			—
4		—		
5	—	—	—	—
6			—	
7	—	—	—	—
8			—	
9		—		
10	—			—
11			—	
12		—		
13	—		—	—

4. In order to match the pilot's remains to the correct family using DNA profiling,
   (A) the majority of the STR bands must match.
   (B) each of the 13 STR bands must match.
   (C) the bands for site 13 must match.
   (D) bands 5 and 7 must match.

5. Based on analysis of the STR sites shown, does the missing pilot belong to any of these three families?
   (A) No, none of the families match.
   (B) Yes, family 1 matches.
   (C) Yes, family 2 matches.
   (D) Yes, family 3 matches.

6. Which of the following mutations would be most likely to have a harmful effect on an organism?
   (A) a deletion of three nucleotides near the middle of a gene
   (B) a single nucleotide deletion in the middle of an intron
   (C) a single nucleotide deletion near the end of the coding sequence
   (D) a single nucleotide insertion downstream of, and close to, the start of the coding sequence

7. Which of the following is NOT true of RNA processing?
   (A) Exons are cut out before mRNA leaves the nucleus.
   (B) Nucleotides may be added at both ends of the RNA.
   (C) Ribozymes may function in RNA splicing.
   (D) RNA splicing can be catalyzed by spliceosomes.

8. What is the basis for the difference in how the leading and lagging strands of DNA molecules are synthesized?
   (A) The origins of replication occur only at the 5' end.
   (B) Helicases and single-strand binding proteins work at the 5' end.
   (C) DNA polymerase can join new nucleotides only to the 3' end of a growing strand.
   (D) DNA ligase works only in the 3' → 5' direction.

Free-Response Question

1. *Genes are located on chromosomes and are the basic unit of heredity that is passed on from parent to child, through generations.*

    (a) Explain how a chromosomal mutation could occur and why mutations are detrimental to the organism in which they take place.

    (b) Explain why it is more common for human males to be color-blind or have hemophilia than females.

    (c) Proper gene dosages are critical to normal development and function. How do we maintain proper gene dosages related to the X chromosome, given that females have two copies of the chromosome and males only one?

Mechanisms of Evolution

Chapter 22: Descent with Modification: A Darwinian View of Life

---

YOU MUST KNOW

- How Lamarck's view of the mechanism of evolution differed from Darwin's.
- Several examples of evidence for evolution.
- The difference between structures that are homologous and those that are analogous, and how this relates to evolution.
- The role of adaptations, variation, time, reproductive success, and heritability in evolution.

***Concept 22.1** The Darwinian revolution challenged traditional views of a young Earth inhabited by unchanging species*

▌ Historical Setting

▪ Scala *naturae* (Aristotle, 384–322 BCE): Life-forms could be arranged on a scale of increasing complexity.

▪ **Old Testament:** Perfect species were individually designed by God.

▪ **Carolus Linnaeus** (1707–1778): Grouped similar species into increasingly general categories reflecting what he considered the pattern of their creation.

- Developed **taxonomy**, the branch of biology dedicated to the naming and classification of all forms of life.

- Developed **binomial nomenclature**, a two-part naming system that includes the organism's genus and species.

▪ **Georges Cuvier** (1769–1832): French geologist opposed to idea of evolution.

- Advocated **catastrophism**, the principle that events in the past occurred suddenly and by different mechanisms than those occurring today.

- This explained boundaries between strata and location of different species.

- **Charles Lyell** (1797–1875): English geologist

  - Developed principle of **uniformitarianism**, the idea that the geologic processes that have shaped the planet have not changed over the course of Earth's history.
  - *Importance*: ***The Earth must be very old.***
  - Lyell's *Principles of Geology* was studied by Darwin during his journeys.

- **Jean-Baptiste de Lamarck** (1744–1829): Developed an early theory of evolution based on two principles:

  - **Use and disuse** is the idea that parts of the body that are used extensively become larger and stronger, while those that are not used deteriorate.
  - **Inheritance of acquired characteristics** assumes that characteristics acquired during an organism's lifetime could be passed on to the next generation. *Example*: A weightlifter's child could be born with a more muscular anatomy.
  - *Importance*: Lamarck recognized that species evolve, though his explanatory mechanism was flawed.

Concept 22.2 *Descent with modification by natural selection explains the adaptations of organisms and the unity and diversity of life*

- **Charles Darwin's** voyage on the HMS *Beagle* from 1831 to 1836 was the impetus for the development of his theory of evolution by natural selection.
- Darwin's mechanism for evolution was natural selection. Recall that Lamarck's mechanism was the inheritance of acquired characteristics.
- **Natural selection** explains how adaptations arise.

  - **Adaptations** are heritable characteristics that enhance organisms' ability to survive and reproduce in specific environments. *Example*: Desert foxes have large ears which radiate heat. Arctic foxes have small ears which conserve body heat.
  - *Overproduction* of offspring leads to *competition* for resources.
  - *Variations* in traits occur. Natural selection is a process in which individuals that have certain *heritable traits* survive and reproduce at a higher rate than other individuals.
  - Over time, natural selection can increase the match between organisms and their environment.
  - If an environment changes, or if individuals move to a new environment, natural selection may result in adaptation to these new conditions, sometimes giving rise to new species in the process.

- **Artificial selection** is the process by which species are modified by humans. *Example*: Selective breeding for milk or meat production; development of dog breeds.
- *Individuals do not evolve.* **Populations** evolve.

Concept 22.3 Evolution is supported by an overwhelming amount of scientific evidence

> **ORGANIZE YOUR THOUGHTS**
>
> EVIDENCE FOR EVOLUTION is seen in the following ways:
>
> 1. Direct observations
> 2. The fossil record
> 3. Homology
> 4. Biogeography

▌ Evidence for Evolution

1. **Direct Observations of Evolutionary Change**

   - Insect populations can rapidly become resistant to pesticides such as DDT.
   - Evolution of drug-resistant viruses and antibiotic-resistant bacteria.

2. **The Fossil Record:** *Fossils provide evidence for the theory of evolution.*

   - Fossils are remains or traces of organisms from the past. They are found in sedimentary rock.
   - **Paleontology** is the study of fossils.
   - Fossils show evolutionary changes have occurred over time and the origin of major new groups of organisms.

3. **Homology and Convergent Evolution**

   - **Homology:** Characteristics in related species can have an underlying similarity even though they have very different functions.
   - **Homologous structures** are anatomical signs of evolution. *Examples*: Forelimbs of mammals that are now used for a variety of purposes, such as flying in bats or swimming in whales, but were present and used in a common ancestor for walking.
   - **Embryonic homologies:** Comparison of early stages of animal development reveals many anatomical homologies in embryos that are not visible in adult organisms. *Examples*: All vertebrate embryos have a post-anal tail and pharyngeal pouches.
   - **Vestigial organs** are structures of marginal, if any, importance to the organism. They are remnants of structures that served important functions in the organisms' ancestors. *Example*: Remnants of the pelvis and leg bones are found in some snakes.
   - **Molecular homologies** are shared characteristics on the molecular level. *Examples*: All life-forms use the same genetic language of DNA and RNA. Amino acid sequences coding for hemoglobin in primate species shows great similarity, thus indicating a common ancestor.
   - **Convergent evolution** explains why distantly related species can resemble one another. Convergent evolution has taken place

when two organisms developed similarities as they adapted to similar environmental challenges—not because they evolved from a common ancestor. The likenesses that result from convergent evolution are considered **analogous** rather than homologous.

- **Similar problem, similar solution** *Examples*:
  - The torpedo shapes of a penguin, dolphin, and shark are the solution to movement through an aqueous environment.
  - Sugar gliders (marsupial mammals) and flying squirrels (eutherian mammals) occupy similar niches in their respective habitats.

> *STUDY TIP Homologous structures* show evidence of common ancestry (whale fin/bat wing).
> *Analogous structures* are similar solutions to similar problems but do *not* indicate close relatedness (bird wing/butterfly wing).

4. **Biogeography:** The geographic distribution of species.

- Species in a discrete geographic area tend to be more closely related to each other than to species in distant geographic areas. *Example*: In South America, desert animals are more closely related to local animals in other habitats than they are to the desert animals of Asia. This reflects evolution, not creation.
- **Continental drift** and the break-up of *Pangaea* can explain the similarity of species on continents that are distant today.
- **Endemic species** are found at a certain geographic location and nowhere else.

  - *Example*: Marine iguanas are endemic to the Galápagos.

- Darwin's theory of evolution through natural selection explains the succession of forms in the fossil record. Transitional fossils have been found that link ancient organisms to modern species, just as Darwin's theory predicts.

> *ORGANIZE YOUR THOUGHTS*
>
> 1. Evolution is change in species over time.
> 2. There is overproduction of offspring, which leads to competition for resources.
> 3. Heritable variations exist within a population.
> 4. These variations can result in differential reproductive success.
> 5. Over generations, this can result in changes in the genetic composition of the population.
>
> *And remember . . . Individuals do not evolve!* **Populations** *evolve.*

Chapter 23: The Evolution of Populations

Concept 23.1 Genetic variation makes evolution possible

- **Microevolution** is change in the allele frequencies of a population over generations. This is evolution on its smallest scale.
- **Mutations** are the only source of *new* genes and *new* alleles.

    - Only mutations in cell lines that produce gametes can be passed to offspring.

- **Point mutations** are changes in one base in a gene. They can have significant impact on phenotype, as in sickle-cell disease.
- **Chromosomal mutations** delete, disrupt, duplicate, or rearrange many loci at once. They are usually harmful.
- However, **most of the genetic variations** within a population are due to the sexual recombination of alleles that already exist in a population.
- Sexual reproduction rearranges alleles into new combinations in every generation. Recall there are *three mechanisms* for this shuffling of alleles:

    - **Crossing over** during prophase I of meiosis.
    - **Independent assortment** of chromosomes during meiosis (2^{23} different combinations possible in the formation of human gametes!)
    - **Fertilization** ($2^{23} \times 2^{23}$ different possible combinations for human sperm and egg)

Concept 23.2 The Hardy-Weinberg equation can be used to test whether a population is evolving

- **Population genetics** is the study of how populations change genetically over time.
- **Population:** A group of individuals of the same species that live in the same area and interbreed, producing fertile offspring.
- **Gene pool:** All of the alleles at all loci in all the members of a population.
- In diploid species, each individual has two alleles for a particular gene, and the individual may be either heterozygous or homozygous.
- If all members of a population are homozygous for the same allele, the allele is said to be **fixed**. Only one allele exists at that particular locus in the population.

▌ The greater the number of fixed alleles, the lower the species' diversity.

▌ The **Hardy-Weinberg principle** is used to describe a population that is *not* evolving. It states that the frequencies of alleles and genes in a population's gene pool will remain constant over the course of generations unless they are acted upon by forces *other* than Mendelian segregation and the recombination of alleles. The population is at **Hardy-Weinberg equilibrium**.

▌ However, it is unlikely that all the conditions for Hardy-Weinberg equilibrium will be met. Allele frequencies change. Populations evolve. This can be tested by applying the **Hardy-Weinberg equation**.

EQUATION FOR HARDY-WEINBERG EQUILIBRIUM

$$p^2 + 2pq + q^2 = 1$$

Where p^2 is equal to the frequency of the homozygous dominant in the population, $2pq$ is equal to the frequency of all the heterozygotes in the population, and q^2 is equal to the frequency of the homozygous recessive in the population.

Consider a gene locus that exists in two allelic forms, *A* and *a*, in a population.

Let p = the frequency of A, the dominant allele

and q = the frequency of a, the recessive allele.

So,
p_2 = AA,
q_2 = aa,
$2pq$ = Aa

If we know the frequency of one of the alleles, we can calculate the frequency of the other allele:

$p = q = 1$, so
$p = 1 - q$
$q = 1 - p$

> **TRY THIS FOR PRACTICE!**
>
> Suppose in a plant population that red flowers (R) are dominant to white flowers (r). In a population of 500 individuals, 25% show the recessive phenotype. How many individuals would you expect to be homozygous dominant and heterozygous for this trait? (**Answer below in bold. Cover this solution and try it first!**)
>
> 1. Let p = frequency of the dominant allele (R) and q = frequency of the recessive allele (r)
> 2. q^2 = frequency of the homozygous recessive = 25% = 0.25. Since q^2 = 0.25, q = 0.5
> 3. Now, $p + q = 1$, so $p = 0.5$
> 4. Homozygous dominant individuals are RR or p^2 = 0.25, and will represent $(0.25)(500)$ = **125 individuals**.
> 5. The heterozygous individuals are calculated from $2pq$ = $(2)(0.5)(0.5)$ = 0.5, and in a population of 500 individuals will be $(0.5)(500)$ = **250 individuals**.

Concept 23.3 Natural selection, genetic drift, and gene flow can alter allele frequencies in a population

▌ **Mutations can alter gene frequency but are rare.**

▌ **The three major factors** that alter allele frequencies and bring about most evolutionary change are *natural selection, genetic drift,* and *gene flow.*

- **Natural selection** results in alleles being passed to the next generation in proportions different from their relative frequencies in the present generation. Individuals with variations that are better suited to their environment tend to produce more offspring than those with variations that are less suited.

- **Genetic drift** is the unpredictable fluctuation in allele frequencies from one generation to the next. The smaller the population, the greater the chance is for genetic drift. This is a *random, nonadaptive* change in allele frequencies. Two examples follow.

  - **Founder effect:** A few individuals become isolated from a larger population and establish a new population whose gene pool is not reflective of the source population.

  - **Bottleneck effect:** A sudden change in the environment (e.g., an earthquake, flood, or fire) drastically reduces the size of a population. The few survivors that pass through the restrictive bottleneck may have a gene pool that no longer reflects the original population's gene pool. *Example*: The population of California condors was reduced to nine individuals. This represents a bottlenecking event.

- **Gene flow** occurs when a population gains or loses alleles by genetic additions to and/or subtractions from the population. This results from the movement of fertile individuals or gametes. Gene flow tends to reduce the genetic differences between populations, thus making populations more similar.

Concept 23.4 Natural selection is the only mechanism that consistently causes adaptive evolution

▋ **Relative fitness** refers to the contribution an organism makes to the gene pool of the next generation relative to the contributions of other members. Fitness does *not* indicate strength or size. It is measured only by reproductive success.

▋ Natural selection acts more directly on the phenotype and indirectly on the genotype and can alter the frequency distribution of heritable traits in three ways (see Figure 6.1).

▪ **Directional selection** *Example*: Large black bears survived periods of extreme cold better than smaller ones, and so became more common during glacial periods.

▪ **Disruptive selection** *Example*: A population has individuals with either large beaks or small beaks, but few with the intermediate beak size. Apparently the intermediate beak size is not efficient in cracking either the large or small seeds that are common.

▪ **Stabilizing selection** *Example*: Birth weights of most humans lie in a narrow range, as those babies who are very large or very small have higher mortality.

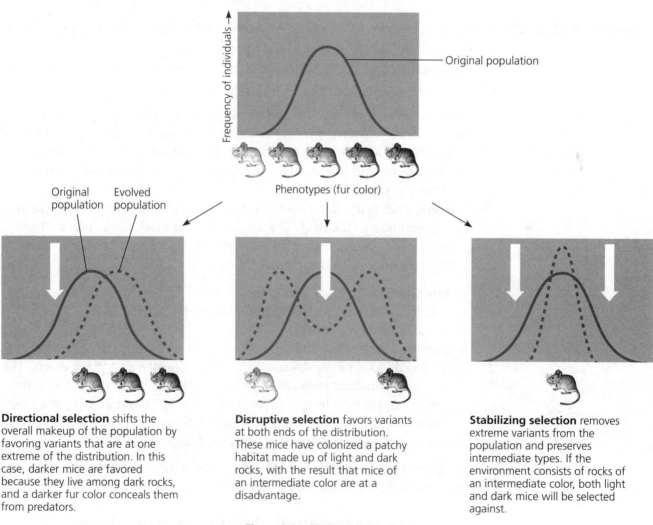

Directional selection shifts the overall makeup of the population by favoring variants that are at one extreme of the distribution. In this case, darker mice are favored because they live among dark rocks, and a darker fur color conceals them from predators.

Disruptive selection favors variants at both ends of the distribution. These mice have colonized a patchy habitat made up of light and dark rocks, with the result that mice of an intermediate color are at a disadvantage.

Stabilizing selection removes extreme variants from the population and preserves intermediate types. If the environment consists of rocks of an intermediate color, both light and dark mice will be selected against.

Figure 6.1   Modes of selection

- **Sexual selection** is a form of natural selection in which individuals with certain inherited characteristics are more likely than other individuals to obtain mates. It can result in **sexual dimorphism**, a difference between the two sexes in secondary sexual characteristics such as differences in size, color, ornamentation, and behavior.
- **How is genetic variation preserved in a population?**

  - **Diploidy:** Because most eukaryotes are diploid, they are capable of hiding genetic variation (recessive alleles) from selection.
  - **Heterozygote advantage:** Individuals who are heterozygous at a certain locus have an advantage for survival. *Example*: In sickle-cell disease, individuals homozygous for normal hemoglobin are more susceptible to malaria, whereas homozygous recessive individuals suffer from the complications of sickle-cell disease. Heterozygotes benefit from protection from malaria and do not have sickle-cell disease, so the mutant allele remains relatively common.

- **Why natural selection cannot produce perfect organisms:**

  - Selection can only edit existing variations.
  - Evolution is limited by historical constraints.
  - Adaptations are often compromises.
  - Chance, natural selection, and the environment interact.

Chapter 24: The Origin of Species

YOU MUST KNOW

- The biological concept of species.
- Prezygotic and postzygotic barriers that maintain reproductive isolation in natural populations.
- How allopatric and sympatric speciation are similar and different.
- How an autopolyploid or an allopolyploid chromosomal change can lead to sympatric speciation.
- How punctuated equilibrium and gradualism describe two different tempos of speciation.

Concept 24.1 *The biological species concept emphasizes reproductive isolation*

- **Speciation** is the process by which new species arise.
- **Microevolution** is change in the genetic makeup of a population from generation to generation. It refers to adaptations that are confined to a single gene pool.
- **Macroevolution** refers to evolutionary change above the species level, such as the appearance of feathers and other such novelties, used to define higher taxa.
- The **biological species concept** defines a species as a population or group of populations whose members have the potential to interbreed in nature and

produce viable, fertile offspring but are unable to produce viable, fertile offspring with members of other populations.

- **Reproductive isolation** is defined as the existence of biological barriers that impede members of two species from producing viable, fertile hybrids.
- **Prezygotic and postzygotic** are two types of barriers that prevent members of different species from producing offspring that can also successfully reproduce. Examples of prezygotic barriers, those that prevent mating or hinder fertilization, include the following:

    - **Habitat isolation:** Two species can live in the same geographic area but not in the same habitat; this will prevent them from mating.
    - **Behavioral isolation:** Some species use certain signals or types of behavior to attract mates, and these signals are unique to their species. Members of other species do not respond to the signals; thus, mating does not occur.
    - **Temporal isolation:** Species may breed at different times of day, different seasons, or different years, and this can prevent them from mating.
    - **Mechanical isolation:** Species may be anatomically incompatible.
    - **Gametic isolation:** Even if the gametes of two species do meet, they might be unable to fuse to form a zygote.

Examples of postzygotic barriers, those that prevent a fertilized egg from developing into a fertile adult, include the following:

- **Reduced hybrid viability:** When a zygote *is* formed, genetic incompatibility may cause development to cease.
- **Reduced hybrid fertility:** Even if the two species produce a viable offspring, reproductive isolation is still occurring if the offspring is sterile and can't reproduce.
- **Hybrid breakdown:** Sometimes two species mate and produce viable, fertile hybrids; however, when the hybrids mate, their offspring are weak or sterile.

Concept 24.2 *Speciation can take place with or without geographic separation*

- The two main types of speciation are **allopatric speciation**, in which a population forms a new species because it is geographically isolated from the parent population, and **sympatric speciation**, in which a small part of a population becomes a new population without being geographically separated from the parent population. It can result from part of the population switching to a new habitat, food source, or other resource.
- Some **geologic events or processes** that can fragment a population include the emergence of a mountain range, the formation of a land bridge, or evaporation in a large lake that produces several small lakes.
- Small, newly isolated populations undergo allopatric speciation more frequently because they are more likely to have their gene pools significantly altered. Speciation is confirmed when individuals from the new population are unable to mate successfully with individuals from the parent population.

- One mechanism that can lead to **sympatric speciation** in plants is the formation of **autopolyploid** plants through nondisjunction in meiosis. For example, these plants may have a 4*n* chromosome number instead of the normal 2*n* number. They cannot breed with diploid members and produce fertile offspring.
- **Adaptive radiation** occurs when many new species arise from a single common ancestor. Adaptive radiation typically occurs when a few organisms make their way to new, distant areas or when environmental changes cause numerous extinctions, opening up ecological niches for the survivors.

Concept 24.4 Speciation can occur rapidly or slowly and can result from changes in few or many genes

- **Gradualism** proposes that species descended from a common ancestor and gradually diverge more and more in morphology as they acquire unique adaptations.
- **Punctuated equilibrium** is a term used to describe periods of apparent stasis punctuated by sudden change observed in the fossil record.

TIP FROM THE READERS

Remember this: Individuals do not evolve! They do not "struggle to survive." They cannot change their genetic makeup in response to a catastrophe. The individual lives or dies. Those that live reproduce and pass on adaptive heritable variations. INDIVIDUALS DO NOT EVOLVE! ONLY POPULATIONS CAN EVOLVE!

Chapter 25: The History of Life on Earth

YOU MUST KNOW

- A scientific hypothesis about the origin of life on Earth.
- The age of the Earth and when prokaryotic and eukaryotic life emerged.
- Characteristics of the early planet and its atmosphere.
- How Miller and Urey tested the Oparin-Haldane hypothesis and what they learned.
- Methods used to date fossils and rocks and how fossil evidence contributes to our understanding of changes in life on Earth.
- Evidence for endosymbiosis.
- How continental drift can explain the current distribution of species (biogeography).
- How extinction events open habitats that may result in adaptive radiation.

Concept 25.1 Conditions on early Earth made the origin of life possible

- Current theory about how life arose consists of four main stages:

    1. Small organic molecules were synthesized.

2. These small molecules joined into macromolecules, such as proteins and nucleic acids.
3. All these molecules were packaged into **protocells**, membrane-containing droplets, whose internal chemistry differed from that of the external environment.
4. Self-replicating molecules emerged that made inheritance possible.

▌ Earth was formed about **4.6 billion years ago**, and life on Earth emerged about **3.8 to 3.9 billion years ago**. For the first three-quarters of Earth's history, all of its living organisms were microscopic and primarily unicellular.

▌ Hypothetical early conditions of Earth have been simulated in laboratories, and organic molecules have been produced.

  ▪ **Oparin** and **Haldane** hypothesized that the early atmosphere, thick with water vapor, nitrogen, carbon dioxide, methane, ammonia, hydrogen, and hydrogen sulfide, provided with energy from lightning and UV radiation, could have formed organic compounds, a primitive "soup" from which life arose.

  ▪ **Miller** and **Urey** tested this hypothesis and produced a variety of amino acids.

▌ It is hypothesized that **self-replicating RNA** (not DNA) was the **first genetic material**. RNA, which plays a central role in protein synthesis, can also carry out a number of enzyme-like catalytic functions. These RNA catalysts are called **ribozymes**.

Concept 25.2 *The fossil record documents the history of life*

▌ The **fossil record** is the sequence in which fossils appear in the layers of sedimentary rock that constitute Earth's surface. **Paleontologists** study the fossil record. Fossils, which may be remnants of dead organisms or impressions they left behind, are most often found in sedimentary rock formed from layers of minerals settling out of water. The fossil record is incomplete because it favors organisms that existed for a long time, were relatively abundant and widespread, and had shells or hard bony skeletons.

▌ Rocks and fossils are dated several ways:

  ▪ **Relative dating** uses the order of rock strata to determine the relative age of fossils.

  ▪ **Radiometric dating** uses the decay of radioactive isotopes to determine the age of the rocks or fossils. It is based on the rate of decay, or **half-life** of the isotope.

Concept 25.3 *Key events in life's history include the origins of single-celled and multicelled organisms and the colonization of land*

▌ The earliest living organisms were **prokaryotes**.

▌ About 2.7 billion years ago, **oxygen** began to accumulate in Earth's atmosphere as a result of photosynthesis.

▌ **Eukaryotes** appeared about 2.1 billion years ago.

- The **endosymbiotic hypothesis** proposes that mitochondria and plastids (chloroplasts) were formerly small prokaryotes that began living within larger cells.
- **Evidence** for this hypothesis includes

  - Both organelles have enzymes and transport systems homologous to those found in the plasma membranes of living prokaryotes.

  - Both replicate by a splitting process similar to prokaryotes.

  - Both contain a single, circular DNA molecule, not associated with histone proteins.

  - Both have their own ribosomes which can translate their DNA into proteins.

- **Multicellular eukaryotes** evolved about 1.2 billion years ago.
- The **colonization of land** occurred about 500 million years ago, when **plants, fungi, and animals** began to appear on Earth.

Concept 25.4 *The rise and fall of groups of organisms reflect differences in speciation and extinction rates*

- **Continental drift** is the movement of Earth's continents on great plates that float on the hot, underlying mantle. The San Andreas Fault marks where two plates are sliding past each other. Where plates have collided, mountains are uplifted.
- Continental drift can help explain the disjunct geographic distribution of certain species, such as a fossil freshwater reptile found in both Brazil and Ghana in west Africa, today widely separated by ocean.
- Continental drift can explain why no eutherian (placental) mammals are indigenous to Australia.
- **Mass extinctions**, loss of large numbers of species in a short period, have resulted from global environmental changes that have caused the rate of extinction to increase dramatically.
- By removing large numbers of species, a mass extinction can drastically alter a complex ecological community. Evolutionary lineages disappear forever. *Example*: The dinosaurs were lost in a mass extinction 65 million years ago.
- **Adaptive radiations** are periods of evolutionary change in which groups of organisms form many new species whose adaptations allow them to fill different ecological niches. *Example*: The Galápagos finch species are the result of an adaptive radiation.

Concept 25.5 *Major changes in body form can result from changes in the sequences and regulation of developmental genes*

- Evolutionary novelty can arise when structures that originally played one role gradually acquire a different one. Structures that evolve in one context but become co-opted for another function are sometimes called **exaptations**. For example, it is possible that feathers of modern birds were co-opted for flight after functioning in some other capacity, such as thermoregulation.
- **Evo-devo** is a field of study in which evolutionary biology and developmental biology converge. This field is illuminating how slight genetic divergences can be magnified into major morphological differences between species.

- **Heterochrony** is an evolutionary change in the rate or timing of developmental events. Changing relative rates of growth even slightly can change the adult form of organisms substantially, thus contributing to the potential for evolutionary change.
- **Homeotic genes** are master regulatory genes that determine the location and organization of body parts.
- *Hox* genes are one class of homeotic genes. Changes in *Hox* genes and in the genes that regulate them can have a profound effect on morphology, thus contributing to the potential for evolutionary change. An example is seen in the variable expression of a *Hox* gene in a snake limb bud and a chicken leg bud, resulting in no legs in the snake, and a skeletal extension in the chicken.

Concept 25.6 *Evolution is not goal oriented*

- Evolution is like tinkering—a process in which new forms arise by the slight modification of existing forms. Even large changes, like the ones that produced the first mammals, can result from the modification of existing structures or existing developmental genes.

Level 1: Knowledge/Comprehension Questions

1. The condition in which there are barriers to inbreeding between individuals of the same species separated by a portion of a mountain range is referred to as
   (A) minute variations.
   (B) geographic isolation.
   (C) infertility.
   (D) a cline.
   (E) differential breeding capacity.

2. Which of the following statements best expresses the concept of punctuated equilibrium?
   (A) Minute changes in the genome of individuals eventually lead to the evolution of a population.
   (B) The five conditions of Hardy-Weinberg equilibrium will prevent populations from evolving quickly.
   (C) Evolution occurs in rapid bursts of change alternating with long periods in which species remain relatively unchanged.
   (D) Profound change over the course of geologic history is the result of an accumulation of slow, continuous processes.
   (E) When two species compete for a single resource in the same environment, one of them will gradually become extinct.

3. In a particular bird species, individuals with average-sized wings survive severe storms more successfully than other birds in the same population with longer or shorter wings. This illustrates
   (A) the founder effect.
   (B) stabilizing selection.
   (C) artificial selection.
   (D) gene flow.
   (E) diversifying selection.

4. All of the following statements are part of Darwin's theory of evolution EXCEPT
   (A) the most prominent contribution to evolution is made by acquiring genetic change during the life span of the organism.
   (B) natural selection is the force behind evolution.
   (C) natural selection occurs as a result of the differing reproductive success of individuals in a population.
   (D) a product of evolution is the adaptation of a population of organisms to their environment.
   (E) more individuals are born in a population than will survive to reproduce.

5. In a certain group of rabbits, the presence of yellow fur is the result of a homozygous recessive condition in the biochemical pathway producing hair pigment. If the frequency of the allele for this condition is 0.10, which of the following is closest to the frequency of the dominant allele in this population? (Assume that the population is in Hardy-Weinberg equilibrium.)
   (A) 0.01
   (B) 0.20
   (C) 0.40
   (D) 0.90
   (E) 1.0

Directions: The group of questions below consists of five lettered choices followed by a list of numbered phrases or sentences. For each numbered phrase or sentence, select the one choice that is most closely related to it. Each choice may be used once, more than once, or not at all.

Questions 6–10
   (A) Artificial selection
   (B) Homology
   (C) Gene pool
   (D) The founder effect
   (E) The bottleneck effect

6. Leads to new species with certain traits desired by humans

7. Can result in a new island population with a limited gene pool

8. One result of evolution from a common ancestor

9. A result of drastic reduction in population size due, for example, to a sudden change in the environment

10. Constitutes all of the alleles in a population

11. All of the following are examples of prezygotic barriers EXCEPT
   (A) habitat isolation.
   (B) behavioral isolation.
   (c) temporal isolation.
   (d) mechanical isolation.
   (E) hybrid breakdown.

12. Species that are found only in one particular geographic location are said to be
   (A) behaviorally evolved.
   (B) endemic.
   (C) speciated.
   (D) undergoing behavioral isolation.
   (E) undergoing mechanical isolation.

13. The allele that causes sickle-cell disease is found with greater frequency in Africa, where malaria is more of a threat, than in the United States. Which genetic phenomenon most likely contributes to the difference in frequency?
   (A) heterozygote advantage
   (B) heterozygote protection theory
   (C) balanced polymorphism
   (D) frequency-dependent selection
   (E) neutral variation

14. In a population of squirrels, the allele that causes bushy tail (B) is dominant, while the allele that causes bald tail is recessive (b). If 64% of the squirrels have a bushy tail, what is the frequency of the dominant allele?
   (A) 0.8
   (B) 0.6
   (C) 0.4
   (D) 0.36
   (E) 0.2

15. Which of the following factors would NOT contribute to allopatric speciation?
    (A) A population becomes geographically isolated from the parent population.
    (B) The separated population is small, and genetic drift occurs.
    (C) The isolated population is exposed to different selection pressures than the ancestral population.
    (D) Gene flow between the two populations is minimal or does not occur.
    (E) The different environments of the two populations create different mutations.

16. A marsupial living in Australia has evolved to eat tree leaves, be diurnal, and raise its young until they are of reproductive age. A grazing placental mammal has also evolved to eat tree leaves, be diurnal, and raise its young until they are of reproductive age. This is an example of which of the following types of evolution?
    (A) divergent evolution
    (B) species-specific evolution
    (C) convergent evolution
    (D) neutral evolution
    (E) sibling evolution

17. Which of the following can lead to sympatric speciation?
    (A) migration of a small number of individuals
    (B) natural disaster that cuts off contact between members of a population
    (C) a newly formed river separating segments of a population
    (D) autopolyploidy
    (E) bottleneck effect

18. Mitochondria and plastids contain DNA and ribosomes and make some, but not all, of their proteins. Some of their proteins are coded for by nuclear DNA and produced in the cytoplasm. What may explain this division of labor?
    (A) Over the course of evolution, some of the original endosymbiont's genes were transferred to the host cell's nucleus.
    (B) The host cell's genome always included genes for making mitochondrial and plastid proteins.
    (C) These organelles do not have sufficient resources to make all their proteins and rely on the help of the cell.
    (D) Some mitochondria and plastid genes were contributed by early bacterial prokaryotes that shared genes with other primitive cells.
    (E) The smaller prokaryotic ribosomes in these organelles cannot produce the eukaryotic proteins required for their functions.

19. Banded iron formations in marine sediments provide evidence of
    (A) the first prokaryotes around 3.5 billion years ago.
    (B) oxidized iron layers in terrestrial rocks.
    (C) the accumulation of oxygen in the seas from the photosynthesis of cyanobacteria.
    (D) the evolution of photosynthetic archaea near deep-sea vents.
    (E) the crashing of meteorites onto Earth, possibly transporting abiotically produced organic molecules from space.

20. Which of the following constitutes the smallest unit capable of evolution?
    (A) individual
    (B) group
    (C) population
    (D) clade
    (E) community

1. Mothers and teachers have often said they need another pair of eyes on the backs of their heads. And another pair of hands would come in handy in many situations. You can imagine that these traits would have been advantageous to our early hunter-gatherer ancestors as well. According to sound evolutionary reasoning, what is the most likely explanation for why humans do not have these traits?
   (A) Because they actually would not be beneficial to the fitness of individuals who possessed them. Natural selection always produces the most beneficial traits for a particular organism in a particular environment.
   (B) Because every time they have arisen before, the individual mutants bearing these traits have been killed by chance events. Chance and natural selection interact.
   (C) Because these variations have probably never appeared in a healthy human. As tetrapods we are pretty much stuck with a four-limbed, two-eyed body plan; natural selection can only edit existing variations.
   (D) Because humans are a relatively young species. If we stick around and adapt for long enough, it is inevitable that the required adaptations will arise.

2. According to this figure, which pair of organisms shares the most recent common ancestor?

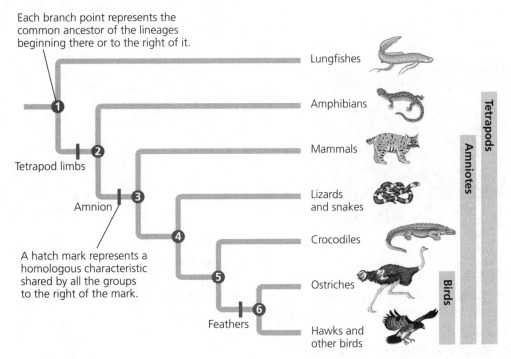

Each branch point represents the common ancestor of the lineages beginning there or to the right of it.

Tetrapod limbs

Amnion

A hatch mark represents a homologous characteristic shared by all the groups to the right of the mark.

Feathers

Lungfishes

Amphibians

Mammals

Lizards and snakes

Crocodiles

Ostriches

Hawks and other birds

Tetrapods

Amniotes

Birds

   (A) lungfish and amphibian
   (B) amphibian and lizard
   (C) mammal and crocodile
   (D) lizard and ostrich

3. According to the figure on the previous page, the group most closely related to lizards and snakes are
   (A) lungfishes.
   (B) amphibians.
   (C) mammals.
   (D) hawks and other birds.

4. Which statement best describes the mode of selection depicted in the figure?

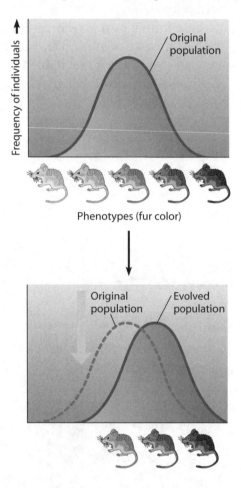

Phenotypes (fur color)

   (A) stabilizing selection, changing the average color of the population over time
   (B) directional selection, favoring the average individual
   (C) directional selection, changing the average color of the population over time
   (D) disruptive selection, favoring the average individual

5. Which species of wheat shown is polyploid? The uppercase letters represent not genes but *sets of chromosomes*.

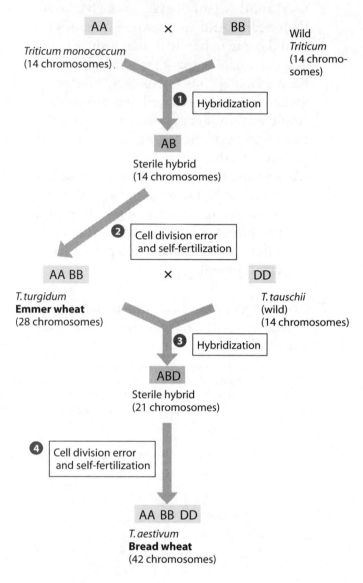

   (A) *T. monococcum*
   (B) the AB sterile hybrid
   (C) *T. turgidun*
   (D) *T. tauschii*

6. Which butterfly has changed gradually but significantly from its ancestor through microevolutionary events that were not part of a speciation event?

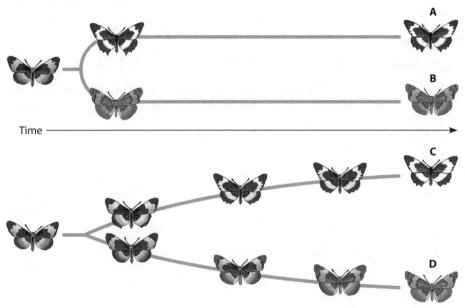

(A) butterfly A
(B) butterfly B
(C) butterfly C
(D) butterfly D

After reading the paragraphs, answer the question(s) that follow.

In 2004, scientists announced the discovery of the fossil remains of some extremely short early humans on the Indonesian island of Flores. The new species has been named *Homo floresiensis*. One hypothesis is that *H. floresiensis* evolved from *Homo erectus*, another early human species. How did a population of *H. erectus* become isolated on this remote island? Early humans constructed boats and rafts, so perhaps they were blown far off course by strong winds during a storm.

H. erectus averaged almost 6 feet in height, but the remains show that adults of *H. floresiensis* were only about 3 feet tall. It is hypothesized that limited resources on this hot and humid island (only 31 square miles) exerted selection pressure and succeeding generations began to shrink in size. Small bodies require less food, use less energy, and are easier to cool than larger bodies. Evolution of small size in similar circumstances has been observed in many other species, but never before in humans. This find demonstrates that evolutionary forces operate on humans in the same way as on all other organisms.

7. The evolution of *Homo floresiensis* is an example of
   (A) sympatric speciation.
   (B) allopatric speciation.
   (C) adaptive radiation.
   (D) hybridization.

8. If *H. floresiensis* were reunited with *H. erectus* at a much later date, but the two populations could no longer interbreed, it would be correct to conclude that *H. floresiensis*

(A) is no longer fertile as a species.
(B) had been isolated for more than 50,000 years.
(C) has become less fit than *H. erectus*.
(D) had evolved reproductive barriers.

Free-Response Question

1. *Microevolution is the change in the gene pool from one generation to the next.*

(a) **Describe** three ways in which microevolution can take place.
(b) **Describe** the difference between microevolution and macroevolution.

The Evolutionary History of Biological Diversity

Chapter 26: Phylogeny and the Tree of Life

YOU MUST KNOW

- The taxonomic categories and how they indicate relatedness.
- How systematics is used to develop phylogenetic trees.
- The three domains of life including their similarities and their differences.

Concept 26.1 Phylogenies show evolutionary relationships

> **TIP FROM THE READERS**
> This topic includes a broad look at taxonomic groups. Because of advances in DNA sequencing, classification is in a state of flux.

▌ **Phylogeny** is the evolutionary history of a species or a group of related species. It is constructed by using evidence from **systematics**, a discipline that focuses on classifying organisms and their evolutionary relationships. Its tools include fossils, morphology, genes, and molecular evidence.

▌ **Taxonomy** is an ordered division of organisms into categories based on a set of characteristics used to assess similarities and differences.

▌ **Binomial nomenclature** uses a two-part naming system that consists of the **genus** to which the species belongs, as well as the organisms' **species** within the genus, such as *Canis familiaris*, the scientific name of the common dog. This system was developed by **Carolus Linnaeus**.

▌ The hierarchical classification of organisms consists of the following levels, beginning with the most general or inclusive: **domain, kingdom, phylum, class, order, family, genus,** and **species**. Each categorization at any level is called a **taxon**.

▍ Systematists use branching diagrams called **phylogenetic trees** to depict hypotheses about evolutionary relationships. The branches of such trees reflect the hierarchical classifications of groups nested within more inclusive groups. (See Figure 7.1.)

▍ Notice on Figure 7.1 that two branch points are numbered. Each number represents a *common ancestor* of the two branches. Can you mark the point where all carnivores shared a common ancestor?

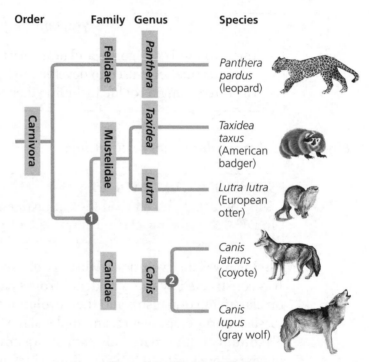

Figure 7.1   Phylogenetic tree

Concept 26.2   Phylogenies are inferred from morphological and molecular data

▍ **Homologous structures** are similarities due to shared ancestry, such as the bones of a whale's flipper and a tiger's paw.

▍ **Convergent evolution** has taken place when two organisms developed similarities as they adapted to similar environmental challenges—not because they evolved from a common ancestor. *Example*: The streamlined bodies of a tuna and a dolphin show convergent evolution.

▍ The likenesses that result from convergent evolution are considered **analogous** rather than homologous. They do not indicate relatedness, but rather similar solutions to similar problems. *Example*: The wing of a butterfly is analogous to the wing of a bat. Both are adaptations for flight.

- **Molecular systematics** uses DNA and other molecular data to determine evolutionary relationships. The more alike the DNA sequences of two organisms, the more closely related they are evolutionarily.

Concept 26.3 *Shared characters are used to construct phylogenetic trees*

- A **cladogram** depicts patterns of shared characteristics among taxa and forms the basis of a **phylogenetic tree**.
- A **clade**, within a tree, is defined as a group of species that includes an ancestral species and all its descendants.
- **Shared derived characters** are used to construct cladograms. They are evolutionary novelties unique to a particular clade. For example, hair is a shared derived character of mammals.
- Study Figure 7.2 to understand why a clade is *monophyletic*, and what is meant by paraphyletic and polyphyletic groupings.

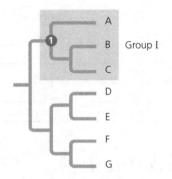

(a) Monophyletic group (clade). Group I, consisting of three species (A, B, C) and their common ancestor ❶, is a clade, also called a monophyletic group. A monophyletic group consists of an ancestral species and *all* of its descendants.

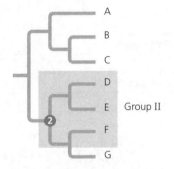

(b) Paraphyletic group. Group II is paraphyletic, meaning that it consists of an ancestral species ❷ and some of its descendants (species D, E, F) but not all of them (missing species G).

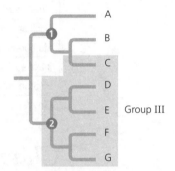

(c) Polyphyletic group. Group III is polyphyletic, meaning that it includes descendants of two or more common ancestors. In this case, species D, E, F, and G share common ancestor ❷, but species C has a different ancestor: ❶.

Figure 7.2   Cladograms

Concept 26.4 *An organism's evolutionary history is documented in its genome*

- The rate of evolution of DNA sequences varies from one part of the genome to another; therefore, comparing these different sequences helps us to investigate relationships between groups of organisms that diverged a long time ago.

  - DNA that codes for ribosomal RNA changes relatively slowly and is useful for investigating relationships between taxa that diverged hundreds of millions of years ago.
  - DNA that codes for mitochondrial DNA evolves rapidly and can be used to explore recent evolutionary events.

Concept 26.5 *Molecular clocks help track evolutionary time*

- **Molecular clocks** are methods used to measure the absolute time of evolutionary change based on the observation that some genes and other regions of the genome appear to evolve at constant rates.

Concept 26.6 **New information continues to revise our understanding of the tree of life**

▌ Taxonomy is in flux! When your authors were in high school, we were taught there were two kingdoms, plants and animals; then in our college courses, we were introduced to five kingdoms: Monera, Protista, Plantae, Fungi, and Animalia.

▌ Now biologists have adopted a **three-domain** system, which consists of the domains Bacteria, Archaea, and Eukarya. This system arose from the finding that there are two distinct lineages of prokaryotes.

▌ The domains Bacteria and Archaea contain *prokaryotic* organisms, and Eukarya contains *eukaryotic* organisms. As we gain more tools for analysis, earlier ideas about evolutionary relatedness are changed, and so taxonomy, too, continues to evolve. That said, let's look at the three domains. Principal differences between the groups are simplified and presented below.

A Comparison of the Three Domains of Life

Characteristic	Bacteria	Archaea	Eukarya
Nuclear envelope	no	no	yes
Membrane-enclosed organelles	no	no	yes
Introns	no	yes	yes
Histone proteins associated with DNA	no	yes	yes
Circular chromosome	yes	yes	no

Chapter 27: Bacteria and Archaea

YOU MUST KNOW

- The key ways in which prokaryotes differ from eukaryotes with respect to genome, membrane-bound organelles, size, and reproduction.
- Mechanisms that contribute to genetic diversity in prokaryotes, including transformation, conjugation, transduction, and mutation.

Concept 27.1 **Structural and functional adaptations contribute to prokaryotic success**

▌ Life is divided into three domains: **Archaea, Bacteria, and Eukarya**. Both domain Bacteria and domain Archaea are made up of prokaryotes.

▌ The most common shapes of prokaryotes are spheres, rods, and helices; most are about 1–5 μm in size. This is perhaps 1/10 the size of a typical eukaryotic cell.

▌ As you read the following sections, study Figure 7.3. Prokaryotes have *no true nuclei* or internal compartmentalization. The DNA is concentrated in a *nucleoid region* and has little associated protein. The small genome consists of a single circular chromosome. Prokaryotes reproduce through an asexual process called **binary fission**, and have short generation times.

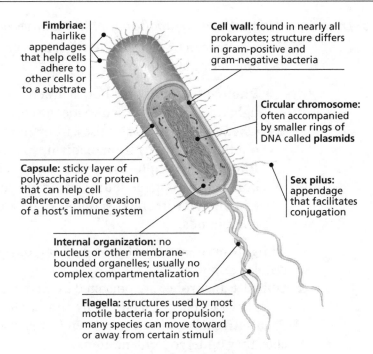

Figure 7.3 Prokaryote

Within the figure:

Fimbriae: hairlike appendages that help cells adhere to other cells or to a substrate

Cell wall: found in nearly all prokaryotes; structure differs in gram-positive and gram-negative bacteria

Circular chromosome: often accompanied by smaller rings of DNA called **plasmids**

Capsule: sticky layer of polysaccharide or protein that can help cell adherence and/or evasion of a host's immune system

Sex pilus: appendage that facilitates conjugation

Internal organization: no nucleus or other membrane-bounded organelles; usually no complex compartmentalization

Flagella: structures used by most motile bacteria for propulsion; many species can move toward or away from certain stimuli

▌ In addition to their one major chromosome, prokaryotic cells may also possess smaller, circular, self-replicating pieces of DNA called **plasmids**.

▌ Outside their cell membranes, most prokaryotes possess a cell wall that contains **peptidoglycans**. **Gram-positive** bacteria have simpler walls with more peptidoglycans, whereas **gram-negative** cells have walls that are structurally more complex.

▌ Prokaryotes use appendages called **pili** that adhere to each other or to surrounding surfaces. About half of the prokaryotes are **motile**, because they possess whiplike **flagella**.

　　▪ Because the flagellum of a bacterium is structurally different from the eukaryotic flagellum, this is another example of analogous structures.

Concept 27.2 Rapid reproduction, mutation, and genetic recombination promote genetic diversity in prokaryotes

▌ It is important for you to know that there are many ways for genetic information to be transferred between bacterial cells. Be able to clearly explain the three mechanisms that follow.

▌ Three mechanisms by which bacteria can transfer genetic material between each other are

　　▪ **Transformation**, in which a prokaryote takes up DNA from its environment.

　　▪ **Transduction**, in which viruses transfer genes between prokaryotes.

　　▪ **Conjugation**, in which genes are directly transferred from one prokaryote to another as they are temporarily joined by a "mating bridge."

▌ The major source of genetic variation in prokaryotes is **mutation.**

Concept 27.3 Diverse nutritional and metabolic adaptations have evolved in prokaryotes

▌ Prokaryotes can be placed in four groups according to how they take in carbon and how they obtain energy:

■ **Photoautotrophs** are photosynthetic, and they use the power of sunlight to convert carbon dioxide into organic compounds.

■ **Chemoautotrophs** also use carbon dioxide as their source of carbon, but they get energy from oxidizing inorganic substances.

■ **Photoheterotrophs** use light to make ATP but must obtain their carbon from an outside source already fixed in organic compounds.

■ **Chemoheterotrophs** get both carbon and energy from organic compounds.

▌ **Obligate aerobes** cannot grow without oxygen because they need oxygen for cellular respiration.

▌ **Obligate anaerobes** are poisoned by oxygen. Some use fermentation, whereas others extract chemical energy by another form of anaerobic respiration.

▌ **Facultative anaerobes** use oxygen if it is available; when oxygen is not available, they undergo fermentation.

▌ Some prokaryotes can use atmospheric nitrogen as a direct source of nitrogen in a process called **nitrogen fixation**. They convert N_2 to NH_4^+. You might want to review the nitrogen cycle from Chapter 55.

Concept 27.4 Molecular systematics is illuminating prokaryotic phylogeny

▌ The first prokaryotes that were classified in the domain Archaea are known as **extremophiles** and live in extreme environments such as geysers:

■ **Extreme halophiles** live in saline environments (highly concentrated with salt).

■ **Extreme thermophiles** live in very hot environments.

▌ Other archaea do not live in extreme environments. **Methanogens** use carbon dioxide to oxidize H_2 and produce methane as a waste product.

Concept 27.5 Prokaryotes play crucial roles in the biosphere

▌ Many prokaryotes are **decomposers,** breaking down dead corpses, vegetation, and waste products.

▌ Many prokaryotes are **symbiotic,** meaning that they form relationships with other species:

■ **Mutualism:** Both symbiotic organisms benefit.

■ **Commensalism:** One organism benefits, whereas the other is neither helped nor harmed.

■ **Parasitism:** One organism benefits at the expense of the other.

Concept 27.6 Prokaryotes have both beneficial and harmful impacts on humans

▌ Some prokaryotes are **pathogenic** and cause illness by producing poisons.

▌ **Antibiotics** are chemicals that can kill prokaryotes. They are *not* effective against viruses. Many bacterial plasmids contain resistance genes to different antibiotics.

▌ Prokaryotes are used by humans in many diverse ways:

■ **Bioremediation**, removing pollutants from soil, air, or water. This includes treating sewage, cleaning up oil spills, and precipitating radio-active materials.

■ Symbionts in the gut, manufacturing vitamins, and digesting foods.

■ Gene cloning and producing transgenic organisms.

■ Production of cheese and yogurt and other products.

Chapter 28: Protists

WHAT'S IMPORTANT TO KNOW?

The new Curriculum Framework does not include specific information that you must know, but we suggest that you work through the short discussion of protists included here.

Concept 28.1 Most eukaryotes are single-celled organisms

▌ **Protist** is now a term used to refer to eukaryotes that are neither plants, animals, nor fungi. Biologists no longer consider Protista a kingdom because it is *paraphyletic*. Refer back to Figure 7.2 to review what this means. Some protists are more closely related to plants, fungi, or animals than to other protists.

▌ Protists vary in structure and function more than any other group of eukaryotes. Here are some general commonalities, but even these are not true for all groups:

■ Most are unicellular.

■ Most use aerobic metabolism and have mitochondria.

▌ According to current theory, mitochondria and chloroplasts evolved through **endosymbiosis**. They were originally unicellular organisms engulfed by other cells that ultimately became organelles in the host cell.

▌ Protists can be divided into three categories: photosynthetic (plantlike) algae, ingestive (animal-like) protozoans, and absorptive (fungus-like) organisms.

▌ Most protists are aquatic and are important constituents of plankton. Many other protists live as symbionts in other organisms.

▌ Protists are such a diverse group that their classification continues to undergo revision. Despite this, you may recognize these protists:

■ *Giardia intestinalis* (causes "hiker's diarrhea"; always treat your water!)

■ *Trichomonas vaginalis* (sexually transmitted infection)

- Trypanosoma species (sleeping sickness and Chagas' disease)
- *Euglena* (remember seeing the tiny flagellated green cell with a red eye-spot in Bio. I?)
- Dinoflagellates (blooms cause "red tides"; many are bioluminescent)
- *Plasmodium* (causative agent of malaria)
- Ciliates (*Paramecium* and *Stentor* are examples; micro and macronuclei)
- Amoebas (move by pseudopodia)
- Diatoms (two-part glass-like wall made of silica)
- Golden algae (have yellow and brown carotenoids)
- Brown algae (kelp)
- Oomycetes (water molds and their relatives; include causative agent of potato blight)
- Red algae (multicellular; some found at great depths; sushi wraps)
- Green algae (*Clamydomonas, Ulva, Volvox*; this group is the closest relative of land plants)
- Slime molds (often seen as flat, yellowish oozes in damp mulch)

Chapter 29: Plant Diversity I: How Plants Colonized Land

WHAT'S IMPORTANT TO KNOW?

Plant evolution is not an explicit topic of the new Curriculum Framework, but your teacher may choose to introduce you to this material. Here are what we consider to be the main points of this chapter:

- Why land plants are thought to have evolved from green algae.
- Disadvantages and advantages of life on land.
- That plants have a unique life cycle termed *alternation of generations* with a gametophyte generation and a sporophyte generation.
- The role of antheridia and archegonia in gametophytes.
- The major characteristics of bryophytes.
- The major characteristics of seedless vascular plants.

Concept 29.1 Land plants evolved from green algae

- Land plants evolved from green algae more than 500 million years ago. Plants have enabled other life-forms to survive on land. Plants supply oxygen and are the ultimate provider of most of the food eaten or absorbed by animals and fungi.
- The evolution of land plants from the green algae group known as the *charophytes* includes the following five lines of evidence:

    1. They produce cellulose for cell walls in the same, unique fashion.
    2. The peroxisomes of these two groups, but no others, have enzymes that reduce the effects of photorespiration.

3. The structure of their sperm is similar.
4. They both produce cell plates in the same way during cell division.
5. Genetic evidence including analysis of nuclear and chloroplast genes indicates the two groups are closely related.

▍ Movement onto land had both advantages and challenges:

- ▪ Advantages included increased sunlight unfiltered by water, more carbon dioxide in the atmosphere than the water, soils rich in nutrients, and fewer predators.
- ▪ Challenges included a lack of water, desiccation, and a lack of structural support against gravity.

▍ All land plants have a life cycle that consists of two multicellular stages, called **alternation of generations**.

▍ The two stages are the **gametophyte** stage (in which the plant cells are haploid) and the **sporophyte** stage (in which the plant cells are diploid).

- ▪ In the gametophyte stage, the gametes are produced. During fertilization, egg and sperm fuse to form a diploid zygote—the sporophyte—which divides mitotically. The sporophyte produces spores by meiosis.
- ▪ The zygote develops within the tissues of the female parent, deriving nutrients from it. For this reason land plants are sometimes referred to as **embryophytes**.

▍ Another key feature of plants is the production of gametes in multicellular organs called **gametangia**. The female gametangia are the **archegonia**, which produce a single egg. The male gametangia are the **antheridia**, which produce many sperm.

Concept 29.2 *Mosses and other nonvascular plants have life cycles dominated by gametophytes*

▍ Bryophytes include three phyla: mosses, liverworts, and hornworts.

▍ The bryophytes are considered **nonvascular** (no xylem or phloem tissue), even though some mosses do have simple vascular tissue. The lack of vascular tissue accounts for the small size of bryophytes.

▍ Unlike vascular plants, in all three bryophyte phyla the gametophytes are the dominant stage of the life cycle. The archegonia and antheridia are found on the gametophytes. Bryophytes require water for the sperm to swim to the egg during fertilization. Follow Figure 7.4 for the basic steps to the life cycle of a typical moss.

▍ Bryophytes have the smallest, simplest sporophytes of all plant groups. Even though the sporophytes are photosynthetic when young, they must absorb water, sugars, and other nutrients from parental gametophytes. Notice from Figure 7.4 that meiosis occurs as the mechanism from sporophyte to gametophyte.

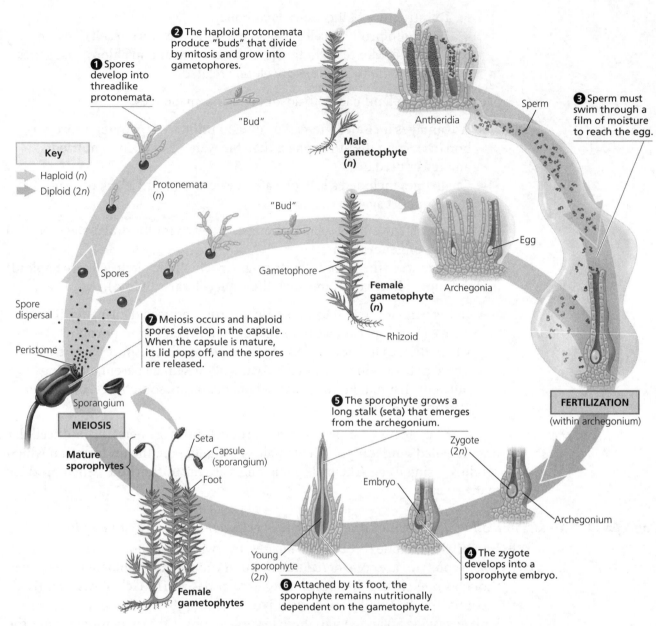

1 Spores develop into threadlike protonemata.

2 The haploid protonemata produce "buds" that divide by mitosis and grow into gametophores.

3 Sperm must swim through a film of moisture to reach the egg.

Sperm

Antheridia

Male gametophyte (n)

"Bud"

Key

→ Haploid (n)
→ Diploid (2n)

Protonemata (n)

"Bud"

Gametophore

Spores

Egg

Archegonia

Female gametophyte (n)

Rhizoid

Spore dispersal

7 Meiosis occurs and haploid spores develop in the capsule. When the capsule is mature, its lid pops off, and the spores are released.

Peristome

MEIOSIS

Sporangium

Mature sporophytes {

Seta
Capsule (sporangium)
Foot

5 The sporophyte grows a long stalk (seta) that emerges from the archegonium.

Zygote (2n)

Embryo

FERTILIZATION

(within archegonium)

Archegonium

4 The zygote develops into a sporophyte embryo.

Young sporophyte (2n)

6 Attached by its foot, the sporophyte remains nutritionally dependent on the gametophyte.

Female gametophytes

Figure 7.4   Life cycle of a typical moss

Concept 29.3 *Ferns and other seedless vascular plants were the first plants to grow tall*

▌ Besides ferns, this group includes two other plants you might commonly recognize, the club mosses and horsetails.

▌ The evolution of vascular tissues allowed vascular plants to grow taller than bryophytes, gaining access to sunlight. Ferns and other seedless vascular plants, however, still require a film of water for the sperm to reach the egg.

▌ In this group, the **sporophyte** is the dominant stage. Meiosis produces haploid **spores** and occurs with the **sporangia**. The haploid spores may grow into gametophytes, where swimming sperm fertilize eggs, yielding the diploid zygote, which will grow into the sporophyte generation. Through this

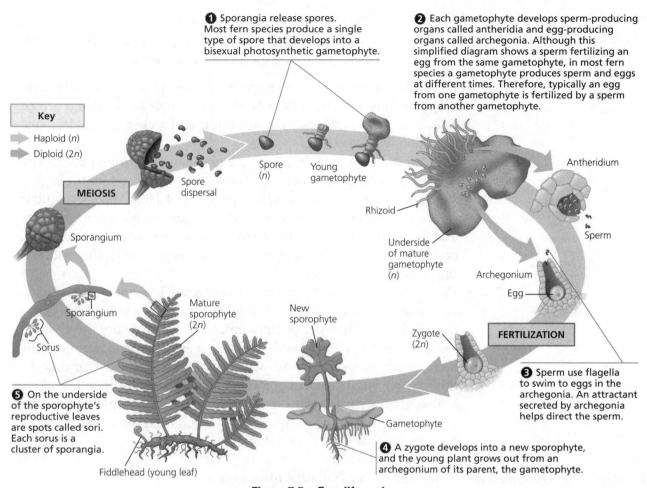

① Sporangia release spores. Most fern species produce a single type of spore that develops into a bisexual photosynthetic gametophyte.

② Each gametophyte develops sperm-producing organs called antheridia and egg-producing organs called archegonia. Although this simplified diagram shows a sperm fertilizing an egg from the same gametophyte, in most fern species a gametophyte produces sperm and eggs at different times. Therefore, typically an egg from one gametophyte is fertilized by a sperm from another gametophyte.

Key
→ Haploid (*n*)
→ Diploid (*2n*)

MEIOSIS

Sporangium

Sporangium

Sorus

⑤ On the underside of the sporophyte's reproductive leaves are spots called sori. Each sorus is a cluster of sporangia.

Fiddlehead (young leaf)

Spore dispersal

Spore (*n*)

Young gametophyte

Mature sporophyte (*2n*)

New sporophyte

Rhizoid

Underside of mature gametophyte (*n*)

Antheridium

Sperm

Archegonium

Egg

Zygote (*2n*)

FERTILIZATION

Gametophyte

③ Sperm use flagella to swim to eggs in the archegonia. An attractant secreted by archegonia helps direct the sperm.

④ A zygote develops into a new sporophyte, and the young plant grows out from an archegonium of its parent, the gametophyte.

Figure 7.5   Fern life cycle

alternation between generations, the life cycle name originated. Use Figure 7.5 above to follow the key steps to a typical fern life cycle.

▌ Seedless vascular plants formed forests of tall plants in the Carboniferous period, eventually forming deposits of coal that we use for fuel today.

Chapter 30: Plant Diversity II: The Evolution of Seed Plants

WHAT'S IMPORTANT TO KNOW?
Plant evolution is not an explicit topic of the new Curriculum Framework, but your teacher may choose to introduce you to this material. Here are what we consider to be the main points of this chapter:

- Key adaptations to life on land unique to seed plants.
- The evolutionary significance of seeds and pollen.
- The role of flowers and fruits in angiosperm reproduction.
- The role of stamens and carpels in angiosperm reproduction.

Concept 30.1 *Seeds and pollen grains are key adaptations for life on land*

- **Seeds** are plant embryos packaged with a food supply in a protective coat.
- Five crucial adaptations led to the success of seed plants:

  - *Reduced gametophytes.* Gametophytes in seed plants are mostly microscopic and entirely dependent on the sporophyte for food and protection. This protects the delicate antheridia and archegonia, increasing reproductive success.
  - *Heterospory.* **Heterospory** means the production of two types of spores. **Megaspores** produce female gametophytes, which produce the eggs. **Microspores** produce male gametophytes, which contain sperm nuclei.
  - *Ovules and the production of eggs.* The megasporangium, megaspore, and the protective tissue around them make an ovule. The ovule increases protection of the egg and the developing zygote, thus increasing reproductive fitness.
  - *Pollen and production of sperm.* A pollen grain is a male gametophyte, containing two sperm nuclei. The pollen grain has a waterproof coating, allowing for transfer by the wind. Until pollen, water was required for sperm transfer. *The evolution of pollen was a key adaptation to land.*
  - *Seeds.* Seeds have several advantages over spores. Seeds are multicellular, with several layers of protective tissue, safeguarding the embryo. Unlike spores, seeds have a supply of stored energy, which allows the seed to wait for good germination conditions and use stored energy to finance the early growth of the embryo.

Concept 30.2 *Gymnosperms bear "naked" seeds, typically on cones*

- Gymnosperms are plants that have "naked" seeds that are not enclosed in ovaries. Their seeds are often exposed on modified leaves that form cones. To compare, angiosperms (flowering plants) have seeds enclosed in fruits, which are mature ovaries. Gymnosperms do not have fruits.
- Four phyla of plants are considered gymnosperms, but the most ecologically significant group is the conifers (*Coniferophyta*), which include pines, spruces, firs, and redwoods.
- Use Figure 7.6 to review the life cycle of a pine. Notice the evolutionary advances listed in Figure 7.6 and the important role they play in the life cycle.

Concept 30.3 *The reproductive adaptations of angiosperms include flowers and fruits*

- **Angiosperms** are seed plants that produce the reproductive structures called flowers and fruits. Today, angiosperms account for about 90% of all plant species.
- The major reproductive adaptation of the angiosperm is the **flower**, which consists of four floral organs: sepals, petals, stamens, and carpels. Use Figure 7.7 to review the major parts of the flower.

  - **Stamens** are the male reproductive structures, producing microspores in the anthers that develop into pollen grains.
  - **Carpels** are the female reproductive structures, producing megaspores and their products—female gametophytes with eggs.

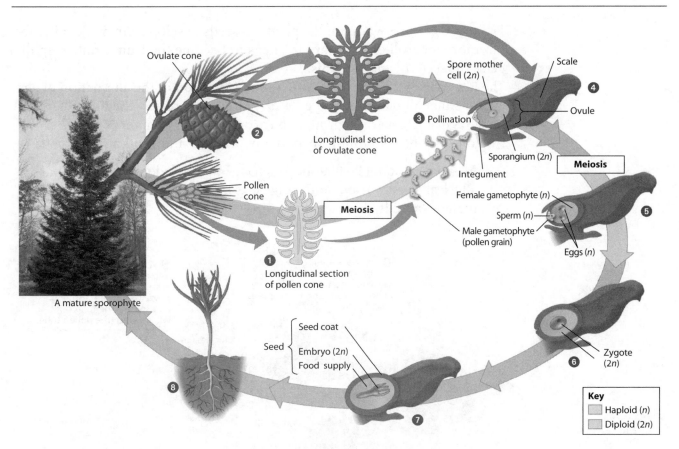

Figure 7.6   Pine life cycle

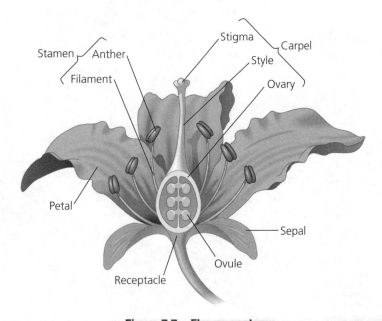

Figure 7.7   Flower anatomy

■ **Fruits** are mature ovaries of the plant. As seeds develop from ovules after fertilization, the wall of the ovary thickens to become the fruit. Fruits help disperse the seeds of angiosperms.

■ The life cycle of the angiosperm is a refined version of the alternation of generations that all plants undergo. Use Figure 7.8 to review the life cycle. More specifics of angiosperm reproduction are covered in Chapter 38.

■ Angiosperms have traditionally been divided into monocots and eudicots:

  ■ **Monocots** (about 70,000 species) have one cotyledon in the seed, parallel leaf venation, and flowering parts in multiples of threes. Examples include orchids, lilies, and grasses.

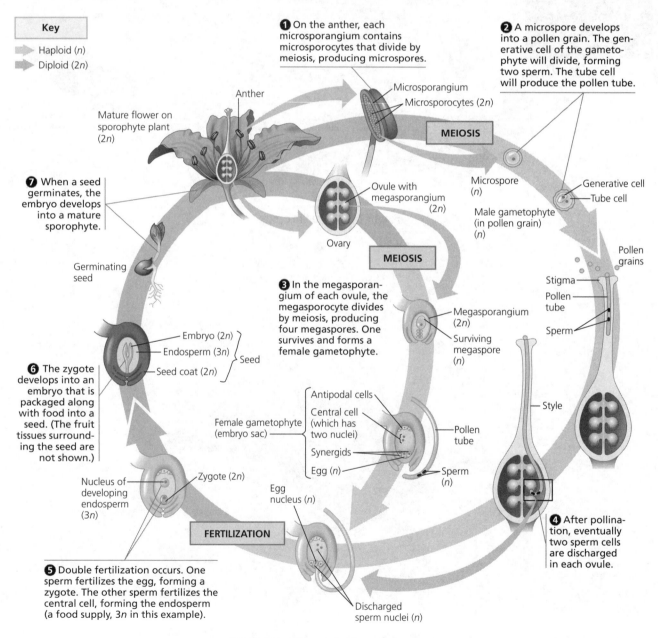

Figure 7.8   Life cycle of an angiosperm

- **Eudicots** (about 170,000 species) have two cotyledons in the seed, net leaf veination, and flowering parts usually in multiples of fours or fives. Examples include roses, peas, beans, and oaks.

Chapter 31: Fungi

WHAT'S IMPORTANT TO KNOW?

The study of fungi is not an explicit topic of the new Curriculum Framework, but your teacher may choose to introduce you to this material and select several illustrative examples from this group. Here is a summary of what may be most useful to know in this chapter:

- The characteristics of fungi.
- Important ecological roles of fungi in mycorrhizal associations, and as decomposers and parasitic plant pathogens.

Concept 31.1 Fungi are heterotrophs that feed by absorption

▌ Fungi are eukaryotes with these characteristics:

- **Multicellular heterotrophs** that obtain nutrients by **absorption**. Fungi secrete hydrolytic enzymes, digest food outside their bodies, and absorb the small molecules.
- The cell walls of fungi are made of *chitin*.
- Bodies composed of filaments called **hyphae** that are entwined to form a mass, the **mycelium**.
- Most fungi are multicellular with hyphae divided into cells by cross-walls called **septa**. *Coenocytic* fungi lack septa and consist of a continuous cytoplasmic mass containing hundreds of nuclei.
- Fungi reproduce by spores.
- Modes of nutrition include decomposers, parasites, and mutualists.

Concept 31.5 Fungi play key roles in nutrient cycling, ecological interactions, and human welfare

▌ Fungi are important decomposers of organic material, including cellulose and lignin.

▌ **Mycorrhizal fungi** are found in association with plant roots and may improve delivery of minerals to the plant, while being supplied with organic nutrients. This is a fine example of mutualism.

▌ **Lichens** are symbiotic associations of photosynthetic microorganisms (algae) embedded in a network of fungal hyphae. They are very hardy organisms that are pioneers on rock and soil surfaces.

▌ Thirty percent of known species of fungi are parasites. Many are plant pathogens (e.g., Dutch elm disease, chestnut blight, dogwood anthracnose, wheat rust, and ergot). Some infect animals (e.g., ringworm, athlete's foot, and *Candida*).

Yeasts are unicellular ascomycetes that figure prominently in molecular biology and biotechnology. Current research includes work into the genes involved in Parkinson's disease and Huntington's disease by examining homologous genes in yeast (*Saccharomyces cerevisiae*).

Chapter 32: An Overview of Animal Diversity

> **WHAT'S IMPORTANT TO KNOW?**
> Animal diversity is not an explicit topic of the new Curriculum Framework, but your teacher may choose to introduce you to this material. Here are what we consider to be some main points of this chapter:
>
> - The characteristics of animals.
> - The stages of animal development.
> - How to sort the animal phyla based on symmetry, development of a body cavity, and the fate of the blastopore.

Concept 32.1 Animals are multicellular, heterotrophic eukaryotes with tissues that develop from embryonic layers

❚ Animals have the following characteristics:

  ▪ They are multicellular heterotrophs.
  ▪ Most have muscle and nervous tissue.
  ▪ Most reproduce sexually, with a flagellated sperm and a large egg uniting to form a diploid *zygote*. The diploid stage dominates the life cycle.

❚ Some animals have **larvae**, an immature form distinct from the adult stage that will undergo **metamorphosis**.

❚ Animals share *Hox* genes, a unique homeobox-containing family of genes that plays important roles in development.

Concept 32.3 Animals can be characterized by "body plans"

❚ **No symmetry:** the sponges.
❚ **Radial symmetry** occurs in jellyfish and other organisms, in which any cut through the central axis of the organism would produce mirror images.
❚ **Bilateral symmetry** occurs in lobsters, humans, and many other organisms. These animals have a right side and a left side, and a single cut would divide the animal into two mirror image halves. There is also a *dorsal* (back) side, *ventral* (belly) side, and *anterior* (head) and *posterior* (tail) ends.
❚ **Cephalization** is the concentration of sensory equipment at one end (usually the anterior, or head end) of the organism.
❚ **Acoelomates**, such as flatworms, have no cavities between their alimentary canal and the outer wall of their bodies.

- **Pseudocoelomates** are triploblastic animals (animals with three tissue layers) with a cavity formed from the mesoderm and endoderm.
- **Coelomates** possess a **true coelom**. This is a body cavity filled with fluid, and this space separates an animal's digestive tract from the outer body wall. The coelom forms from tissue derived from mesoderm only.
- *Functions of the body cavity* (see Figure 7.9) include the following:
  - It cushions suspended organs.
  - It acts as a hydrostatic skeleton.
  - It enables internal organs to grow and move independently.

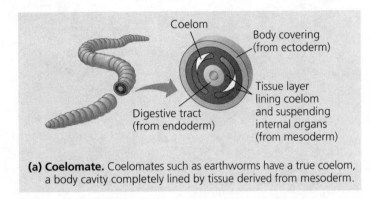

(a) Coelomate. Coelomates such as earthworms have a true coelom, a body cavity completely lined by tissue derived from mesoderm.

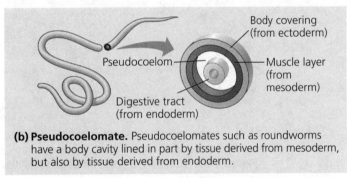

(b) Pseudocoelomate. Pseudocoelomates such as roundworms have a body cavity lined in part by tissue derived from mesoderm, but also by tissue derived from endoderm.

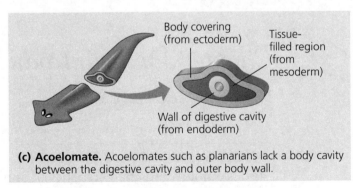

(c) Acoelomate. Acoelomates such as planarians lack a body cavity between the digestive cavity and outer body wall.

Figure 7.9   Body cavities of triploblastic animals

▌ **Protostomes** and **deuterostomes** differ in three major ways. Study Figure 7.10 to summarize these.

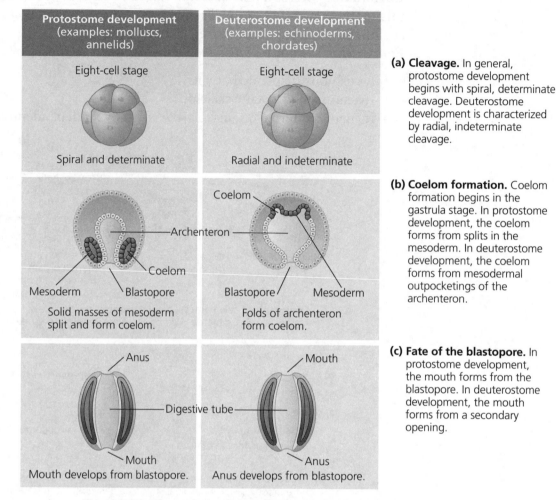

Key
- Ectoderm
- Mesoderm
- Endoderm

Protostome development (examples: molluscs, annelids)

Eight-cell stage

Spiral and determinate

Mesoderm | Coelom | Blastopore

Solid masses of mesoderm split and form coelom.

Anus

Digestive tube

Mouth

Mouth develops from blastopore.

Deuterostome development (examples: echinoderms, chordates)

Eight-cell stage

Radial and indeterminate

Coelom | Archenteron

Blastopore | Mesoderm

Folds of archenteron form coelom.

Mouth

Digestive tube

Anus

Anus develops from blastopore.

(a) Cleavage. In general, protostome development begins with spiral, determinate cleavage. Deuterostome development is characterized by radial, indeterminate cleavage.

(b) Coelom formation. Coelom formation begins in the gastrula stage. In protostome development, the coelom forms from splits in the mesoderm. In deuterostome development, the coelom forms from mesodermal outpocketings of the archenteron.

(c) Fate of the blastopore. In protostome development, the mouth forms from the blastopore. In deuterostome development, the mouth forms from a secondary opening.

Figure 7.10   Comparison of protostome and deuterostome

Chapter 33: An Introduction to Invertebrates

WHAT'S IMPORTANT TO KNOW?
The taxonomy and characteristics of animals are not part of the new Curriculum Framework and will not be covered on the exam. Your teacher may choose to have you learn about strategies in different groups for managing physiological challenges, and the material that follows may help you organize your knowledge of the invertebrates.

- **Phylum Porifera** (pore-bearing animals) includes the sponges.
- Sponges are *sessile* (anchored to the substrate) and lack true tissues. They are *diploblastic* (their body consists of only two layers of cells).
- Sponges have no nerves or muscles. The body of a sponge looks like a sac with holes in it. Water is drawn through the pores into the **spongocoel** and flows out through the **osculum** through the movement of flagellated **choanocytes**.
- **Spicules** comprise a skeletal framework.
- The **Eumetazoans** (animals with true tissues)
- **Radially symmetrical animals**

  - Members of **phylum Cnidaria** exist in *polyp* (vase-like; think of a sea anemone) and *medusa* form (think of a jellyfish). Examples of cnidarians are hydras, jellyfish, and corals.
  - They have radial symmetry, a central digestive compartment known as a gastrovascular cavity, and *cnidocytes* (cells that function in defense and the capture of prey).

- **Bilaterally symmetrical animals**

  - **Acoelomates** (animals without a body cavity)

    - **Phylum Platyhelminthes** (flatworms) *Examples*: planarians, tapeworms, and flukes

      - Flattened bodies with *cephalization* (sense organs at the anterior end)
      - Excretion by *flame bulbs* and *protonephridia*
      - No specialized organs for gas exchange or circulation
      - Gastrovascular cavity with a single opening

  - **Pseudocoelomates**

    - **Phylum Nematoda** (roundworms) *Examples*: *Ascaris*, *C. elegans*, *Trichinella spiralis*, pinworms, and hookworms

      - Cylindrical bodies, with a tough *cuticle*
      - Complete alimentary canal; no circulatory system

  - **Coelomates**

    - **Phylum Mollusca** (soft-bodied animals) *Examples*: slugs, clams, snails, squids, and octopuses

      - Muscular *foot* for movement, a *visceral mass* containing most of the organs, and a *mantle*, which secretes a shell.
      - Open circulatory system. This means that the circulatory fluid (hemolymph) is not always contained within vessels but sometimes circulates through body sinuses collectively called the *hemocoel*.
      - Excretion through *nephridia*.
      - Clades include Polyplacophora (chitons), Gastropoda (snails and slugs), Bivalvia (clams and oysters), and Cephalopoda (squids and octopuses).

- **Phylum Annelida** (segmented worms) *Examples*: earthworms, leeches
  - Internal and external segmentation
  - Excretion by *metanephridia* in each segment
  - Closed digestive system with specialized regions (the crop, gizzard, esophagus, and intestine)
  - Brainlike *central ganglia* with a ventral nerve cord
  - Closed circulatory system

- **Phylum Arthropoda** (jointed-legged animals) *Examples*: insects, arachnids, millipedes, centipedes, and crustaceans
  - *Exoskeleton* of *chitin*, which must be *molted* to grow
  - Jointed appendages
  - Open circulatory system
  - Various organs for gas exchange including *gills, book lungs,* and *tracheal systems*
  - Ventral nerve cords
  - Insects undergo metamorphosis during development:
    - **Incomplete metamorphosis:** egg, nymph, adult (e.g., grasshoppers)
    - **Complete metamorphosis:** egg, larva, pupa, adult (e.g., butterflies)

- **Deuterostomia**
  The clade Deuterostomia contains a diverse array of organisms, from sea stars to chordates. All have radial cleavage and share common developmental processes. There are two main phyla:

- **Phylum Echinodermata** (spiny-skinned animals) *Examples*: starfish, brittle stars, sea urchins, sand dollars, sea lilies, and sea cucumbers
  - Larvae have bilateral symmetry; adults radiate from the center, often as five spokes.
  - Have a thin skin covering an *exoskeleton*.
  - Have a *water vascular system*. This is a network of internal canals that branch into *tube feet* used for moving, feeding, and gas exchange.

- **Phylum Chordata** includes two invertebrate subphyla as well as all vertebrates.

Chapter 34: The Origin and Evolution of Vertebrates

WHAT'S IMPORTANT TO KNOW?
The taxonomy and characteristics of animals are not part of the new Curriculum Framework and will not be covered on the exam. Your teacher may choose to have you learn about strategies in different groups for managing physiological challenges, and the material that follows may help you organize your knowledge of this group.

Concept 34.1 *Chordates have a notochord and a dorsal, hollow nerve cord*

▌ **Vertebrates** derive their name from the **vertebrae,** a series of bones that make up the backbone. In the majority of vertebrates, vertebrae enclose the spinal cord and have assumed the roles of the notochord.

▌ Vertebrates are members of the phylum **Chordata.**

▌ There are four shared derived characteristics of all chordates. Note that some of these are present only during embryonic development (see Figure 7.11).

  ■ A **notochord**—a long, flexible rod that appears during embryonic development between the digestive tube and the dorsal nerve cord. This is *not* the spinal cord or the vertebral column! Students commonly confuse all of these.

  ■ A **dorsal, hollow nerve cord**—formed from a plate of ectoderm that rolls into a hollow tube. Recall that other phyla you have studied, such as annelids and arthropods, had a *ventral* nerve cord.

  ■ **Pharyngeal clefts**—grooves that separate a series of pouches along the sides of the pharynx. In most chordates (but not tetrapods, the critters with four legs), the clefts develop into slits that allow water to enter and exit the mouth without going through the digestive tract.

  ■ A muscular **tail** posterior to the anus.

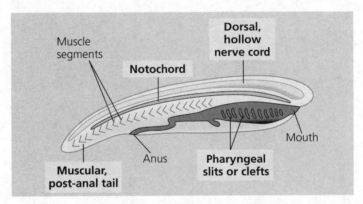

Figure 7.11   Chordate characteristics

Concept 34.3 *Vertebrates are craniates that have a backbone*

As you review this section, use Figure 7.12 to help you organize these groups.

▌ **Lampreys** are the oldest lineage of vertebrates. They are jawless parasitic fish with a skeleton of cartilage.

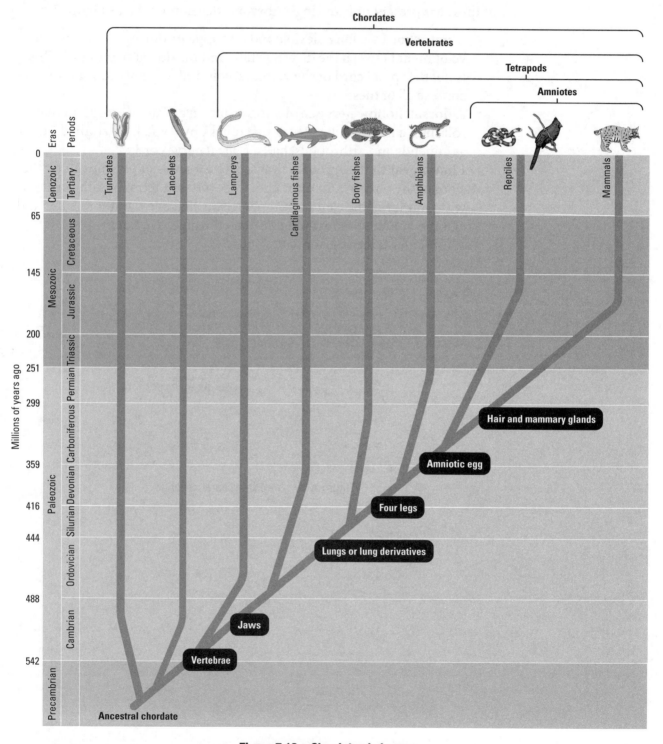

Figure 7.12   Chordate phylogeny

Concept 34.4 *Gnathostomes are vertebrates that have jaws*

▌ The jaws of vertebrates evolved from the modification of skeletal parts that had once supported the pharyngeal (gill) slits.

- **Class Chondrichthyes** members have flexible endoskeletons composed of cartilage, possess streamlined bodies, are denser than water, and will sink if they stop swimming. Some examples are sharks and rays (such as stingrays).
- **Class Osteichthyes** members are the bony fishes; these are the most numerous of all vertebrate groups. They have a bony endoskeleton, are covered in scales, and possess a swim bladder for buoyancy control. Some examples are trout and salmon.

Concept 34.5 *Tetrapods are gnathostomes that have limbs*

▌ **Class Amphibia** members include frogs, toads, salamanders, and newts.

- Gas exchange can occur across their thin, moist skin, though many members have lungs.
- Some, such as frogs, have an aquatic larval stage with gills and metamorphosis to an adult stage with lungs.
- They have external fertilization and external development in an aquatic environment. Eggs lack a shell.
- As larvae, amphibians have a two-chambered heart (one atrium, one ventricle), whereas the adult stage generally has three chambers (two atria, one ventricle).

Concept 34.6 *Amniotes are tetrapods that have a terrestrially adapted egg*

▌ The clade of **amniotes** consists of mammals and reptiles (including birds).

▌ The **amniotic egg** was an important evolutionary development for life on land. Amniotic eggs have a shell that retains water and, thus, can be laid in a dry environment.

▌ Amniotic eggs have four specialized **extraembryonic membranes:**

- The **amnion** encloses a fluid compartment that bathes the embryo and absorbs shock.
- The **chorion**, **allantois**, and **yolk sac** function in gas exchange, waste storage, and transfer of stored nutrients.

▌ **Reptiles** include turtles, lizards and snakes, alligators and crocodiles, birds, and the extinct dinosaurs.

- Modern reptiles have **scales**, containing keratin, which are an adaptation for terrestrial living because they reduce water loss. They obtain oxygen through their lungs, not their skin.
- Reproductive adaptations of internal fertilization and the amniote shelled egg allow reproduction on land. Nitrogenous waste is produced as *uric acid*.

- Most reptiles are *ectothermic*. They regulate body temperature through behavioral adaptations rather than by metabolism.
- Most reptiles have a heart with two atria and a ventricle, which has a partial septum.

Birds all have wings, a body covering of feathers, and many adaptations that facilitate flight.

- Most birds' bodies are constructed for flight, with light, hollow bones; relatively few organs; wings; and feathers.
- **Feathers** and, in some cases, a layer of fat insulate birds and help them maintain internal temperature.
- Birds have a **four-chambered heart** and a high rate of metabolism.
- Birds are **endotherms** and maintain a warm, consistent body temperature.

Concept 34.7 Mammals are amniotes that have hair and produce milk

Mammals share certain characteristics:

- All possess **mammary glands** that produce milk.
- Mammals have a body covering of **hair.**
- They have a four-chambered heart and are **endothermic.**
- Mammals have internal fertilization, and most are born rather than hatched.
- They have **proportionally larger brains** than other vertebrates, and all have **teeth** of differing size and shape.

Mammals can be placed into three groups:

1. **Monotremes** are egg-laying mammals that have hair and produce milk. Examples are platypuses and spiny anteaters.
2. **Marsupials** are born early in development and complete embryonic development in a marsupium (pouch) while nursing. Examples include kangaroos and opossums.
3. **Placental mammals (eutherians)** have a longer period of pregnancy; they complete their development in the uterus. Examples include mice, dogs, and humans.

Humans belong to the order **Primates,** along with monkeys and gorillas. Some characteristics common to all **primates** include opposable thumbs, large brains and short jaws, forward-looking eyes, flat nails, well-developed parental care, and complex social behavior.

Some features of **human evolution** are increased brain volume, shortening of the jaw, bipedal posture, reduced size difference between the sexes, and certain important changes in family structure.

For Additional Review

Compare the land adaptations of plants and animals, including how each manages the uptake of nutrients and water as well as the excretion of wastes.

Level 1: Knowledge/Comprehension Questions

1. Which of the following groups is best characterized as being eukaryotic, multicellular, heterotrophic, and without a cell wall?
   (A) Plantae
   (B) Animalia
   (C) Fungi
   (D) Viruses
   (E) Monera

2. Systematists categorize all living creatures into what three domains?
   (A) Bacteria, Euglena, Eukarya
   (B) Bacteria, Archaea, Eukarya
   (C) Archaea, Plantae, Eukarya
   (D) Protista, Plantae, Eukarya
   (E) Prokaryota, Eukarya, Plantae

3. Which of the following is a symbiotic relationship in which both organisms benefit?
   (A) parasitism
   (B) commensalism
   (C) mutualism
   (D) obligate
   (E) heterotrophic

4. From an evolutionary perspective, mitochondria and chloroplasts are the result of what process?
   (A) endosymbiosis
   (B) conjugation
   (C) transformation
   (D) ectosymbiosis
   (E) compartmentalization

5. All of the following are adaptations of plants to terrestrial life EXCEPT
   (A) stomata.
   (B) a cuticle.
   (C) xylem.
   (D) phloem.
   (E) photosynthesis.

6. Which of these is an organism with symbiotic associations of photosynthetic microorganisms in a network of fungal hyphae?
   (A) mushrooms
   (B) yeast
   (C) lichens
   (D) slime molds
   (E) protozoans

7. Which of these is a mutualistic association of plant roots and fungi?
   (A) mushrooms
   (B) mycorrhizae
   (C) lichens
   (D) hyphae
   (E) sporangia

8. Place the following groups of plants in order beginning with those that first appeared on Earth.
   (A) moss, angiosperms, ferns, gymnosperms
   (B) liverworts, ferns, gymnosperms, angiosperms
   (C) moss, ferns, angiosperms, gymosperms
   (D) moss, liverworts, tracheophytes, bryophytes
   (E) seed plants, cone-bearing plants, bryophytes

9. All of the following terms reflect characteristics of almost all animals EXCEPT
   (A) multicellularity.
   (B) heterotrophic nutrition.
   (C) homeobox genes.
   (D) eukaryotic.
   (E) diurnal.

10. Evolution of which feature enabled vertebrates to reproduce successfully on land?
    (A) the amniotic egg
    (B) quadruped locomotion
    (C) body hair
    (D) opposable thumbs
    (E) the cloaca

11. Based on this tree, which statement is NOT correct?

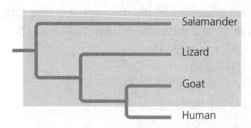

(A) The lineage leading to salamanders was the first to diverge from the other lineages.
(B) Salamanders are a sister group to the group containing lizards, goats, and humans.
(C) Salamanders are as closely related to goats as to humans.
(D) Lizards are more closely related to salamanders than to humans.
(E) The group highlighted by shading is paraphyletic.

12. Genetic variation in bacterial populations cannot result from
(A) transduction.
(B) transformation.
(C) conjugation.
(D) mutation.
(E) meiosis.

13. Shared derived characters are
(A) those that characterize all the species on a branch of a phylogenetic tree.
(B) determined through a computer comparison of paralogous genes in a group of species.
(C) homologous structures that develop during adaptive radiation.
(D) characters found in the outgroup but not in the species under consideration.
(E) used to identify species but not higher taxa.

Level 2: Application/Analysis/Synthesis Questions

1. In a comparison of birds and mammals, having four appendages is
(A) a shared ancestral character.
(B) a shared derived character.
(C) a character useful for distinguishing birds from mammals.
(D) an example of analogy rather than homology.

2. An outgroup is a species from an evolutionary lineage that is known to have diverged before the lineage under study. If you were using cladistics to build a phylogenetic tree of cats, which of the following would be the best outgroup?
(A) lion
(B) domestic cat
(C) wolf
(D) leopard

3. When a bacterial cell with a chromosome-borne F factor conjugates with another bacterium, how is the transmitted donor DNA incorporated into the recipient genome?
(A) It is substituted for the equivalent portion of the recipient chromosome by the process of crossing over.
(B) It circularizes and becomes one of the recipient cell's plasmids.
(C) The genes on the donor DNA of which the recipient does not have a copy are added to the recipient chromosome; the remainder of the donor DNA is degraded.
(D) The donor and recipient DNA are both chopped into segments by restriction enzymes, and a new, composite chromosome is assembled from the fragments.

4. In the 1920s, Frederick Griffith conducted an experiment in which he mixed the dead cells of a bacterial strain that can cause pneumonia with live cells of a bacterial strain that cannot. When he cultured the live cells, some of the daughter colonies proved able to cause pneumonia. Which of the following processes of bacterial DNA transfer does this experiment demonstrate?
   (A) transduction
   (B) conjugation
   (C) transformation
   (D) transposition

Free-Response Question

1. *Systematists are scientists who study evolutionary relationships between organisms; they use scientific evidence to construct hypothetical phylogenies that show these relationships. Systematists have replaced the five-kingdom system with a more accurate three-domain system, containing the Archaea, the Bacteria, and the Eukarya.*

   (a) **Describe** three types of evidence that scientists use to discover evolutionary relationships between organisms.

Plant Form and Function

Chapter 35: Plant Structure, Growth, and Development

WHAT'S IMPORTANT TO KNOW?

While much of plant anatomy is not a part of the new Curriculum Framework and will not be covered on the exam, your teacher might choose to have you learn about plant structures to better understand what you see in a slide of an onion root tip, or to learn more about how plants acquire resources. Some of this chapter will be important to you in understanding the topic of movement of materials through plants, including leaf anatomy. The material that follows will help you focus on some important concepts if your teacher emphasizes plant anatomy.

Concept 35.1 Plants have a hierarchical organization consisting of organs, tissues, and cells

▌ Plants have a **root system** beneath the ground that is a multicellular organ which anchors the plant, absorbs water and minerals, and often stores sugars and starches. Additional structural characteristics of roots include the following:

- *Fibrous roots* are made up of a mat of thin roots that are spread just below the soil's surface. *Taproots* are made up of one thick, vertical root with many lateral roots that come out from it.
- At the tips of the roots vast numbers of tiny *root hairs* increase the surface area enormously, making efficient absorption of water and minerals possible. Plants may also have a symbiotic relationship with fungi at the tips of the roots, termed **mycorrhizae** ("fungus roots"). Mycorrhizae assist in the absorption process and are found in the vast majority of all plants.

▌ Plants have a **shoot system** above the ground that works as a multicellular organ consisting of stems and leaves.

- **Stems** function primarily to display the leaves. Two types of buds help achieve this purpose:
  - A **terminal bud** is located at the top end of the stem where growth usually occurs. Often the terminal bud secretes a hormone that inhibits the growth of axillary buds. This concentrates the growth of the plant upward, toward more light.

- **Axillary buds** are located in the V formed between the leaf and the stem; these buds have the potential to form a branch (or lateral shoot).

  - **Leaves** are the main photosynthetic organ in most plants.

- Plant organs—leaf, stem, and root—are composed of three tissue types:

  - **Dermal tissue** is a single layer of closely packed cells that cover the entire plant and protect it against water loss (a waxy layer termed the **cuticle** in the leaves) and invasion by pathogens like viruses and bacteria.
  - **Vascular tissue** is continuous through the plant and transports materials between the roots and shoots. Vascular tissue is made up of **xylem**, which transports water and minerals up from the roots, and **phloem**, which transports food from the leaves to the other parts of the plant.
  - **Ground tissue** is anything that isn't dermal tissue or vascular tissue. Any ground tissue located inside the vascular tissue is *pith*; any ground tissue outside the vascular tissue is *cortex*. Roles of the ground tissue include photosynthesis and storage.

- **Xylem** cells conduct water and minerals, and are dead at maturity.
- **Phloem cells** conduct sugar and other organic compounds. Phloem is composed of two types of cells, sieve tubes and companion cells, both alive at maturity.

Concept 35.2 Meristems generate cells for primary and secondary growth

- Based on their life cycle, flowering plants can be classified as **annuals** (life cycle completed in one year), **biennials** (life cycle completed in two years), or **perennials** (life cycle continues for many years).
- **Meristems** are perpetually embryonic tissues. Unlike growth in animals, growth in plants occurs only as a result of cell division in the meristems.

  - **Apical meristems** are located at the tips of roots and in buds of shoots, allowing the plant's stems and roots to extend. This is *primary growth*.
  - **Lateral meristems** result in growth, which thickens the shoots and roots. This is termed *secondary growth*.

Concept 35.3 Primary growth lengthens roots and shoots

- Study Figure 8.1 as you follow these descriptions. The **root cap** protects the delicate meristem of the root tip as it pushes through the soil. The **root tip** contains three zones of cells in various stages of growth. It is likely that you have done a lab to investigate mitosis in an onion root tip and should have observed the following:

  1. The **zone of cell division** includes root apical meristem and its derivatives. New root cells are produced in this region, including the cells of the root cap. (This is where you would have seen cells in various stages of the cell cycle when studying onion root tips with a microscope.)

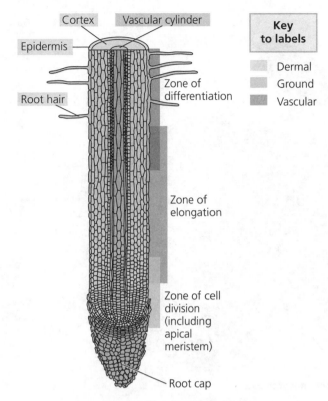

Figure 8.1   Primary growth of a root

2. Above the zone of cell division is the **zone of elongation**, in which cells elongate significantly.
3. In the **zone of differentiation**, the three systems in primary growth become functionally mature.

> **TIP FROM THE READERS**
> As you learn plant anatomy, such as the structure of the leaf, focus on how the structures enhance functions, such as gas exchange, photosynthesis, and reduction of water loss. The Curriculum Framework asks you to integrate knowledge of factual material with function!

▌ Refer to Figure 8.2 as you review this. The epidermis of the underside of the leaf is interrupted by **stomata**, which are small pores flanked by *guard cells*, which open and close the stomata.

▌ In leaves, the ground tissue is sandwiched between the *upper* and *lower epidermis*, in the *mesophyll*. It is made up of *parenchyma* cells, the sites of photosynthesis.

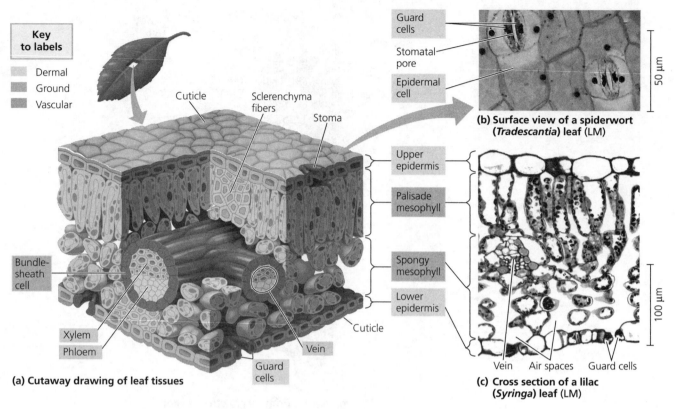

(a) Cutaway drawing of leaf tissues

Key to labels
- Dermal
- Ground
- Vascular

Cuticle

Sclerenchyma fibers

Stoma

Bundle-sheath cell

Xylem

Phloem

Vein

Guard cells

Guard cells

Stomatal pore

Epidermal cell

50 µm

(b) Surface view of a spiderwort (*Tradescantia*) leaf (LM)

Upper epidermis

Palisade mesophyll

Spongy mesophyll

Lower epidermis

Cuticle

Vein   Air spaces   Guard cells

100 µm

(c) Cross section of a lilac (*Syringa*) leaf (LM)

Figure 8.2   Leaf anatomy

Concept 35.4 Secondary growth increases the diameter of stems and roots in woody plants

▐ Two lateral meristems take part in plant growth. The **vascular cambium** produces secondary xylem (wood). The **cork cambium** produces a tough covering that replaces epidermis early in secondary growth.

▐ **Bark** is all the tissues outside the vascular cambium. Bark includes the phloem derived from the vascular cambium, the cork cambium, and the tissues derived from the cork cambium.

Chapter 36: Resource Acquisition and Transport in Vascular Plants

YOU MUST KNOW

- The role of passive transport, active transport, and cotransport in plant transport.
- The role of diffusion, active transport, and bulk flow in the movement of water and nutrients in plants.
- How the transpiration cohesion-tension mechanism explains water movement in plants.
- How pressure flow explains translocation.

Concept 36.1 Adaptations for acquiring resources were key steps in the evolution of vascular plants

▮ Study Figure 8.3 where you will see an overview of resource acquisition and transport. Take your time with the image, as it has much important information.

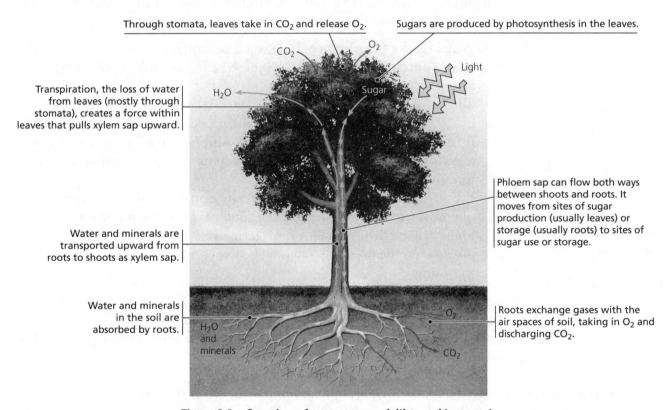

Through stomata, leaves take in CO_2 and release O_2.

Sugars are produced by photosynthesis in the leaves.

Transpiration, the loss of water from leaves (mostly through stomata), creates a force within leaves that pulls xylem sap upward.

Phloem sap can flow both ways between shoots and roots. It moves from sites of sugar production (usually leaves) or storage (usually roots) to sites of sugar use or storage.

Water and minerals are transported upward from roots to shoots as xylem sap.

Water and minerals in the soil are absorbed by roots.

Roots exchange gases with the air spaces of soil, taking in O_2 and discharging CO_2.

CO_2 O_2

Light

H_2O Sugar

H_2O and minerals

O_2

CO_2

Figure 8.3   Overview of resource acquisition and transport

Concept 36.2 Different mechanisms transport substances over short or long distances

▮ Transport begins with the movement of water and solutes across a cell membrane.

  ▪ Solutes diffuse down their electrochemical gradients (recall from Chapter 7 that electrochemical gradients are the combined effects of the concentration gradient of the solute and the voltage or charge differential across the membrane).

- If no energy is required to move a substance across the membrane, then the movement is termed *passive transport*. Diffusion is an example of passive transport.
- If energy is required to move solutes across the membrane, it is termed *active transport*. As most solutes cannot move across the phospholipid barrier of the membrane, a **transport protein** is required. The most important transport protein in plants is the **proton pump**.
- A proton pump creates an electrochemical gradient by using the energy of ATP to pump hydrogen ions across the membrane. This potential energy can then be used in the process of **cotransport**—the coupling of the steep gradient of one solute (hydrogen in our example) with a solute like sucrose. The drop in potential energy experienced by the hydrogen ion pays for the transport of the sucrose.

- The uptake of water across cell membranes occurs through *osmosis*, the passive transport of water across a membrane.

  - Water moves from areas of high water potential to low water potential. **Water potential** includes the combined effects of solute concentration and physical pressure.
  - The water potential equation is $\psi = \psi_s + \psi_p$, where ψ is water potential, ψ_s is solute potential, and ψ_p is the pressure potential.
  - By definition the ψ_s of pure water is 0. Adding solutes to pure water always lowers water potential. The solute potential of a solution is therefore always negative.
  - Pressure potential is the physical pressure on a solution. An example of positive ψ_p occurs when the cell contents press the plasma membrane against the cell wall, a force termed *turgor pressure*. If the cell loses water, the pressure potential becomes more negative, resulting in wilting.

  - **Aquaporins** are the transport proteins (channels) in the plant plasma membrane specifically designed for the passage of water.
  - **Bulk flow** is the movement of water through the plant from regions of high pressure to regions of low pressure.

Concept 36.3 Transpiration drives the transport of water and minerals from roots to shoots via the xylem

- Water and minerals from the soil enter the plant through the root epidermis, cross the cortex, pass into the vascular cylinder, and then flow up the xylem. Use Figure 8.4 to chart the movement of water and minerals out of the soil, across the root, and up through the xylem to the rest of the plant.

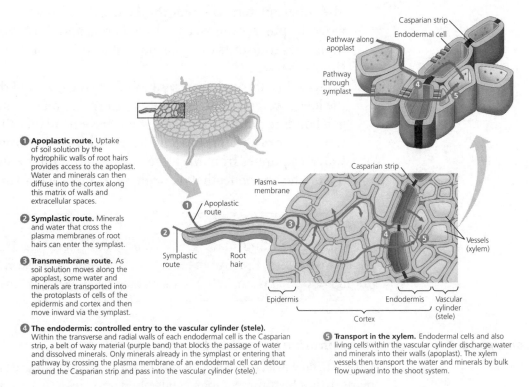

① **Apoplastic route.** Uptake of soil solution by the hydrophilic walls of root hairs provides access to the apoplast. Water and minerals can then diffuse into the cortex along this matrix of walls and extracellular spaces.

② **Symplastic route.** Minerals and water that cross the plasma membranes of root hairs can enter the symplast.

③ **Transmembrane route.** As soil solution moves along the apoplast, some water and minerals are transported into the protoplasts of cells of the epidermis and cortex and then move inward via the symplast.

④ **The endodermis: controlled entry to the vascular cylinder (stele).** Within the transverse and radial walls of each endodermal cell is the Casparian strip, a belt of waxy material (purple band) that blocks the passage of water and dissolved minerals. Only minerals already in the symplast or entering that pathway by crossing the plasma membrane of an endodermal cell can detour around the Casparian strip and pass into the vascular cylinder (stele).

⑤ **Transport in the xylem.** Endodermal cells and also living cells within the vascular cylinder discharge water and minerals into their walls (apoplast). The xylem vessels then transport the water and minerals by bulk flow upward into the shoot system.

Figure 8.4   Lateral transport of minerals and water in roots

- The movement between cells is termed the *apoplastic* route. The *symplastic* route occurs only after the solution crosses a plasma membrane.
- Water can move inward by either route until reaching the innermost layer of the cortex, the **endodermis**, labeled number 4. The endodermal cells are sealed one to the other by the *Casparian strip*, a belt of waxy material that blocks the passage of water and dissolved materials. This is a critical control point for materials moving into the plant, because at the Casparian strip the soil solution must cross the plasma membrane. The plasma membrane determines what can cross into the xylem tissue and gain entrance to the rest of the plant.

- Once in the root xylem, water and minerals are transported long distances—to the rest of the plant—by bulk flow. The water and minerals, termed *xylem sap*, flow out of the root and up through the shoot, eventually exiting the plant primarily through the leaves.

- **Transpiration** is the loss of water vapor from the leaves and other parts of the plant that are in contact with air. It plays a key role in the movement of water from the roots.

- Two mechanisms influence how water is moved up through the plant:

  - **Root pressure** occurs when water diffusing in from the root cortex generates a positive pressure that *pushes* sap up. Root pressure does not have the force to push water to the tops of trees.

- The **cohesion-tension hypothesis** describes how transpiration provides the pull for the ascent of xylem sap, and the cohesion of water molecules transmits this pull along the entire length of the xylem from shoots to roots.
- Water is lost through transpiration from the leaves of the plant due to the lower water potential of the air. The *cohesion* of water due to hydrogen bonding plus the *adhesion* of water to the plant cell walls enables the water to form a water column. Water is drawn up through the xylem as water evaporates from the leaves, each evaporating water molecule pulling on the one beneath it through the attraction of *hydrogen bonds*.

> **TIP FROM THE READERS**
> Hydrogen bonding plays a key role in cohesion-tension mechanisms. Be able to explain the importance of cohesion, adhesion (water hydrogen bonded to the xylem walls), and surface tension in this mechanism.

Concept 36.4 *The rate of transpiration is regulated by stomata*

- Large surface area increases photosynthesis but also increases water loss by the plant through stomata. Guard cells open and close the stomata, controlling the amount of water lost by transpiration, but also the amount of carbon dioxide available from the atmosphere for photosynthesis.
- Guard cells control the size of the stomata opening by changing shape, widening or closing the gap between them. When the guard cells take up K^+ from the surrounding cells, this decreases water potential in the guard cells, causing them to take up water. The guard cells then swell and buckle, increasing the size of the pore between them. When the guard cells lose K^+, the cells then lose water, become less bowed, and the pore closes.
- Guard cells are stimulated to open by the presence of light, loss of carbon dioxide in the leaf, and by normal circadian rhythms. Circadian rhythms are part of the plant's internal clock mechanism and cycle with intervals of about 24 hours. Even plants kept in the dark will open their stomata as dawn approaches.

Concept 36.5 *Sugars are transported from sources to sinks via the phloem*

- Phloem transports organic products of photosynthesis from the leaves throughout the plant, a process called **translocation**. The mechanism for translocation is **pressure flow**.
- Sieve tubes, a specialized cell type in phloem tissue, always carry sugars from a sugar source to a sugar sink. A **sugar source** is an organ that is a net producer of sugar, such as the leaves. A **sugar sink** is an organ that is a net consumer or storer of sugar, such as a fruit, or roots during the summer. Follow Figure 8.5 while noting the key steps.

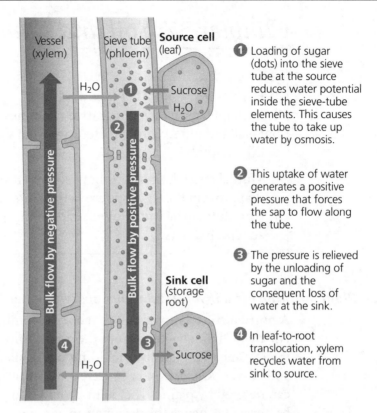

Figure 8.5   Pressure flow in a sieve tube

① Loading of sugar (dots) into the sieve tube at the source reduces water potential inside the sieve-tube elements. This causes the tube to take up water by osmosis.

② This uptake of water generates a positive pressure that forces the sap to flow along the tube.

③ The pressure is relieved by the unloading of sugar and the consequent loss of water at the sink.

④ In leaf-to-root translocation, xylem recycles water from sink to source.

1. Sucrose is loaded into the sieve tubes at the sugar source. Proton pumps are used to create an electrochemical gradient that is utilized to load sucrose. This decreases water potential and causes the uptake of water, creating positive pressure.
2. The pressure is relieved at the sugar sink by the unloading of sucrose followed by the loss of water. In leaf-to-root translocation, xylem recycles the water back to the sugar source. Translocation via pressure flow is a second example of bulk flow.

Concept 36.6 The symplast is highly dynamic

▮ The symplast, or network of living phloem cells that connects all parts of the plant, is dynamic and interconnected.

- Plasmodesmata allow for the movement of informational molecules like RNAs and proteins that coordinate development between cells.
- In some plants the phloem carries out rapid, long-distance electrical signaling. This may lead to changes in gene transcription, respiration, photosynthesis, and other cellular functions in widely spaced organs. This is a nerve-like function, allowing for swift communication.

Chapter 37: Soil and Plant Nutrition

> **WHAT'S IMPORTANT TO KNOW?**
>
> This chapter provides illustrative examples that your teacher might select, but is not required content. Here are what we consider to be the main points of this chapter:
>
> - Mutualistic relationships between plant roots and the bacteria and fungi that grow in the rhizosphere help plants acquire important nutrients.
> - Nonmutualistic nutritional adaptations in plants enable plant survival in adverse habitats.

Concept 37.3 Plant nutrition often involves relationships with other organisms

▮ A *mutualistic* relationship occurs between nitrogen-fixing bacteria and plants. Nitrogen-fixing bacteria from the genus *Rhizobium* live in the root nodules of the legume plant family, including plants like peas, soybeans, alfalfa, peanuts, and clover. *Rhizobium* bacteria can fix atmospheric nitrogen into a form that can be used by plants. The plant provides food into the root nodule where the bacteria live, hence the designation as a mutualistic relationship.

▮ Bacteria also play a critical role in the nitrogen cycle, discussed in Chapter 55. Ammonia is recycled in the soil by bacteria into forms that can be absorbed and used by plants.

▮ *Mycorrhizae* are another example of mutualistic relationships with roots, this time between the roots and fungi in the soil. In mycorrhizae, the fungus benefits from a steady supply of sugar donated by the host plant. In return, the fungus increases the surface area for water uptake, selectively absorbs minerals that are taken up by the plant, and secretes substances that stimulate root growth and antibiotics that protect the plant from invading bacteria.

▮ Plants also form symbiotic relationships that are not mutualistic:

   ▪ **Parasitic plants**, such as mistletoe, are not photosynthetic and rely on other plants for their nutrients. They tap into the host plant's vascular system.

   ▪ **Epiphytes** are not parasitic but just grow on the surfaces of other plants instead of the soil. Many orchids grow as epiphytes.

   ▪ **Carnivorous plants** are photosynthetic, but they get some nitrogen and other minerals by digesting small animals. They are commonly found in nitrogen-poor soil, such as in bogs.

Chapter 38: Angiosperm Reproduction and Biotechnology

WHAT'S IMPORTANT TO KNOW?

Most of this chapter's information is not required knowledge, but below are what we consider to be the main points:

- The relationship between seed and fruit.
- How temperature and moisture determine seed germination.
- How different modes of plant reproduction affect their genetic diversity.

Concept 38.1 *Flowers, double fertilization, and fruits are unique features of the angiosperm life cycle*

▌ **Pollination** is the transfer of pollen from an anther to a stigma.

▌ The ripe ovary develops into the **fruit**. The ovules within the ovaries develop into seeds. The fruit protects the enclosed seeds and aids in their dispersal by wind or animals.

▌ As the seed matures, it enters dormancy, in which it has a low metabolic rate and its growth and development are suspended. The seed resumes growth when there are suitable environmental conditions for germination. If you did a laboratory activity on cellular respiration, such as AP Investigation 6, you will know that dry seeds have a very low rate of respiration because they are dormant, not dead.

Concept 38.2 *Flowering plants reproduce sexually, asexually, or both*

▌ Asexual reproduction, or **vegetative reproduction**, produces clones. Fragmentation is an example, in which pieces of the parent plant break off to form new individuals that are exact genetic replicas of the parent.

▌ While some flowers self-fertilize, others have methods to prevent self-fertilization and maximize genetic variation. One of these is **self-incompatibility**, in which a plant rejects its own pollen or that of a closely related plant, thus ensuring cross-pollination.

▌ Agriculture uses several techniques of artificial vegetative reproduction such as grafting, cuttings, and test-tube cloning.

Concept 38.3 *Humans modify crops by breeding and genetic engineering*

▌ Humans have intervened in the reproduction and genetic makeup of plants for thousands of years through **artificial selection**.

▌ The conversion of plant material to sugars, which can be fermented to form alcohols and distilled to yield **biofuels**, is currently under study.

▌ **Genetically modified organisms** are engineered to express a gene from another species. *Examples*: "Golden Rice," engineered to include large amounts of vitamin A; and *Bt* corn, engineered to contain a toxin that kills specific crop pests. There is some debate over the creation of these GM (genetically modified) crops due to fear of human allergies and possible effects on nontarget organisms, among other concerns.

Chapter 39: Plant Responses to Internal and External Signals

YOU MUST KNOW

- The three steps to a signal transduction pathway.
- The role of auxins in plants.
- How phototropism and photoperiodism use changes in the environment to modify plant growth and behavior.
- How plants respond to attacks by herbivores and pathogens.

Concept 39.1 Signal transduction pathways link signal reception to response

> **TIP FROM THE READERS**
> Essays on the AP Biology Exam often cover ideas in different units in the textbook. This concept brings together the general ideas on cell communication from Chapter 11 with the specific results of cell communication in plants. This would be an excellent topic for an essay question.

▌ Let's begin with a review from Chapter 11. Signal transduction pathways involve three steps:

1. **Reception:** Cell signals are detected by receptors that undergo changes in shape in response to a specific stimulus. Two common plasma membrane receptors are *G protein-coupled receptors* and *tyrosine kinase receptors.*
2. **Transduction:** Transduction is a multistep pathway that amplifies the signal. This allows a small number of signal molecules to produce a large cellular response.
3. **Response:** Cellular response is primarily accomplished by two mechanisms: (1) increasing or decreasing mRNA production, or (2) activating existing enzyme molecules.

Concept 39.2 Plant hormones help coordinate growth, development, and responses to stimuli

▌ **Hormones** are defined as chemical messengers that coordinate the different parts of a multicellular organism. They are produced by one part of the body and transported to another.

▌ A **tropism** is a plant growth response that results in the plant growing either toward or away from a stimulus. Tropisms result from hormone production.

▪ **Phototropism** is the growth of a shoot in a certain direction in response to light. **Positive phototropism** is the growth of a plant toward light; **negative phototropism** is growth of a plant away from light.

▌ Following is a survey of the most important actions of hormones.

1. The natural auxin in plants is *indoleacetic acid*, usually abbreviated as *IAA*. Auxins have many functions and play key roles in phototropisms and gravitropisms.

   - **Auxins** stimulate elongation of cells within young developing shoots. Auxins produced in the apical meristems activate proton pumps in the plasma membrane, which results in a lower pH (acidification of the cell wall). This weakens the cell wall, allowing turgor pressure to expand the cell wall, resulting in cell elongation.
   - Synthetic auxins are often used as herbicides. Monocots, like grasses, can quickly inactivate synthetic auxins, but eudicots cannot. Thus, the high concentrations of auxins kill broadleaf weeds (dicots), whereas grasses (monocots like turf grass or corn) are not harmed.

2. **Cytokinins** play an essential role in cell division and differentiation. These hormones stimulate cytokinesis or cell division. When the proper ratio of cytokinins to auxins exists, cytokinins stimulate cell division and the pathways to cell differentiation.

3. **Gibberellins** work in concert with auxins to stimulate stem elongation. Gibberellins help loosen cell walls, allowing expansion of cells and therefore of stems. Many dwarf varieties of plants do not produce working gibberellins—including Mendel's dwarf pea variety. Gibberellins are also used to signal the seed to break dormancy and germinate.

4. In general, **abscisic acid** slows growth, often acting antagonistically to the other hormones mentioned. For example, abscisic acid promotes seed dormancy, thus keeping seeds from germinating too quickly. When leaves are under water stress, it is abscisic acid that signals the stomata to close, saving water.

5. **Ethylene** is unusual as a hormone because it is a gas. Ethylene plays a critical role in programmed cell death, or apoptosis. The shedding of leaves and the death of an annual after flowering are examples. Ethylene also promotes the ripening of fruits. Ethylene triggers ripening, and ripening triggers more ethylene. This rare positive feedback loop can rapidly promote the ripening of fruit. Because of ethylene, one rotten apple *can* spoil the bunch!

Concept 39.3 Responses to light are critical for plant success

- Plants can detect not only the presence of light, but also its direction, intensity, and wavelength. Action spectra (introduced in Chapter 10 on photosynthesis) reveal that red and blue light are the most important colors in plant responses to light.

- *Blue-light photoreceptors* initiate a number of plant responses to light including phototropisms and the light-induced opening of stomata.
- Light receptors termed **phytochromes** absorb mostly red light.

  - Phytochromes exist in two isomer forms, P_r and P_{fr}, that can switch back and forth depending on the wavelength of light in greatest supply. P_{fr} is the form of phytochrome that triggers many of a plant's developmental responses to light. The plant produces phytochrome in the P_r form, but upon illumination, the P_{fr} level increases by rapid conversion from P_r. The relative amounts of the two pigments provide a baseline for measuring the amount of sunlight in a day.
  - **Circadian rhythms** are physiological cycles that have a frequency of about 24 hours and that are not paced by a known environmental variable. In plants, the surge of P_{fr} at dawn resets the biological clock. The combination of a phytochrome system and a biological clock allows the plant to accurately assess the amount of daylight or darkness and hence the time of year.

- A physiological response to a photoperiod (the relative lengths of night and day), such as flowering, is called **photoperiodism**. Photoperiodism controls when plants will flower. When this was first discovered, scientists thought the critical factor was the length of the day. It was later determined that the length of the night is the actual critical factor. The old terminology, however, is still used, so take note of that in the following categories:

  - Study the response of a short-day plant, such as the chrysanthemum shown in Figure 8.6, to day length. Notice that if an amount of *uninterrupted* darkness does not exceed the critical period, the plant will not flower.
  - **Short-day plants** flower in early spring or fall. Short-day plants are actually long-night plants; that is, what the plant measures is the length of the night.
  - **Long-day plants** flower only if a period of continuous darkness was shorter than a critical period. They often flower in the late spring or early summer. Long-day plants are actually short-night plants.

- **Day-neutral plants** can flower in days of any length.

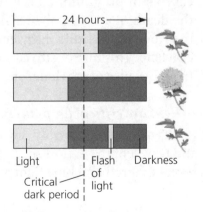

Figure 8.6   Short-day (long-night) plant

Concept 39.4 *Plants respond to a wide variety of stimuli other than light*

▌ What environmental cue causes the shoot of a young seedling to grow up and the root to grow down? **Gravitropism** is a plant's response to gravity. Roots show **positive gravitropism** and grow toward the source of gravity, whereas shoots show **negative gravitropism** and grow away from gravity.

▪ The hormone **auxin** plays a key role in gravitropism in both roots and stems. In the young root gravity indirectly causes a high concentration of auxin on the root's lower side. *High* concentrations of auxins inhibit cell elongation, causing the lower side to grow more slowly, whereas more rapid elongation of cells on the upper side causes the root to curve as it grows.

▌ **Thigmotropism** is directional growth in a plant as a response to a touch. Vines display thigmotropism when their tendrils coil around supports.

▌ Plants have various responses to stresses. In times of drought, the guard cells lose turgor. This causes the stomata to close; young leaves will stop growing, and they will roll into a shape that slows transpiration rates. Also, deep roots continue to grow, whereas those near the surface (where there isn't much water) do not grow very quickly.

Concept 39.5 *Plants respond to attacks by herbivores and pathogens*

▌ Some physical defenses plants have against predators (herbivores) are thorns and chemicals such as bitter or poisonous compounds. Some plants produce airborne attractants to recruit other animals to kill the herbivores. (Remember reading how parasitoid wasps lay their eggs on caterpillars, and the emerging wasp larva eats the caterpillar from the inside out?)

▌ The first line of defense against viruses for a plant (as for humans) is the epidermal layer.

For Additional Review

Describe the plant structures that make plants ideally suited for trapping and processing the sun's energy, and for the process of carbon fixation.

Level 1: Knowledge/Comprehension Questions

1. Which of the following processes is responsible for the bending of the stem of a plant toward a light source?
   (A) The amount of chlorophyll produced on the side facing the light increases.
   (B) The rate of cell division on the side facing the light increases.
   (C) The rate of cell division on the side away from the light increases.
   (D) The cells on the side of the stem facing the light elongate.
   (E) The cells on the side of the stem away from the light elongate.

2. The driving force for the movement of materials in the xylem of plants is
   (A) gravity.
   (B) root pressure.
   (C) transpiration.
   (D) the difference in osmotic pressure between the source and the sink.
   (E) osmosis.

3. Hydrogen bonding plays a particularly important role in which plant process?
   (A) pressure-flow hypothesis
   (B) mutualistic relationships
   (C) attraction of the sperm and egg
   (D) the transpiration-cohesion-tension mechanism
   (E) attraction of the pollen tube to the female gametophyte

4. In plants, translocation occurs as a result of
   (A) a difference in water potential between a sugar source and a sugar sink.
   (B) transpiration.
   (C) cohesion-adhesion.
   (D) active transport by guard cells.
   (E) active transport by tracheid and vessel elements.

5. In a mesophyll cell of a leaf, the synthesis of ATP takes place in the mitochondria and which of the following other cell organelles?
   (A) chloroplasts
   (B) Golgi apparatus
   (C) nucleus
   (D) ribosomes
   (E) lysosomes

6. All of the following enhance the uptake of water by a plant's roots EXCEPT
   (A) root hairs.
   (B) the large surface area of cortical cells.
   (C) mycorrhizae.
   (D) the attraction of water and dissolved minerals to root hairs.
   (E) gravitational force.

7. The waxy barrier found in the endodermal wall of a root that prevents the passage of unwanted materials into the vascular tissue is called the
   (A) pericycle.
   (B) ghostly strip.
   (C) Casparian strip.
   (D) cortex.
   (E) epidermis.

8. All of the following contribute to the closing of stomata during the day EXCEPT
   (A) water deficiency.
   (B) wilting.
   (C) high temperatures.
   (D) excessive rainfall.
   (E) excessive transpiration.

9. How does the sperm of an angiosperm reach the egg?
   (A) via the pollen tube that grows from the pollen grain through the carpel tissues to the ovule
   (B) via the pollen tube that grows from the ovule to reach the pollen grain on the stigma
   (C) usually via an insect, which places sperm in the ovary while probing for nectar
   (D) by actively swimming down through the style to the egg
   (E) via raindrops of a dew film that allows the sperm to swim from the male plant to the female plant

10. A seed develops from
    (A) an ovum.
    (B) a pollen grain.
    (C) an ovule.
    (D) an ovary.
    (E) an embryo.

11. A fruit is
    (A) a mature ovary.
    (B) a mature ovule.
    (C) a seed plus its integuments.
    (D) a fused carpel.
    (E) an enlarged embryo sac.

12. In a flowering plant, sperm are produced by meiosis in the
    (A) petals.
    (B) ovaries.
    (C) sepals.
    (D) anthers.
    (E) stigma.

13. Which vascular tissue in plants is responsible for carrying sugars down from the leaves to the rest of the plant?
    (A) xylem
    (B) phloem
    (C) dermal tissue
    (D) tracheids
    (E) vessel elements

Level 2: Application/Analysis/Synthesis Questions

1. In order to flower, short-day plants require a period of
   (A) light greater than a critical period.
   (B) darkness greater than a critical period.
   (C) light less than a critical period.
   (D) darkness less than a critical period.

2. If a long-day plant has a critical length of 9 hours, which 24-hour cycle would prevent flowering?
   (A) 16 hours light/8 hours dark
   (B) 14 hours light/10 hours dark
   (C) 15.5 hours light/8.5 hours dark
   (D) 4 hours light/8 hours dark/4 hours light/ 8 hours dark

3. Which of the following does NOT occur in a signal transduction pathway?
   (A) stimulation of the receptor by a relay molecule
   (B) expression of specific genes
   (C) activation of protein kinases
   (D) phosphorylation of transcription factors

4. In systemic acquired resistance, salicylic acid probably
   (A) destroys pathogens directly.
   (B) activates defenses throughout the plant before infection spreads.
   (C) closes stomata, thus preventing the entry of pathogens.
   (D) activates heat-shock proteins.

5. The figure shows key elements of Went's experiment. Which plant shows auxin stimulating elongation in the left side of the plant only?

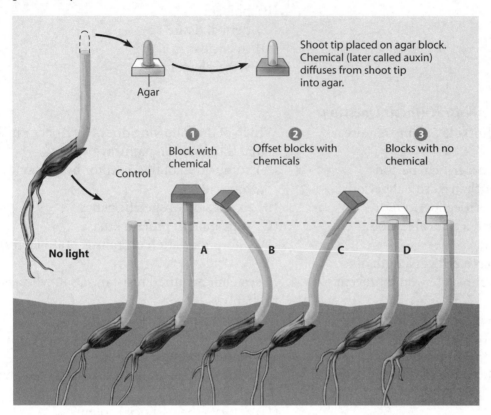

(A) plant A
(B) plant B
(C) plant C
(D) plant D

6. Which of the following plant responses is affected by photoperiod?
(A) gravitropism
(B) apical dominance
(C) onset of dormancy
(D) cell division

7. Which of the following statements about germination is *false*?
(A) The germination of a seed represents the beginning of life.
(B) Germination usually begins when a seed takes up water.
(C) A hydrated seed expands, rupturing its seed coat.
(D) Germination usually takes place after a period of dormancy.

Free-Response Question

1. *Photoperiodism and the control of flowering is a plant behavior essential to survival.*

   (a) **Describe** the importance of the process to the plant's life cycle.
   (b) **Explain** the mechanisms by which it is controlled, including triggers and responses.

Animal Form and Function

Chapter 40: Basic Principles of Animal Form and Function

YOU MUST KNOW

- The importance of homeostasis and examples.
- How feedback systems control homeostasis.
- One example of positive feedback and one example of negative feedback.

Concept 40.1 Animal form and function are correlated at all levels of organization

▌ **Tissues** are groups of cells that have a common structure and function. Tissues are further organized into functional units called **organs**. Groups of organs that work together make up **organ systems** (e.g., the digestive, circulatory, and excretory systems).

▌ For animal survival tissues, organs, and organ systems must act in a coordinated manner. Two major organ systems specialize in control and coordination:

　■ In the **endocrine system**, chemical signals called **hormones** are released into the bloodstream and are broadcast throughout the body. Different hormones cause specific effects, but only in cells with specific receptors for the released hormone.

　■ In the **nervous system, neurons** transmit information between specific locations. Only three types of cells receive nerve impulses: neurons, muscle cells, and endocrine cells.

Concept 40.2 Feedback control maintains the internal environment in many animals

▌ In **homeostasis** animals maintain a relatively constant internal environment, even when the external environment changes significantly.

TIP FROM THE READERS
Homeostasis is a fundamental concept in biology. You should concentrate on knowing how any change in a system affects it, from the level of cells through an ecosystem. This topic offers many good examples of positive and negative feedback and how small changes can affect a system.

- Homeostatic control systems function by having a **set point** (like a body temperature to maintain), **sensors** to detect any stimulus above or below the set point, and a physiological **response** that helps return the body to its set point.
- In **negative-feedback systems**, the animal responds to the stimulus in a way that reduces the stimulus. For example, in response to exercise, the body temperature rises, which initiates sweating to cool the body. *More gets you less.*
- In **positive-feedback systems**, a change in some variable triggers mechanisms that amplify rather than reverse the change. For example, during childbirth, pressure of the baby's head against receptors near the opening of the uterus stimulates greater uterine contractions, which cause greater pressure against the uterine opening, which heightens the contractions, and so forth. *More gets you more.*

Concept 40.3 *Homeostatic processes for thermoregulation involve form, function, and behavior*

- **Thermoregulation** refers to how animals maintain their internal temperature within a tolerable range. It may involve *physiological changes* such as panting, sweating, shivering, or *behaviors* such as burrowing and sunning.
- **Endotherms** (such as mammals and birds) are warmed mostly by heat generated by metabolism. **Ectotherms** (such as most invertebrates, fishes, amphibians, and reptiles) generate relatively little metabolic heat, gaining most of their heat from external sources.
- In many birds and mammals, reduction of heat loss relies on **countercurrent exchange**. Heat transfer involves antiparallel arrangement of blood vessels such that warm blood from the core of the animal, en route to the extremities, transfers heat to colder blood returning from the extremities. Heat that would have been lost to the environment is conserved in the blood returning to the core of the animal.

Concept 40.4 *Energy requirements are related to animal size, activity, and environment*

- An animal's **metabolic rate** is the total amount of energy it uses in a unit of time. Metabolic rates are generally higher for endotherms than for ectotherms.
- Under similar conditions and for animals of the same size, the **basal metabolic rate** of endotherms is substantially higher than the **standard metabolic rate** of ectotherms.
- Metabolic rate is inversely related to body size among similar animals. For example, elephants have a slower metabolic rate, whereas mice have a very fast metabolic rate.
- **Torpor** involves a decrease in metabolic rate, conserving energy during environmental extremes. Animals may enter torpor in winter (*hibernation*), summer (*estivation*), or during sleep periods (*daily torpor*, such as seen in hummingbirds).

Chapter 41: Animal Nutrition

WHAT'S IMPORTANT TO KNOW?
This chapter provides illustrative examples that your teacher might select, but is not required content.

Concept 41.1 An animal's diet must supply chemical energy, organic molecules, and essential nutrients

▌ The **essential nutrients** required by an animal include both minerals and preassembled organic molecules that the animal cannot produce from raw materials.

> ***STUDY TIP*** If your teacher covers this topic, you might refer to Figure 9.1 and consider questions like, How does the structure fit its function? For example, how do the villi and microvilli of the intestine increase digestion and absorption? How does the epithelial lining of the intestine aid in absorption of nutrients?

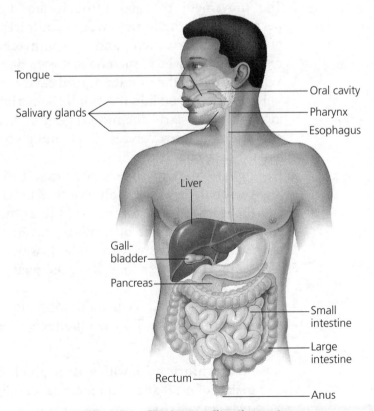

Figure 9.1   The human digestive system

Concept 41.2 **The main stages of food processing are ingestion, digestion, absorption, and elimination**

▌ **Intracellular digestion** occurs within a cell enclosed by a protective membrane. Sponges digest their food this way.

▌ **Extracellular digestion** is carried out by most animals; in this type of digestion, food is broken down outside cells. This process allows the animal to devour much larger sources of food than can be handled using only intracellular digestion.

▌ More complex animals have **complete digestive tracts (alimentary canals)**, which are *one-way* digestive tubes that begin at the mouth and terminate at the anus.

Concept 41.3 **Organs specialized for sequential stages of food processing form the mammalian digestive system**

> *STUDY TIP* You will not be expected to name the enzymes of digestion or know the anatomy of this system. An important idea here is how the large chunks of food you ingest become reduced to monosaccharides, amino acids, fatty acids, and glycerol, and are able to enter your cells. Enzymatic reactions as well as passive and active transport are required knowledge.

▌ The movement of food through the digestive system is controlled by **peristalsis**, the rhythmic waves of contraction by smooth muscle in the walls of the alimentary canal, and by **sphincters**—muscular, ringlike valves that regulate the passage of material between digestive compartments.

▌ When food is in the mouth, or **oral cavity**, a nervous reflex occurs that causes saliva to be secreted into the mouth. Saliva lubricates the food to facilitate swallowing. It also starts chemical digestion, because saliva contains the enzyme **amylase**, which hydrolyzes starch and glycogen into smaller polysaccharides and the disaccharide maltose.

▌ During chewing, food is shaped into a ball called a **bolus**. During swallowing, the bolus enters the **pharynx**—a junction that opens to the esophagus and the trachea (generally referred to as the throat). During swallowing, the **epiglottis**, a flap made of cartilage, covers the trachea. This diverts the food down the esophagus away from the airway.

▌ The **esophagus** moves food from the pharynx down to the stomach through peristalsis.

▌ The stomach's functions include storing food and secreting the digestive fluid termed gastric juice. Two components of **gastric juice** carry out chemical digestion:

▪ **Hydrochloric acid**, with a pH of about 2, breaks down the extracellular matrix of meat and plant materials, and it also kills most of the bacteria ingested with food.

- **Pepsin** is an enzyme in gastric juice that begins to hydrolyze proteins into smaller polypeptides. Pepsin is secreted in an inactive form called *pepsinogen*, which is activated by hydrochloric acid in the stomach. The inactive form protects the cells that produce the protein-digesting enzyme from self-digestion. For further protection, a thick **mucus** is produced by the lining of the stomach.

- The result of digestion in the stomach is a substance called **chyme**. Chyme is shunted from the stomach into the small intestine via the **pyloric sphincter**.

- The first section of the **small intestine** is known as the duodenum. The **duodenum** is the major site of chemical digestion. In the duodenum, chyme mixes with secretions from the pancreas and the liver. The pancreas releases an alkaline fluid rich in bicarbonate, which acts as a buffer against acidic contents from the stomach.

- **Bile** is made in the liver and stored in the gallbladder. Bile emulsifies fats; that is, bile coats fat droplets turning large fat droplets into small fat droplets, which are easier to digest.

- Chemical digestion in the duodenum can be summarized as follows:

  - **Carbohydrates:** The breakdown of starch and glycogen began with salivary amylase in the mouth. In the small intestine, pancreatic **amylases** break starch, glycogen, and small polysaccharides into the disaccharide maltose. The breakdown of maltose and other disaccharides (sucrose and lactose) into their monomers occurs at the wall of the duodenal epithelium.

  - **Proteins:** Pepsin begins the breakdown of proteins in the stomach. In the duodenum, **trypsin** and **chymotrypsin** break polypeptides into smaller chains. **Dipeptidases, carboxypeptidase**, and **aminopeptidase** break apart polypeptides into amino acids.

  - **Nucleic acids:** The breakdown of nucleic acids starts with the hydrolysis of DNA and RNA to their respective nucleotides. The nucleotides are then broken down to nitrogenous bases, sugars, and phosphate groups. Most of the enzymes responsible for nucleic acid digestion enter the duodenum from the pancreas.

  - **Fats:** Chemical digestion of fats begins in the small intestine, where fats are emulsified by bile secreted by the liver. The enzyme **lipase**, which is produced in the pancreas, hydrolyzes the small fat droplets into fatty acids and glycerol.

- The epithelial lining of the small intestine has folds called **villi**, which in turn bear projections called **microvilli**—both of which radically increase the surface area available for absorption.

- In each villus are capillaries for the absorption of monomers, including monosaccharides and amino acids, and a lymph vessel, termed a **lacteal**, which absorbs small fatty acids. Passive facilitated diffusion and active transport are used to move monomers across the intestinal membrane and into blood vessels.

- The capillaries and veins that drain the nutrients away from the villi flow into the **hepatic portal vein**, a blood vessel that goes to the liver. The liver then regulates the distribution of nutrients to the body.

▌ Hormones are chemical messengers that travel to target tissues through the bloodstream. Hormones help to coordinate the digestive process as follows:

  ▪ **Gastrin** is produced by the stomach and increases the production of gastric juices.
  ▪ **Secretin** and **cholecystokinin (CCK)**, secreted by the walls of the duodenum, increase the flow of digestive juices from the pancreas and gallbladder.

▌ The **large intestine**, also called the **colon**, is connected to the small intestine by a sphincter. The point of the connection is the site of the **cecum**, a small pouch with an extension called the **appendix**.

▌ The main functions of the large intestine are to compact waste and reabsorb water.

▌ The large intestine includes a rich flora of mostly harmless bacteria, including *Escherichia coli*. Some of these bacteria are important symbionts, producing several vitamins including B vitamins and vitamin K. The presence of *E. coli* in lakes and streams is a useful indicator of contamination by untreated sewage.

▌ At the end of the colon is the **rectum**, where feces are stored until they are eliminated.

Concept 41.4 *Evolutionary adaptations of vertebrate digestive systems correlate with diet*

▌ A mammal's **dentition** is generally correlated with its diet. In particular, mammals have specialized dentition that best enables them to ingest their food.

▌ Herbivores generally have longer alimentary canals than carnivores, reflecting the longer time needed to digest vegetation. Much of the chemical energy in herbivore diets comes from the cellulose of plant cell walls. Many vertebrates (as well as termites) house large populations of symbiotic bacteria and protists whose enzymes actually digest the cellulose.

Concept 41.5 *Feedback circuits regulate digestion, energy storage, and appetite*

▌ Vertebrates store excess calories as *glycogen* in the liver and muscles, and as fat.

▌ Overnourishment can lead to obesity. Several hormones regulate appetite, including *leptin,* which suppresses appetite.

Chapter 42: Circulation and Gas Exchange

WHAT'S IMPORTANT TO KNOW?

This chapter provides illustrative examples that your teacher might select, but is not required content. Here are what we consider to be the main points of this chapter:

- RBCs demonstrate the relationship of structure to function.
- The general characteristics of a respiratory surface.
- How O_2 and CO_2 are transported in the blood.
- The components of blood pressure and how it is measured.
- The roles of diet, blood pressure, and genetics in cardiovascular disease.
- The pathway a molecule of oxygen takes from the air until it is delivered by a red blood cell to the tissues.

Concept 42.1 Circulatory systems link exchange surfaces with cells throughout the body

- Exchange of gases, nutrients, and wastes occurs at the cellular level. Since diffusion is rapid only across small distances, natural selection has led to two general solutions.

  - Body size and shape that keep many or all cells in direct contact with the environment, such as with sponges. Cnidarians and flatworms possess a **gastrovascular cavity** that serves both in digestion and distribution of substances throughout the body.
  - Larger animals have a **circulatory system** that moves fluid to the tissues and cells for exchange.

- A *circulatory system* has three components: **blood** (a circulatory fluid), **vessels** (tubes through which blood moves), and a **heart** (a structure that pumps the blood).

- There are three main types of blood vessels:

  - **Arteries** carry blood *away* from the heart and branch into smaller **arterioles**. Their walls are relatively thick and include a significant amount of smooth muscle. The *pulse* is felt in an artery.
  - **Capillaries** are microscopic vessels that are composed of only a single layer of cells, the *endothelium*, on a basement membrane. All diffusion occurs here.
  - **Veins** carry the blood back to the heart. They have valves to prevent backflow.

- **Atria** are heart chambers that receive blood and convey it to **ventricles**, which pump blood.

Concept 42.2 **Coordinated cycles of heart contraction drive double circulation in mammals**

▌ **Heart rate** is the rate of contraction per minute, and the **stroke volume** is the amount of blood pumped by the left ventricle during each contraction.

▌ An **atrioventricular (AV) valve** between each atrium and ventricle prevents the backflow of blood into the atria; there are also two **semilunar valves**—one located at the entrance to the pulmonary artery and the second at the entrance to the aorta—that prevent backflow of blood into the ventricles.

▌ The **sinoatrial (SA) node** is the pacemaker of the heart. It is located in the upper wall of the right atrium. It generates electrical impulses that set the rate at which cardiac muscle cells contract.

▌ The **AV node**, located in the lower wall of the right atrium, delays the impulses from the SA node to allow the atria to completely empty before the ventricles contract. The AV node generates electrical impulses that cause the ventricles to contract.

▌ Heart rate is regulated by at least three factors. The sympathetic nerves accelerate heart rate, and the parasympathetic nerves slow it down. Hormones such as epinephrine increase heart rate, as does an increase in body temperature.

Concept 42.3 **Patterns of blood pressure and flow reflect the structure and arrangement of blood vessels**

▌ **Blood pressure** is measured with a *sphygmomanometer*.

▪ *Short-term regulation of blood pressure* occurs when the smooth muscle of the arterioles contracts or relaxes, due to changes in activity or hormonal signals.

▪ *Long-term regulation of blood pressure* is accomplished by changes in blood volume due to the rennin-angiotensin-aldosterone system. (See Chapter 44.)

▌ Blood moves through the veins and venules propelled by rhythmic contractions of smooth muscle, and it is squeezed by contraction of skeletal muscles during exercise. Valves prevent backflow.

Concept 42.4 **Blood components function in exchange, transport, and defense**

▌ **Plasma** is mostly water, but it also contains ions, electrolytes, and plasma proteins. It transports nutrients, metabolic wastes, gases, and hormones. In addition, blood plasma carries

▪ **Red blood cells (erythrocytes or RBCs)**, which transport oxygen via *hemoglobin* (an iron-containing protein).
▪ **White blood cells (leukocytes or WBCs)**, which are part of the immune system.
▪ **Platelets**, which are fragments of cells responsible for blood clotting.

■ Red blood cells are *biconcave disks*. This shape increases surface area to enhance oxygen transport. Each RBC contains about 250 million molecules of hemoglobin, and each hemoglobin molecule can bind up to four molecules of oxygen.

■ RBCs lack nuclei, which increases space for hemoglobin.

■ RBCs lack mitochondria, so the oxygen they carry is not consumed.

Concept 42.5 Gas exchange occurs across specialized respiratory surfaces

■ **Gas exchange**, or **respiration**, is the uptake of molecular oxygen (O_2) from the environment and the discharge of carbon dioxide (CO_2) to the environment.

■ The **respiratory medium** is the source of the O_2. It is air for terrestrial animals and water for most aquatic animals.

■ The **respiratory surface** is the part of an animal's body where gases are exchanged with the surrounding environment. It can be the body wall, the skin, gills, tracheae, or lungs.

  ■ General characteristics of respiratory surfaces include

    ■ Must be moist.
    ■ Favorable surface area/volume ratio. Respiratory surfaces are often extensively folded or branched. (Think structure/function.)
    ■ Closely associated with the vascular system of larger animals.

■ **Gills** are respiratory organs in aquatic animals. Water flows through them, and blood flowing through capillaries within the wall of the gill picks up oxygen from the water. Blood flows in a direction opposite to the flow of water. This is called **countercurrent exchange**, and it maximizes the absorption of oxygen.

■ Countercurrent exchange mechanisms allow for more diffusion to occur than would otherwise be possible. To quickly review, you will see this mechanism again in the flow of filtrate in loops of Henle in the kidney and in the capillary beds of the feet of many aquatic birds (temperature regulation, Chapter 40).

■ Identify the structures of the mammalian respiratory system in Figure 9.2. Note that the *pharynx* is the common passage of the nasal and oral cavities.

■ The **larynx** (voice box) is the upper part of the respiratory tract. It is a tube with cartilage-reinforced walls that leads to the trachea (windpipe). The *epiglottis* is a flap of cartilage that keeps food and drink from entering the trachea.

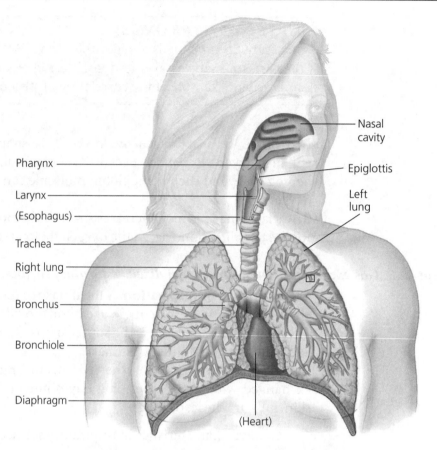

Figure 9.2 Mammalian respiratory system

▌ The **trachea** divides into two bronchi, each of which leads to a lung. *Cilia* and *mucus-producing cells* line the trachea, and their action keeps particulate matter from reaching the lungs. The trachea has *C-shaped cartilage rings*, which keep it from collapsing, much like the reinforcements in a vacuum cleaner's hose.

▌ In the lungs, the **bronchi** branch into **bronchioles**.

▌ The **alveoli** are air sacs clustered at the ends of bronchioles. They are thin, moist, have a large surface area (in a human, the size of a tennis court if spread out!), and are associated with capillary beds. Here, O_2 diffuses into the blood by passing through the simple squamous membrane of an alveolus and through the simple squamous membrane of a capillary, into the blood, and attaches to a hemoglobin molecule in a red blood cell.

> **STUDY TIP** Refer again to the characteristics of a respiratory surface and note how the mammalian lung fulfills the requirements.

Concept 42.6 Breathing ventilates the lungs

▌ **Breathing** is the inhalation and exhalation of air that ventilates lungs.

▌ In mammals, breathing involves movement of the **diaphragm**—a dome-shaped muscle separating the thoracic cavity from the abdominal cavity. Lung volume increases when the rib muscles and diaphragm contract, pressure within the lungs decreases, and air flows into the lungs. During exhalation, lung volume is decreased as the diaphragm relaxes and moves up, and pressure within the lungs increases.

- Breathing is under control of regions located in the medulla and the pons. It is influenced by pH, which is an indirect indication of blood CO_2 levels.

- Increased metabolic activity lowers pH by increasing the concentration of CO_2 in the blood. CO_2 reacts with water to form carbonic acid, which dissociates into a bicarbonate ion and a hydrogen ion. This chemical reaction is catalyzed by **carbonic anhydrase**, an enzyme found in RBCs. As blood pH drops, the rate and depth of respiration will increase.

Concept 42.7 *Adaptations for gas exchange include pigments that bind and transport gases*

- **Hemoglobin** is the respiratory pigment found in almost all vertebrates. It consists of four subunits, each of which has a *heme group* with an embedded iron atom. The iron atom binds O_2, so each hemoglobin can carry four oxygen molecules.

- A lowering of the pH in blood lowers the affinity of hemoglobin for oxygen, and oxygen dissociates. This is called the **Bohr shift**.

- CO_2 is most commonly carried in the blood in the form of bicarbonate ions (70%). Less commonly, it is transported via hemoglobin (23%) and in solution in the blood plasma (7%).

> **STUDY TIP** Know how oxygen and carbon dioxide are transported in the blood!

Chapter 43: The Immune System

> ### YOU MUST KNOW
> - Several elements of an innate immune response.
> - The differences between B and T cells relative to their activation and actions.
> - How antigens are recognized by immune system cells.
> - The differences in humoral and cell-mediated immunity.
> - Why helper T cells are central to immune responses.

> **STUDY TIP** The immune system is one of three human systems that is explicitly included in the Curriculum Framework. As you work through this chapter, pay particular attention to the ways cells communicate with each other to maintain defense.

Concept 43.1 **In innate immunity, recognition and response rely on traits common to groups of pathogens**

▌ **Innate immune responses** include barrier defenses as well as defenses to combat pathogens that enter the body. Innate immune responses are the same whether or not the pathogen has been encountered previously.

  ▪ **Barrier defenses** include skin and the mucous membranes that cover the surface and line the openings of the animal body. These provide a physical barrier and also produce secretions that result in a skin pH from 3 to 5, and the antimicrobial **lysozyme** found in saliva, mucous secretions, and tears.

  ▪ **Cellular innate defenses** combat pathogens that get through the skin—for example, in a cut. They include phagocytic white blood cells and antimicrobial proteins.

  ▪ Phagocytic white blood cells recognize microbes using **toll-like receptors or TLRs**. Toll-like receptors recognize fragments of molecules characteristic of a particular type of pathogen. For example, a TLR might recognize a polysaccharide found on the surface of many bacteria. TLRs increase the efficiency of phagocytosis.

    Phagocytic White Blood Cells

    • **Neutrophils** are white blood cells that ingest and destroy microbes in a process called **phagocytosis**.
    • **Monocytes** are another type of phagocytic leukocyte. They migrate into tissues and develop into **macrophages**, which are even more efficient.
    • **Eosinophils** are leukocytes that defend against parasitic invaders such as worms by positioning themselves near the parasite's wall and discharging hydrolytic enzymes.
    • **Dendritic cells** populate tissues in contact with the environment, where they capture pathogens, display foreign antigens, and start the primary immune response.

    Antimicrobial Proteins

    • **Interferon** proteins provide innate defense against viral infections. They cause cells adjacent to infected cells to produce substances to inhibit viral replication.
    • The **complement system** consists of roughly 30 proteins with a variety of functions that enhance the immune response. One function is to lyse invading cells.

  ▪ A local **inflammatory response** is triggered by damage to tissue by physical injury or the entry of pathogens. It leads to release of numerous chemical signals. For example, **histamines** are released by *mast cells* in response to injury. Histamines trigger the dilation and permeability of nearby capillaries. This aids in delivering clotting agents and phagocytic cells to the injured area. Systemic inflammatory responses include fever and increased production of white blood cells to fight infection.

  - **Natural killer (NK) cells** help recognize and remove diseased cells. *Examples*: cells containing viruses, and cancer cells.

Concept 43.2 *In adaptive immunity, receptors provide pathogen-specific recognition*

  ▌ Vertebrates have two types of lymphocytes: **B lymphocytes (B cells)**, which proliferate in the bone marrow, and **T lymphocytes (T cells)**, which mature in the thymus. They circulate through the blood and lymph, and both recognize particular antigens. All blood cells proliferate from stem cells in the bone marrow.

  ▌ **Antigens** are foreign molecules that elicit a response by lymphocytes. B and T cells recognize them by specific receptors embedded in their plasma membranes.

  ▌ **Antibodies** are soluble proteins secreted by B cells during an immune response.

  ▌ **B** or **T cell activation** occurs when an antigen binds to a B or T cell. B cell activation is enhanced by cytokines. The lymphocyte forms two clones of cells in a process called **clonal selection**. This results in thousands of cells, all specific to this antigen.

    - **Effector cells** combat the antigen.
    - **Memory cells**, which are long-lived, bear receptors for the same antigen, thus allowing them to quickly mount an immune response in subsequent infections.

  ▌ B cell receptors bind intact antigens.

  ▌ T cell receptors bind antigens that are displayed by **antigen-presenting cells (APCs)** on their **MHCs**.

  ▌ **Major histocompatibility complex (MHC) molecules** are proteins that are the product of a group of genes. (Individuals differ in their MHCs. This is a major component of "self.")

  ▌ There are two types of MHCs:

    - **Class I MHCs** are found on almost all cells of the body, except RBCs.
    - **Class II MHCs** are made by some cells of the immune system, including dendritic cells, macrophages, and B cells.

  ▌ The specificity of B and T cells is a result of the shuffling and recombination of several gene segments and results in more than 1 million *different* B cells and 10 million different T cells.

  ▌ Each B or T cell responds to only one antigen.

  ▌ A **primary immune response** occurs when the body is first exposed to an antigen and a lymphocyte is activated.

  ▌ A **secondary immune response** occurs when the same antigen is encountered at a later time. It is faster and of greater magnitude. (Refer to Figure 9.3.)

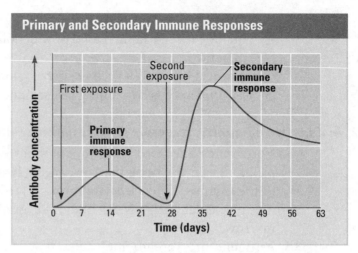

Figure 9.3 Primary and secondary immune responses

Concept 43.3 *Adaptive immunity defends against infection of body fluids and body cells*

- ▌ Acquired immunity has two branches:

  - ▪ **Humoral immune response** involves the activation and clonal selection of effector B cells, which produce antibodies that circulate in the blood.
  - ▪ **Cell-mediated immune response** involves the activation and clonal selection of cytotoxic T cells, which identify and destroy infected cells.

- ▌ **Helper T cells** aid both responses. When activated by interaction with the class II MHC molecule of an APC, they secrete *cytokines* that stimulate and activate both B cells and cytotoxic T cells. Refer to Figure 9.4 to see the central role of helper T cells.

- ▌ **Cytotoxic T cells** bind to class I MHC molecules, displaying antigenic fragments on the surface of infected body cells. Cytotoxic T cells destroy infected body cells.

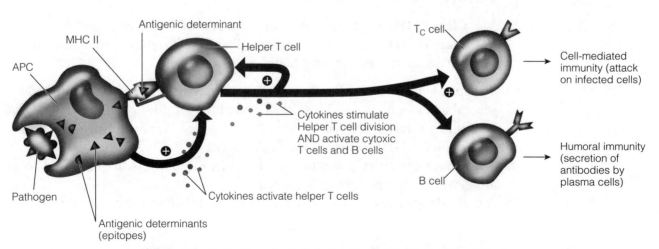

Figure 9.4 Activation of cytotoxic T cells and B cells by helper T cells

> **ORGANIZE YOUR THOUGHTS**
>
> Confused by all the different cell names?
>
> 1. **B cells** make antibodies, which provide humoral immunity. This helps fight pathogens that are circulating in body fluids.
> 2. **Cytotoxic T cells** destroy body cells that are infected by a pathogen or cancer cells.
> 3. **Helper T cells** activate both B and T cells.

▌ Recall that activated B cells produce memory cells as well as plasma cells. The plasma cells secrete antibodies in prodigious numbers. These will circulate in the blood, and bind and destroy the antigen.

▌ Modes of antibody action include

  ▪ **Neutralization:** Antibodies bind the pathogen's surface proteins, which prevents it from entering and infecting cells.
  ▪ **Opsonization:** Results in increased phagocytosis of the antigen.
  ▪ **Lysis:** Caused by activation of the complement system.

▌ **Active immunity** develops naturally in response to an infection; it also develops artificially by immunization (vaccination). In immunization, a non-pathogenic form of a microbe or part of a microbe elicits an immune response resulting in immunological memory for that microbe.

▌ **Passive immunity** occurs when an individual receives antibodies, such as those passed to the fetus across the placenta and to infants via milk.

▌ Certain antigens on red blood cells determine whether a person has **type A, B, AB**, or **O blood**. Because antibodies to nonself blood antigens already exist in the body, transfusion with incompatible blood leads to destruction of the transfused cells and a life-threatening situation for the patient.

▌ **MHC molecules** are responsible for stimulating the rejection of tissue grafts and organ transplants. The chances of successful transplantation are increased if the donor's tissue-bearing MHC molecules closely match the recipient's. The recipient also must take immunosuppressant drugs.

Concept 43.4 Disruptions in immune system function can elicit or exacerbate disease

▌ In localized **allergies** such as hay fever, IgE antibodies produced after first exposure to an allergen attach to receptors on mast cells. The next time the same allergen enters the body, it bonds to mast cell–associated IgE molecules, inducing the cell to release histamine and other mediators that cause vascular changes and typical symptoms.

▌ **Lupus, rheumatoid arthritis, type I diabetes mellitus, and multiple sclerosis** are examples of autoimmune diseases. In each case, the immune system turns against particular molecules of the body, allowing cytotoxic T cells to attack and damage the body's own healthy cells.

■ **HIV** infects helper T cells. Refer again to Figure 9.4. Can you see why people with AIDS are immune suppressed? Note what cell is central in both humoral and cell-mediated immunity.

Chapter 44: Osmoregulation and Excretion

> **WHAT'S IMPORTANT TO KNOW?**
> This topic is not part of the required human systems, but the role of a nephron in maintaining solute and water balance makes it an excellent example of a homeostatic mechanism. We consider the role of osmoregulation in homeostasis to be a main point of this chapter.

Concept 44.1 Osmoregulation balances the uptake and loss of water and solutes

■ **Osmoregulation** is the process by which animals control solute concentrations and balance water gain and loss.

■ Most metabolic wastes must be excreted from the body. One of the most important types is **nitrogenous wastes** from the breakdown of proteins and nucleic acids.

■ **Excretion** includes the removal of nitrogenous wastes from the body.

Concept 44.3 Diverse excretory systems are variations on a tubular theme

■ Most excretory systems produce urine in a four-step process. Focus on the location of each of these processes in the mammalian kidney, which is introduced in the next two concepts and shown in Figure 9.5.

■ **Nephrons** are the functional units of the kidney. Each kidney has approximately 1 million! They are made up of a single long tubule and the **glomerulus**, a ball of capillaries. At one end of the tubule is the **Bowman's capsule**, a C-shaped structure that surrounds the glomerulus.

■ The filtrate flows through the **proximal tubule**, the descending **loop of Henle**, the ascending loop of Henle, and the **distal tubule**. The distal tubule empties into a **collecting duct**, which receives wastes from many nephrons. The filtrate empties into the renal pelvis.

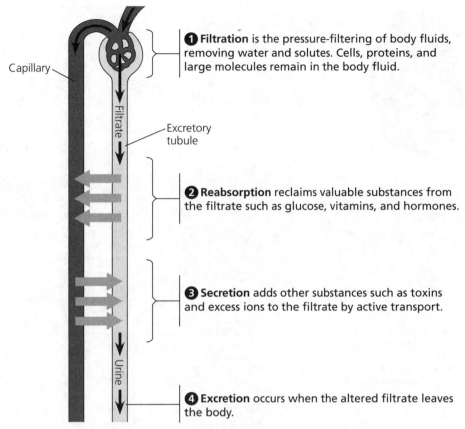

1 Filtration is the pressure-filtering of body fluids, removing water and solutes. Cells, proteins, and large molecules remain in the body fluid.

Capillary

Filtrate

Excretory tubule

2 Reabsorption reclaims valuable substances from the filtrate such as glucose, vitamins, and hormones.

3 Secretion adds other substances such as toxins and excess ions to the filtrate by active transport.

Urine

4 Excretion occurs when the altered filtrate leaves the body.

Figure 9.5   Key functions of excretory systems

Concept 44.4 *The nephron is organized for stepwise processing of blood filtrate*

▌ There are five main steps in the **transformation of blood filtrate to urine**, as shown in Figure 9.6. Study each step in the figure as you read the following descriptions.

1. In the *proximal tubule*, secretion and reabsorption change the volume and composition of the filtrate. The pH of body fluids is controlled, and bicarbonate is absorbed, as are NaCl and water.
2. In the *descending loop of Henle*, reabsorption of water continues. *Aquaporins* are water transport proteins that provide channels for the free movement of water.
3. In the *ascending loop of Henle*, the filtrate loses salt without giving up water and becomes more dilute.
4. In the *distal* tub*ule*, K^+ and NaCl levels are regulated, as is filtrate pH.
5. The *collecting duct* carries the filtrate through the medulla to the renal pelvis, and the filtrate becomes more concentrated by the movement of salt.

▌ Let's go back to the four steps of urine formation introduced earlier. Study Figure 9.5 again, and make specific note of where each step occurs in the mammalian kidney.

▪ **Filtration** occurs in the glomerulus, when filtrate is forced into the Bowman's capsule. Blood cells and proteins do *not* enter the filtrate.

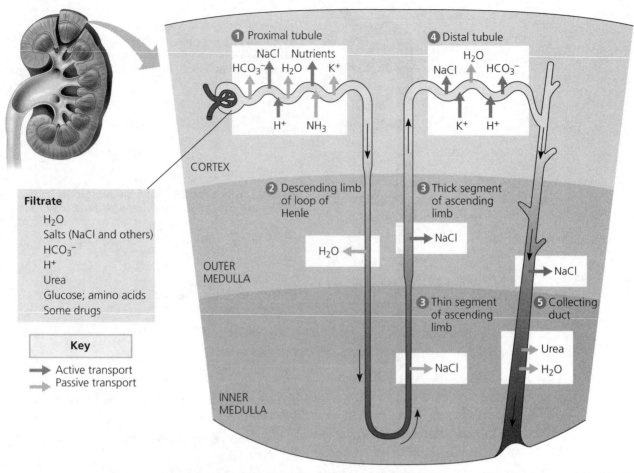

Figure 9.6   The nephron

- **Reabsorption** occurs in the proximal and distal tubules, as well as the loop of Henle.
- **Secretion** occurs in the proximal tubule.
- **Excretion** occurs when the filtrate leaves the body as the collecting tubules carry urine toward the ureters, then to the bladder, then out the urethra.

Note that the flow of filtrate in the loop of Henle is an example of a **countercurrent** system, and allows the kidney to form a concentrated urine with minimal water loss.

Concept 44.5 Hormonal circuits link kidney function, water balance, and blood pressure

Antidiuretic hormone (ADH) is an important hormone in the regulation of water balance. It is produced in the hypothalamus and released from the pituitary gland. It makes the collecting ducts more permeable to water, so more water leaves the filtrate, resulting in more concentrated urine and reduced loss of water from the body.

Also involved in regulation of water balance are **renin**, **angiotensin**, and **aldosterone**. Let's look at what happens if blood pressure or blood volume drops.

1. **Renin**, an enzyme, is released in the kidney. Renin activates angiotensin II.
2. **Angiotensin II** acts as a hormone and causes arterioles to constrict, which will raise blood pressure. It will also cause the adrenal glands to release the hormone **aldosterone**.
3. **Aldosterone** causes the kidney to reabsorb more Na^+, which increases retention of water and blood volume and pressure.

In summary, this pathway in the kidney helps in *long-term blood pressure regulation*. Note the two ways in which angiotensin II lowers blood pressure.

Chapter 45: Hormones and the Endocrine System

YOU MUST KNOW

- How hormones bind to target receptors and trigger specific pathways.
- The secretion, target, action, and regulation of at least two hormones.
- An illustration of both positive and negative feedback in the regulation of homeostasis by hormones.

TIP FROM THE READERS
The endocrine system is one of three human systems that is explicitly included in the Curriculum Framework. Feedback systems control the release of hormones, so you will want to be familiar with a couple of specific hormones and their regulation.

Concept 45.1 *Hormones and other signaling molecules bind to target receptors, triggering specific response pathways*

- The **endocrine system** of an animal is the sum of all its hormone-secreting cells and tissues.
- **Endocrine glands** are *ductless* and secrete hormones directly into body fluids.
- **Hormones** are chemical signals that cause a response in *target cells*.
- **Positive and negative feedback** regulate most endocrine secretion.

Concept 45.2 **Feedback regulation and antagonistic hormone pairs are common in endocrine systems**

▌ Study two mechanisms of hormone action. Recall that these mechanisms were covered in some detail in Chapter 11, Cell Communication.

■ **Cell-surface receptors** bind the hormone, and a *signal transduction pathway* is triggered. A signal transduction pathway consists of a series of molecular events that initiate a response to the signal. *Example*: The binding of epinephrine to liver cells causes a cascade that leads to the conversion of glycogen to glucose.

■ **Intracellular receptors** are bound by hormones that are lipid-soluble. The receptor then acts as a transcription factor, causing a change in gene expression. *Example*: Testosterone and estrogen enter the nuclei of target cells, bind the DNA, and stimulate transcription of certain genes.

▌ Hormones in the body can affect one tissue, a few tissues, or most of the tissues in the body (as with the sex hormones), or they may affect other endocrine glands (these last are referred to as **tropic hormones**).

Concept 45.3 **The hypothalamus and pituitary are central to endocrine regulation**

▌ The **hypothalamus** receives information from nerves throughout the body and from other parts of the brain and then initiates endocrine signals in response.

▌ The **posterior pituitary** is an extension of the hypothalamus that stores and releases these two hormones (which are produced in the hypothalamus):

■ **Oxytocin** causes contraction of the uterine muscles in childbirth and ejection of milk in nursing.

■ **Antidiuretic hormone (ADH)** makes the collecting tubules of the kidney more permeable to water, increasing water retention.

▌ The **anterior pituitary** consists of endocrine cells that synthesize and secrete several hormones. Some of these are *tropic hormones*, which means they stimulate the activity of other endocrine tissues (FSH, LH, TSH, and ACTH).

■ **Follicle-stimulating hormone (FSH)** stimulates development of the ovarian follicles in females and promotes spermatogenesis in males by acting on the cells in the seminiferous tubules.

■ **Luteinizing hormone** triggers ovulation in females and stimulates the production of testosterone by the interstitial cells of the testes.

Concept 45.4 **Endocrine glands respond to diverse stimuli in regulating homeostasis, development, and behavior**

▌ The maintenance of blood calcium level shown in Figure 9.7 is one example of how homeostasis is maintained by *negative feedback*. Remember, in negative feedback, *more gets you less*.

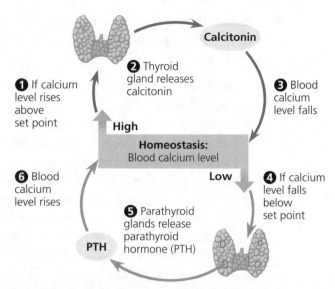

Figure 9.7   Maintenance of glucose homeostasis by insulin and glucagon

---

ORGANIZE YOUR THOUGHTS

Now that you have reviewed the negative-feedback regulation of blood calcium, prepare your own figure to show the regulation of blood glucose by insulin and glucagon. Having trouble? See Figure 45.13 in your text!

▍ While we are on regulation by feedback loops, let's jump ahead to Chapter 46 to use *oxytocin regulation* as an example of *positive feedback*. Oxytocin stimulates uterine contractions and also stimulates production of prostaglandins by the placenta. Prostaglandins stimulate release of more oxytocin and more prostaglandins, stimulating more uterine contractions. Remember, in positive feedback, *more gets you more.*

Chapter 46: Animal Reproduction

WHAT'S IMPORTANT TO KNOW?
This chapter provides illustrative examples that your teacher might select, but is not required content.

Concept 46.1 Both asexual and sexual reproduction occur in the animal kingdom

▍ **Sexual reproduction** is the creation of offspring by the fusion of haploid gametes to form a zygote. The female gamete is the *ovum*, and the male gamete is the *sperm*.

- **Asexual reproduction** is reproduction in which all genes come from one parent; there is no fusion of egg and sperm. There are several modes of asexual reproduction.

  - **Fission** is the separation of a parent into two or more individuals of about the same size.
  - **Budding** occurs when new individuals arise from outgrowths of the parent. This is seen in corals and hydras.
  - **Fragmentation** occurs when an individual breaks into several pieces, all of which then may form complete adults. *Regeneration*, the re-growth of body parts, is a necessary part of fragmentation. This mode of reproduction can be seen in starfish, sponges, and cnidarians.
  - **Parthenogenesis** is the process in which a female produces eggs that develop *without* being fertilized. Male bees are produced this way and are always haploid.

- Why sex? It may result in beneficial gene combinations arising through recombination that speed up adaptation, or the shuffling of genes during sexual reproduction might allow a population to rid itself of sets of harmful genes more readily.

Concept 46.2 Fertilization depends on mechanisms that bring together sperm and eggs of the same species

- **Fertilization** is the union of sperm and egg.
- **External fertilization** occurs when eggs are shed by the female and fertilized by the male outside the female's body, usually in water. The release of gametes must be synchronous and often involves environmental cues or courtship behaviors. Species with external fertilization in general produce very large numbers of gametes.
- **Internal fertilization** occurs when sperm are deposited in the female reproductive tract, and fertilization occurs within the tract. It is an adaptation that allows reproduction in a dry environment. Fewer gametes and fewer zygotes are often produced.
- When fertilized eggs are protected by shells or within the female's body, fewer zygotes are produced.
- **Gonads** are the organs that produce gametes in most animals.

Concept 46.3 Reproductive organs produce and transport gametes

- **Spermatogenesis** is the production of mature sperm cells, and it occurs in the seminiferous tubules. The cells that give rise to sperm are called *spermatogonia*. They undergo meiosis and differentiation eventually to form mature, motile sperm.
- **Oogenesis** is the development of mature ova. **Oogonia** are the cells that develop into ova; they multiply and begin meiosis, but they stop at prophase I

of meiosis I. These egg cells are called **primary oocytes**, which are quiescent until puberty. From puberty onward, FSH periodically stimulates a follicle to grow and its egg cell to complete meiosis I and begin meiosis II. This forms the **secondary oocyte**.

▮ Spermatogenesis differs from oogenesis in several ways:

- ▪ Production of four sperm cells versus a single egg.
- ▪ Spermatogenesis continues throughout a mature male's entire life; oogenesis ends with menopause.
- ▪ Spermatogenesis is continuous after puberty, whereas oogenesis has long interruptions. Recall that in humans, eggs are arrested in prophase I prior to a female's birth! Meiosis is not completed until after fertilization.

Concept 46.4 *The interplay of tropic and sex hormones regulates mammalian reproduction*

▮ Humans and other primates have **menstrual cycles**. Menstruation occurs when the endometrium is shed from the uterus through the cervix and vagina. Other mammals have **estrous cycles** when the vagina is receptive to mating. There are three phases in the menstrual cycle:

1. In the **menstrual flow phase**, most of the endometrium is shed and menstrual bleeding occurs.
2. In the **proliferative phase**, the endometrium begins to regenerate and thicken.
3. In the **secretory phase**, the endometrium continues to thicken, and if an embryo has not implanted in the lining by the end of this phase, menstrual flow occurs.

> **TIP FROM THE READERS**
> Because the hormonal control of menstruation is rich in examples of feedback and illustrates so well the circulation of hormones in the blood affecting target tissues, this is a good process to understand in detail.

▮ The **ovarian cycle** parallels the menstrual flow cycle. This cycling is under hormonal influence, and there is an interplay between pituitary hormones, whose target tissues are in the ovary, and ovarian hormones. Since you have probably just reviewed the chapter on hormones, let's look at what happens in this particular example. Note which hormone levels are increasing and which are decreasing and what these changes effect by studying Figure 9.8.

▮ Notice that the menstrual cycle is affected by two pituitary hormones and two ovarian hormones. We will first look at the pituitary hormones; since their target tissues are in the ovary, they are *gonadotropic hormones:*

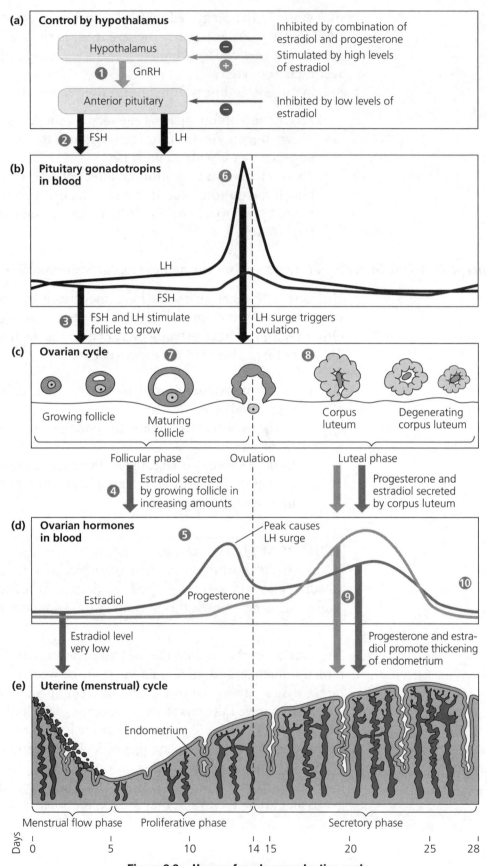

(a) Control by hypothalamus

Hypothalamus

❶ GnRH

Anterior pituitary

Inhibited by combination of estradiol and progesterone
Stimulated by high levels of estradiol

Inhibited by low levels of estradiol

❷ FSH   LH

(b) Pituitary gonadotropins in blood

❻

LH

FSH

❸ FSH and LH stimulate follicle to grow

LH surge triggers ovulation

(c) Ovarian cycle

❼   ❽

Growing follicle   Maturing follicle   Corpus luteum   Degenerating corpus luteum

Follicular phase   Ovulation   Luteal phase

❹ Estradiol secreted by growing follicle in increasing amounts

Progesterone and estradiol secreted by corpus luteum

(d) Ovarian hormones in blood

❺   Peak causes LH surge

Estradiol   Progesterone

❾   ❿

Estradiol level very low

Progesterone and estradiol promote thickening of endometrium

(e) Uterine (menstrual) cycle

Endometrium

Menstrual flow phase   Proliferative phase   Secretory phase

Days   0   5   10   14 15   20   25   28

Figure 9.8   Human female reproductive cycle

- **FSH** is secreted from the anterior pituitary. Its target is the follicle. As its level increases, the follicle enlarges and matures. FSH levels are affected by estradiol levels.
- **LH** is also secreted from the anterior pituitary. It induces final development of the follicle, and a surge of LH triggers ovulation.

▌ Now, take a look at the ovarian hormones:

- **Estradiol** is an estrogen secreted by the ovarian follicle. As the follicle develops, the level of estradiol increases and the endometrium thickens. Estradiol levels affect production of gonadotropins. (Study Figure 9.8.)
- **Progesterone** levels increase markedly after the ovarian follicle ruptures in ovulation and becomes the corpus luteum. High progesterone and estradiol levels promote further endometrial development.

▌ If fertilization of the egg does not occur, the corpus luteum will disintegrate, and the sharp decline in levels of estradiol and progesterone will lead to shedding of the endometrium (menstruation).

▌ Spermatogenesis and androgen levels in males also involve feedback loops with varying FSH, LH, and testosterone levels.

Concept 46.5   In placental mammals, an embryo develops fully within the mother's uterus

▌ **Human chorionic gonadotropin (hCG)** is secreted by the developing embryo and acts like LH to maintain the secretion of progesterone and estrogens by the corpus luteum in early pregnancy. Its high level in the urine is used as the basis of a common early pregnancy test.

Chapter 47: Animal Development

> **WHAT'S IMPORTANT TO KNOW?**
> This chapter provides illustrative examples that your teacher might select, but is not required content.

Concept 47.1   Fertilization and cleavage initiate embryonic development

▌ There are three stages that begin to build the body of most animals. Study Figure 9.9 as you read the accompanying text.

1. **Cleavage** is a period of rapid mitotic cell division that partitions the cytoplasm of the zygote into smaller cells called **blastomeres**, each of which has its own nucleus. Continued cleavage leads to a ball of cells called a **morula**, and then a fluid-filled central cavity called the **blastocoel** forms within the morula to produce a **blastula**.

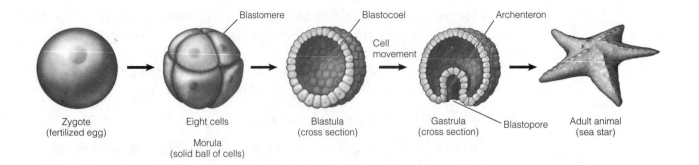

Figure 9.9   Animal development

2. **Gastrulation** is a drastic rearrangement of the cells in the blastula to form a three-layered embryo with a primitive gut (the *archenteron*). Invagination begins at the *blastopore*. In gastrulation, three germ cell layers are produced: the *ectoderm, endoderm*, and *mesoderm*. The chart below gives some of the most important derivatives of each germ layer.

Ectoderm	Mesoderm	Endoderm
• Skin, nails, teeth	• Skeletal, muscular systems	• Epithelial *linings* of digestive, respiratory, excretory tracts
• Lens of eye	• Excretory, circulatory systems	• Liver, pancreas
• Nervous system	• Reproductive system	
	• Blood, bone, and muscle	

3. **Organogenesis** is the development of the three germ layers into the rudiments of organs. Some additional notes about chordate development follow:

- The **notochord** is a stiff dorsal skeletal rod characteristic of all chordate embryos. It forms from mesoderm.
- The **neural plate** forms from ectoderm above the notochord. It curves inward, rolling into a *neural tube* that will become the brain and spinal cord.
- **Neurulation** is the process in which the hollow dorsal nerve chord forms.
- **Somites** are blocks of mesoderm that are serially arranged along the notochord. They are a sign of segmentation.

▍ The **blastocyst** is the mammalian version of a blastula (see Figure 9.10).

  ■ The **inner cell mass** is a group of cells that will develop into the embryo. Its cells are the source of embryonic stem cell lines.
  ■ The **trophoblast** is the outer epithelium of the blastocyst. It initiates formation of the fetal portion of the placenta.

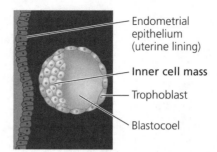

Endometrial
epithelium
(uterine lining)

Inner cell mass

Trophoblast

Blastocoel

Figure 9.10   Blastocyst structure

- The patterns of development are due to a combination of different **cytoplasmic determinants** and **inductive cell signals**. *Cytoplasmic determinants* are chemical signals such as mRNAs and transcription factors that may be parceled out unevenly in early cleavages. *Induction* is an interaction among cells that influences their fate, usually by causing changes in gene expression.
- The dorsal lip of the blastopore is an "organizer" which induces a series of events that result in formation of the notochord and neural tube.
- **Totipotent cells** are capable of developing into all the different cell types of that species. The cells of mammalian embryos remain totipotent until the 16-cell stage.

Chapter 48: Neurons, Synapses, and Signaling

YOU MUST KNOW
• The anatomy of a neuron.
• The mechanisms of impulse transmission in a neuron.
• The process that leads to release of neurotransmitter, and what happens at the synapse.

Concept 48.1 Neuron organization and structure reflect function in information transfer

- The **neuron** is the functional unit of the nervous system (see Figure 9.11). It is composed of a **cell body**, which contains the nucleus and organelles; **dendrites**, which are cell extensions that receive incoming messages from other cells; and **axons**, which transmit messages to other cells.
- Many axons are covered by an insulating fatty **myelin sheath**. This speeds the rate of impulse transmission.
- The **synapse** is a junction between two neurons (or a neuron and a muscle fiber or gland).
- **Neurotransmitters** are chemical messengers released from vesicles in the **synaptic terminals** into the synapse. They will diffuse across the synapse and bind to receptors on the neuron, muscle fiber, or gland across the synapse,

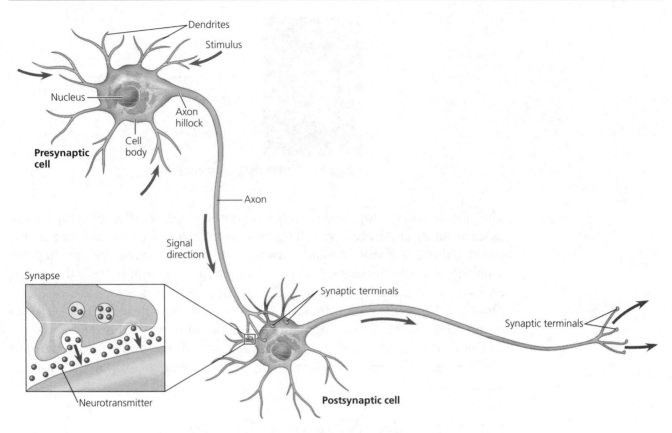

Figure 9.11   Structure of a vertebrate neuron

effecting a change in the second cell. *Examples*: acetylcholine, dopamine, and serotonin.

▌ The **central nervous system (CNS)** consists of the brain and spinal cord, and the **peripheral nervous system (PNS)** consists of the nerves that communicate motor and sensory signals throughout the rest of the body.

▌ **Sensory receptors** collect information about the world outside the body as well as processes inside the body. *Examples*: the rods and cones of the eye; pressure receptors in the skin.

▌ **Sensory neurons** transmit information from eyes and other sensors that detect stimuli to the brain or spinal cord for processing.

▌ **Interneurons** connect sensory and motor neurons or make local connections in the brain and spinal cord.

▌ **Motor neurons** transmit signals to *effectors*, such as muscle cells and glands.

▌ **Nerves** are bundles of neurons. A nerve can contain all motor neurons, all sensory neurons, or be mixed.

Concept 48.2 Ion pumps and ion channels establish the resting potential of a neuron

▌ **Membrane potential** describes the difference in electrical charge across a cell membrane.

▌ The membrane potential of a nerve cell at rest is called its **resting potential**. It exists because of differences in the ionic composition of the extracellular and intracellular fluids across the plasma membrane.

- The concentration of Na⁺ is higher outside the cell, whereas the concentration of K⁺ is higher inside the cell.
- Changes in the membrane potential of a neuron are what give rise to **nerve impulses**. A stimulus changes the permeability of the membrane to Na⁺.

Concept 48.3 Action potentials are the signals conducted by axons

- An **action potential** (nerve impulse) is an *all-or-none response* to depolarization of the membrane of the nerve cell. Carefully study Figure 9.12 to follow what happens in an action potential.

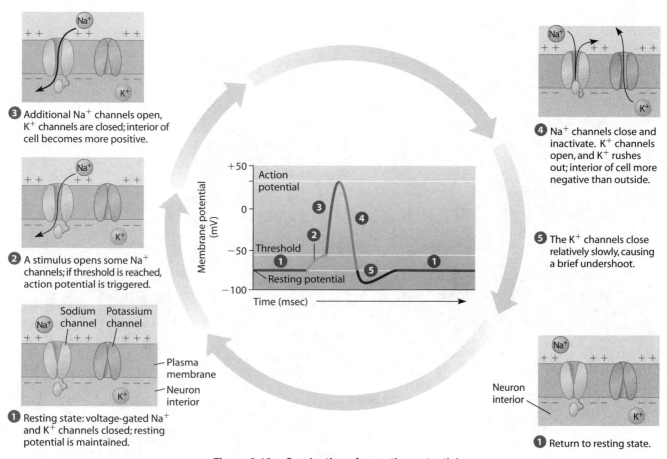

Figure 9.12 Conduction of an action potential

- A stimulus opens voltage-gated Na⁺ channels, allowing Na⁺ to enter the cell and bringing the membrane potential to a positive value.
- In order to generate an action potential, a certain level of depolarization must be achieved, known as the **threshold**.
- The membrane potential is restored to its normal resting value by the inactivation of Na⁺ channels and by opening voltage-gated K⁺ channels, which increases K⁺ leaving the cell.
- A *refractory period* follows the action potential, corresponding to the interval when the Na⁺ channels are inactivated. A new action potential cannot be generated during this time.

- **Saltatory conduction**, which is the jumping of the nerve impulse between *nodes of Ranvier* (areas on the axon not covered by the myelin sheath), speeds up the conduction of the nerve impulse.

Concept 48.4 Neurons communicate with other cells at synapses

- The signal is conducted from the axon of a presynaptic cell to the dendrite of a postsynaptic cell via an **electrical** or **chemical synapse**.
- Study the steps involved in neurotransmitter release in Figure 9.13.

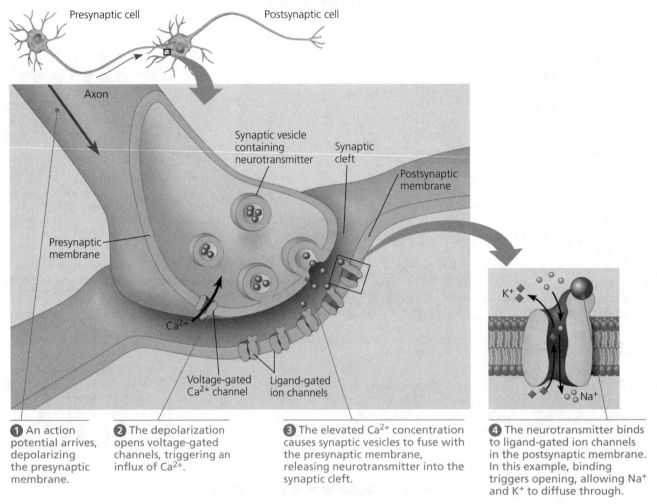

① An action potential arrives, depolarizing the presynaptic membrane.

② The depolarization opens voltage-gated channels, triggering an influx of Ca²⁺.

③ The elevated Ca²⁺ concentration causes synaptic vesicles to fuse with the presynaptic membrane, releasing neurotransmitter into the synaptic cleft.

④ The neurotransmitter binds to ligand-gated ion channels in the postsynaptic membrane. In this example, binding triggers opening, allowing Na⁺ and K⁺ to diffuse through.

Figure 9.13   Synapse

- Neurotransmitters are released by the presynaptic membrane into the *synaptic cleft*. They bind to receptors on the postsynaptic membrane and are then broken down by enzymes, or taken back up into surrounding cells.
- There are two categories of neurotransmitters: *excitatory* and *inhibitory*. Excitatory causes depolarization of the postsynaptic membrane, whereas inhibitory causes hyperpolarization of the postsynaptic membrane.
- **Acetylcholine** is a very common neurotransmitter; it can be inhibitory or excitatory. It is released by neurons at the neuromuscular junction. Other common **neurotransmitters** are epinephrine, norepinephrine, dopamine, and serotonin.

Chapter 49: Nervous Systems

Concept 49.1 Nervous systems consist of circuits of neurons and supporting cells

▌ A **reflex** is a simple automatic nerve circuit in response to a stimulus. Let's use a protective reflex, such as jerking your finger off a flame, as an example:

  ▪ The *stimulus* is detected by a *receptor* in the skin, conveyed via a *sensory neuron* to an *interneuron* in the spinal cord, which synapses with a *motor neuron*, which will cause the *effector*, a muscle cell, to contract.

▌ Note that at its simplest level, conscious thought is not required in a reflex.

▌ **Cerebrospinal fluid** circulates through a central canal in the spinal cord and the ventricles of the brain, bathing cells with nutrients and carrying away wastes. It also cushions the brain and spinal cord.

▌ **Gray matter** consists of mainly neuron cell bodies and unmyelinated axons.

▌ **White matter** is white because of the myelin sheaths about the axons.

▌ **Glia** are cells that support neurons. Three important kinds of glia are **astrocytes**, which provide support for neurons; **oligodendrocytes**, which form myelin sheaths in the CNS; and **Schwann cells**, which form myelin sheaths in the PNS.

▌ Study Figure 9.14 as you review the organization of the nervous system.

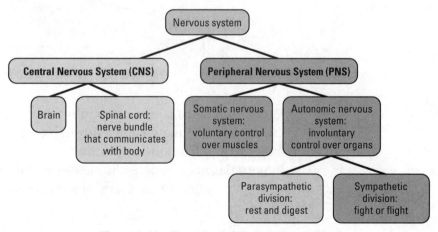

Figure 9.14   Branches of the nervous system

- The **central nervous system (CNS)** is the brain and spinal cord.
- The **peripheral nervous system (PNS)** consists of paired cranial and spinal nerves and associated ganglia. It is divided into

  - The *motor (somatic) nervous system*, which carries signals to skeletal muscles. It is a voluntary system.
  - The *autonomic nervous system*, which regulates the primarily automatic, visceral functions of smooth and cardiac muscles. This is the involuntary system.

- The **autonomic nervous system** transmits signals that regulate the internal environment by controlling smooth and cardiac muscle, including those in the gastrointestinal, cardiovascular, excretory, and endocrine systems. Its divisions are as follows:

  - The **sympathetic division**, which, when activated, causes the heart to beat faster and adrenaline to be secreted (with all its effects).
  - The **parasympathetic division**, which has the opposite effect when activated, slowing heartbeat and digestion.

Concept 49.2 *The vertebrate brain is regionally specialized*

- The **brainstem** is made up of the medulla oblongata, pons, and midbrain. The brainstem controls homeostatic functions such as breathing rate, conducts sensory and motor signals between the spinal cord and higher brain centers, and regulates arousal and sleep. (See Figure 9.15.)

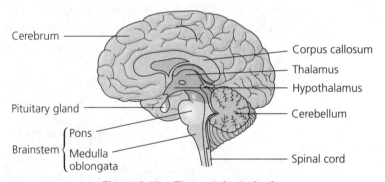

Figure 9.15   The vertebrate brain

- The **cerebellum** helps coordinate motor, perceptual, and cognitive functions.
- The **thalamus** is the main center through which sensory and motor information passes to and from the cerebrum.
- The **hypothalamus** regulates homeostasis; basic survival behaviors such as feeding, fighting, fleeing, and reproducing; thermostat, appestat, thirst center, and circadian rhythms.

> **STUDY TIP** Different regions of the brain have different functions. You are not expected to know each region and its function, but should have a firm grasp of an illustrative example.

- The **cerebrum** has two hemispheres, each with a covering of gray matter over white matter. Information processing is centered here, and this region is particularly extensive in mammals.
- The **cerebral cortex** controls voluntary movement and cognitive functions.
- The **corpus callosum** is a thick band of axons that enables communication between the right and left cortices.

Chapter 50: Sensory and Motor Mechanisms

> **WHAT'S IMPORTANT TO KNOW?**
> Big Idea 3 includes the idea that living systems respond to information essential to life processes. The sense organs allow detection of this information. Keep this in mind as you study the types of receptors and sense organs. Here are what we consider to be the main points of this chapter:
>
> - The location and function of several types of sensory receptors.
> - How skeletal muscle contracts.
> - Cellular events that lead to muscle contraction.

Concept 50.1 **Sensory receptors transduce stimulus energy and transmit signals to the central nervous system**

- **Mechanoreceptors** are receptors stimulated by physical stimuli, such as pressure, touch, stretch, motion, or sound.
- **Thermoreceptors** detect heat or cold and help maintain body temperature.
- **Chemoreceptors** transmit information about solute concentration in a solution. Taste and smell receptors are two types of chemoreceptors.
- **Electromagnetic receptors** detect various forms of electromagnetic energy such as visible light (*photoreceptors*), electricity, and magnetism.
- **Pain receptors** respond to excess heat, pressure, or specific classes of chemicals released from damaged or inflamed tissues.
- **Reception** occurs when a receptor detects a stimulus. **Perception** occurs in the brain as this information is processed. As an example, when you view an optical illusion in which a figure seems to change, what is actually changing is your perception of the object.

Concept 50.5 **The physical interaction of protein filaments is required for muscle function**

> **WHAT'S IMPORTANT TO KNOW?**
> The mechanism of skeletal muscle contraction is not required content, but is a nice illustrative example of a number of cellular activities.

- **Skeletal muscle** is attached to bones and responsible for the movement of bones. It consists of long fibers, each of which is a single muscle cell.
- Skeletal muscle is *striated* (striped in appearance).
- Each muscle fiber is a bundle of **myofibrils**, which in turn are composed of two kinds of myofilaments: **thin filaments** and **thick filaments**.
  - The thin filaments are *actin* and a regulatory protein.
  - The thick filaments are *myosin*.
- The **sarcomere** is the basic contractile unit of the muscle. Study the sarcomere in Figure 9.16 as you review these components.

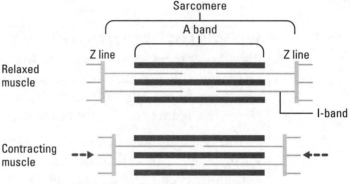

Figure 9.16   Sections of a sarcomere

  - **Z lines** make up the border of sarcomeres. Actin is attached here.
  - The **I band** is the area near the end of the sarcomere where only the thin actin filaments are located.
  - The **A band** is the entire length of the thick myosin filaments.
- During **muscle contraction**, the length of the sarcomere is reduced. Actin filaments slide over the myosin.
- The **sliding-filament model** states that the thick and thin filaments slide past each other so that their degree of overlap increases.
- Study Figure 9.17 as you review what happens during depolarization of a muscle fiber.
- A **motor neuron** will cause a muscle fiber to contract when its depolarization causes the neurotransmitter *acetylcholine* to be released into the synapse of the *neuromuscular junction.*
- As acetylcholine binds to receptors on the muscle fiber, ion channels open on the muscle cell membrane. This triggers an *action potential* (wave of depolarization) in the muscle cell.
  - The action potential spreads along **T tubules (transverse tubules)** to the **sarcoplasmic reticulum**.
  - This depolarization causes the sarcoplasmic reticulum to release **calcium ions**.
  - The calcium ions bind to **troponin** and cause it to move, exposing the myosin-binding sites on actin.

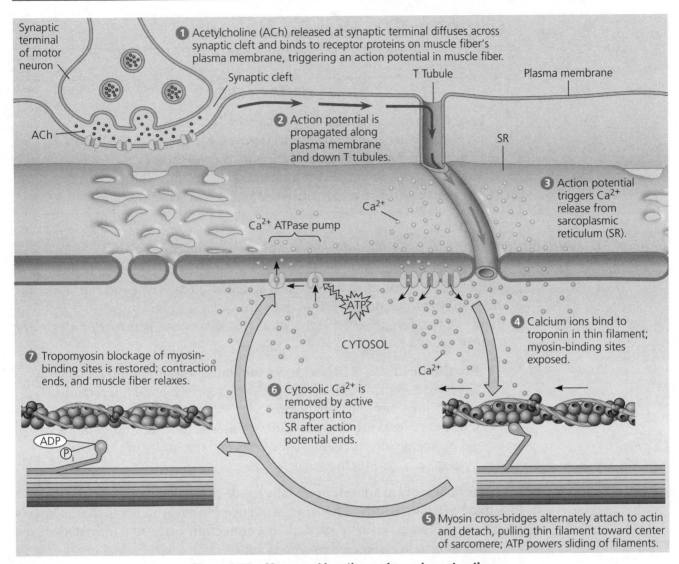

Figure 9.17 Myoneural junction and muscle contraction

- Actin and myosin now interact as myosin heads attach to the actin filaments, after being phosphorylated by ATP.
- The muscle contracts, with actin filaments sliding over myosin.

TIP FROM THE READERS

The mechanism of impulse transmission in a neuron (Concepts 48.3 and 48.4 and the mechanism of muscle contraction (Concept 50.5) both involve cell signaling, specific activities of many organelles, and therefore integrate many areas of your course.

Chapter 51: Animal Behavior

```
┌─────────────────────────────────────────────────────────────┐
│                       YOU MUST KNOW                          │
│  • How behaviors are the result of natural selection.        │
│  • How innate behavior and various types of learning increase fitness. │
│  • How organisms use communication to increase fitness.      │
│  • Various forms of animal communication.                    │
│  • The role of altruism and inclusive fitness in kin selection. │
└─────────────────────────────────────────────────────────────┘
```

Concept 51.1 Discrete sensory inputs can stimulate both simple and complex behaviors

▮ **Behavior** is what an animal does and how it does it. Behavior is a result of genetic and environmental factors, is essential for survival and reproduction, and is subject to natural selection over time.

▮ There are two fundamental levels of analysis, *proximate* and *ultimate*, in the study of behavior.

    ▪ **Proximate** causes of behavior are the "how" questions and include the effects of heredity on behavior, genetic-environmental interactions, and sensory-motor mechanisms.

    ▪ **Ultimate** causes are the "why" questions and include studies of the origin of a behavior, its change over time, and the utility of the behavior in terms of reproductive success.

▮ **Innate behaviors** are developmentally fixed. They are unlearned behaviors.

▮ A **fixed action pattern (FAP)** is a sequence of unlearned acts that is largely unchangeable and usually carried to completion once it is initiated. Fixed action patterns are triggered by *sign stimuli*.

▮ A classic example of a FAP was noted by *Niko Tinbergen* in male stickleback fish, which attack red objects. The red object is the sign stimulus; the attack is the FAP.

▮ A **kinesis** is a simple change in activity in response to a stimulus, whereas a taxis is an automatic movement toward or away from a stimulus. For example, the movement of pill bugs toward a moist habitat is a kinesis.

▮ **Migration** is a complex behavior seen in a wide variety of animals. Navigation may be by detection of the earth's magnetic field or visual cues.

▮ **Circadian rhythms** are those that occur on a daily cycle. Other rhythms occur over longer periods and can be triggered by differing day lengths or lunar cycles.

▮ A **signal** is a behavior that causes a change in the behavior of another individual and is the basis for animal communication. Examples of signals follow:

    ▪ **Pheromones** are chemical signals that are emitted by members of one species that affect other members of the species.

    ▪ **Visual signals** such as the warning flash of white of a mockingbird's wing.

    ▪ **Auditory signals** such as the screech of a blue jay or song of a warbler.

Concept 51.2 Learning establishes specific links between experience and behavior

■ **Learning** is the modification of behavior based on specific experiences.

■ **Imprinting** is a combination of learned and innate components that are limited to a *sensitive period* in an organism's life and is generally irreversible. Recall *Konrad Lorenz*'s demonstration of imprinting in greylag geese. When hatchlings spent their first few hours with him, they followed him as though he were their mother!

■ **Associative learning** is the ability of many animals to associate one feature of their environment with another feature. **Classical conditioning** involves learning to associate certain stimuli with reward or punishment. **Operant conditioning** occurs as an animal learns to associate one of its behaviors with a reward or punishment.

■ Twin studies in humans indicate that both environment and genetics contribute significantly to behaviors.

■ Behavior can be directed by genes. For example, a single gene appears to control the courtship ritual in fruit flies.

Concept 51.3 Selection for individual survival and reproductive success can explain most behaviors

■ **Foraging behavior** includes not only eating, but also mechanisms used in searching for, recognizing, and capturing food.

■ **The optimal foraging model** proposes that it is a compromise between the benefits of nutrition and the cost of obtaining food.

■ **Mating systems** vary between species. The needs of the young are important constraints in the development of these systems. Various systems are

    ■ **Promiscuous** with no strong pair-bonds.
    ■ **Monogamous** with one male/one female.
    ■ **Polygamous** with one individual mating with several others.

■ **Agonistic behaviors** are often ritualized contests that determine which competitor gains access to a resource, such as food or mates.

Concept 51.4 Inclusive fitness can account for the evolution of behavior, including altruism

■ **Altruism** occurs when animals behave in ways that reduce their individual fitness but increase the fitness of other individuals in the population. *Example*: A blue jay giving an alarm call attracts attention to its location.

■ **Inclusive fitness** is the total effect an individual has on proliferating its genes by producing its own offspring and by providing aid that enables other close relatives to produce offspring. The natural selection that favors this kind of altruistic behavior by enhancing reproductive success of relatives is called **kin selection**.

■ Just checking to see if you are really awake as you read all this! Here is one of our favorite quotes on the topic, from J. B. S. Haldane: "I would lay down my life for two brothers or eight cousins." Can you explain what is meant?

For Additional Review

Consider how the immune, digestive, nervous, circulatory, and respiratory systems and the senses all contribute to homeostasis in animals. In doing so, connect the stimuli that engage these systems with the way the systems respond to those stimuli.

1. Which of the following is required in ALL living things in order for gas exchange to occur?
   (A) lungs
   (B) gills
   (C) moist membranes
   (D) blood
   (E) lymph

2. In animals, all of the following are associated with embryonic development EXCEPT
   (A) gastrulation.
   (B) cleavage.
   (C) depolarization.
   (D) organogenesis.
   (E) cell migration.

3. Which of the following is most likely to result in a release of epinephrine (adrenaline) from the adrenal glands?
   (A) falling asleep in front of the TV
   (B) watching a checkers tournament
   (C) doing yoga
   (D) fleeing from a harmful situation
   (E) being in the kitchen while dinner is being cooked

4. Oxygen is transported in human blood by which type of cell?
   (A) erythrocytes
   (B) leukocytes
   (C) phagocytes
   (D) B cells
   (E) platelets

5. The proximal tubules in the kidney reabsorb most of which of the following compounds?
   (A) H^+
   (B) proteins
   (C) urea
   (D) HCO_3^-
   (E) glucose

6. Salivary amylase, an enzyme secreted in saliva, begins the breakdown of which substance?
   (A) starches
   (B) proteins
   (C) fatty acids
   (D) nucleic acids
   (E) polypeptides

Directions: The following group of questions consists of five lettered choices followed by a list of numbered phrases or sentences. For each numbered phrase or sentence, select the one choice that is most closely related to it. Each choice may be used once, more than once, or not at all.

Questions 7–10
   (A) Ovary
   (B) Thyroid gland
   (C) Posterior pituitary gland
   (D) Adrenal medulla
   (E) Anterior pituitary gland

7. Releases at least six different hormones, including several tropic hormones

8. Releases hormones that stimulate the mammary gland cells and contraction of the uterus

9. Releases hormones that stimulate growth of the uterine lining and promote the development of female secondary sex characteristics

10. Releases hormones that stimulate and maintain metabolic processes

11. Which of the following is an example of negative feedback?
    (A) The movement of sodium across a membrane through a transport protein and the movement of potassium in the opposite direction through the same transport
    (B) The pressure of the baby's head against the uterine wall during childbirth, stimulating uterine contractions, causing greater pressure against the uterine wall, which produces still more contractions
    (C) The growth of a population of bacteria in a petri dish until it has used all its nutrients and its subsequent decline
    (D) A heating system in which the heat is turned off when the temperature exceeds a certain point and is turned on when the temperature falls below a certain point
    (E) The progress of a chemical reaction until equilibrium is reached and then the cycling back and forth of reactant to product

12. Which of the following mechanisms does NOT help prevent the gastric juice from digesting the stomach lining?
    (A) Pepsin is stored and secreted as pepsinogen.
    (B) Mucus lines the inside surface of the stomach.
    (C) Mitosis generates enough new cells to replace the stomach lining every few days.
    (D) Gastric juice is not secreted continuously.
    (E) Pepsinogen is activated by pepsin.

13. Pepsin in the stomach is primarily responsible for the breakdown of which type of molecule?
    (A) starches
    (B) proteins
    (C) lipids
    (D) nucleic acids
    (E) glycogens

14. Which of the following structures is primarily responsible for reabsorbing water from the alimentary canal?
    (A) small intestine
    (B) stomach
    (C) pancreas
    (D) large intestine
    (E) cecum

15. In the mammalian heart, the sinoatrial (SA) node is responsible for which of the following functions?
    (A) delaying the nerve impulse to the walls of the ventricle
    (B) controlling the atrioventricular valve
    (C) controlling the semilunar valves
    (D) setting the rate and timing of cardiac muscle contraction
    (E) monitoring stroke volume

16. Fluid lost from the capillaries and not reabsorbed is returned to the blood via
    (A) the venous system.
    (B) the arteriole system.
    (C) the lymphatic system.
    (D) capillary beds.
    (E) the hemolymph system.

17. In the blood, carbon dioxide is primarily transported in what way?
    (A) by hemoglobin
    (B) by hemocyanin
    (C) as carbon monoxide
    (D) as bicarbonate
    (E) in erythrocytes

18. All of the following are barrier defenses against infectious agents EXCEPT
    (A) skin.
    (B) nasal membranes.
    (C) saliva.
    (D) mucous secretions.
    (E) phagocytes.

19. An immune response to a specific antigen generates the production of which type of cell that launches an attack the next time that same antigen infects the body?
    (A) effector cells
    (B) memory cells
    (C) T cells
    (D) B cells
    (E) antibodies

20. The ball of capillaries that is associated with the nephron and associated with filtration in the kidney is
    (A) the Bowman's capsule.
    (B) the loop of Henle.
    (C) the proximal tubule.
    (D) the glomerulus.
    (E) the distal tubule.

21. Muscle cell contraction occurs via
    (A) contraction of the A band.
    (B) contraction of the I band.
    (C) contraction of the Z lines.
    (D) the sliding of the thin filaments by the thick filaments.
    (E) the contraction of the sarcoplasmic reticulum.

22. The succession of rapid cell division that follows fertilization is called
    (A) gastrulation.
    (B) cleavage.
    (C) morulation.
    (D) involution.
    (E) polarization.

23. The circuit of a sensory neuron, the spinal cord, a motor neuron, and an effector cell constitutes a
    (A) presynaptic sequence.
    (B) reflex arc.
    (C) nerve circuit.
    (D) nerve impulse.
    (E) saltatory conduction system.

24. Which of the following is released into the synaptic cleft and acts as an intercellular messenger?
    (A) sodium
    (B) chloride
    (C) neurotransmitter
    (D) action potential
    (E) voltage gradient

25. The regulation of the internal environment in animals is referred to as
    (A) equilibrium.
    (B) stasis.
    (C) homeostasis.
    (D) regulation.
    (E) feedback.

26. Fertilization—the fusion of egg and sperm cell—results in which of the following?
    (A) embryo
    (B) zygote
    (C) gamete
    (D) ovum
    (E) follicle

27. Fixed action patterns (FAPs) are instigated by which of the following?
    (A) mating behavior
    (B) ritual behavior
    (C) innate stimulus
    (D) sign stimulus
    (E) action potential

28. One morning, a woman who usually feeds her two cats in the morning passes by the food bowl without putting food in it. The cats usually run over to the bowl as she approaches it, but after four mornings of her passing the bowl without putting food in it, the cats no longer run over to the bowl. This is an example of
    (A) maturation.
    (B) imprinting.
    (C) habituation.
    (D) foraging.
    (E) sensitivity.

29. Pavlov's dogs learned to salivate when they heard the ring of a particular bell; this is an example of
    (A) classical conditioning.
    (B) operant conditioning.
    (C) sensitivity.
    (D) imprinting.
    (E) maturation.

30. The phenomenon in which young ducks follow their mother in a line is a result of which of the following?
    (A) habituation
    (B) imprinting
    (C) maturation
    (D) foraging
    (E) conditioning

31. Altruism exists in populations because
    (A) it deprives members of the species of territory and results in agonistic behavior.
    (B) it can result in the passing on of the altruistic member's genes.
    (C) it can result in the overall success of the ecosystem.
    (D) it can result in a bond between the altruistic member and the recipient of the altruism, and the recipient might later reciprocate the altruism.
    (E) it can result in the maximizing of the altruistic member's genetic representation in a population, if the altruistic member's behavior is directed toward a close relative.

Use this figure to answer the next question.

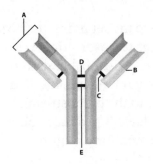

32. Where is the antigen-binding site of this antibody?
    (A) site A
    (B) site B
    (C) site C
    (D) site D
    (E) site E

Use this figure to answer the next question.

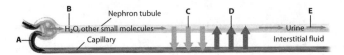

33. Which arrow in this schematic view of the nephron shows reabsorption?
    (A) arrow A
    (B) arrow B
    (C) arrow C
    (D) arrow D
    (E) arrow E

34. What type of cell acts as an intermediary between humoral and cell-mediated immunity?
    (A) plasma cell
    (B) cytotoxic T cell
    (C) B cell
    (D) helper T cell
    (E) macrophage

35. What occurs during gastrulation?
    (A) A solid embryo is changed into a hollow morula.
    (B) A solid blastula is changed into a hollow embryo that has four tissue layers.
    (C) A hollow blastula is changed into a hollow embryo that has three tissue layers.
    (D) A neural tube is created by invagination of the ectoderm.
    (E) A notochord is created by invagination of the ectoderm.

Use this figure to answer the next two questions.

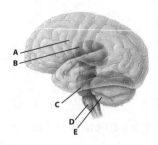

36. Which part of this diagram of the human brain depicts the brain region that coordinates muscular functions such as walking and tying shoelaces?
    (A) part A
    (B) part B
    (C) part C
    (D) part D
    (E) part E

37. Which structure in the diagram of the human brain is the center for the control of the involuntary centers for breathing, heartbeat, respiration, and digestion?
    (A) part A
    (B) part B
    (C) part C
    (D) part D
    (E) part E

38. The autonomic nervous system
    (A) integrates sensory inputs to the brain.
    (B) carries signals to and from skeletal muscles.
    (C) regulates the internal environment of the body.
    (D) is mostly voluntary.
    (E) is part of the central nervous system.

39. Once the threshold potential is reached,
    (A) the membrane potential is positive.
    (B) K^+ channels open.
    (C) Na^+ channels close.
    (D) an action potential is inevitable.
    (E) the interior of the cell becomes negative with respect to the outside.

40. During transmission across a typical chemical synapse,
    (A) neurotransmitter molecules are stored in the synaptic knob.
    (B) action potentials trigger chemical changes that make the synaptic vesicles fuse with each other.
    (C) vesicles containing neurotransmitters diffuse to the receiving cell's plasma membrane.
    (D) neurotransmitter molecules bind to receptors in the receiving cell's plasma membrane.
    (E) the binding of neurotransmitters to receptors initiates exocytosis.

Level 2: Application/Analysis/Synthesis Questions

After reading the paragraph, answer the question(s) that follow.

Under normal conditions, blood sugar levels are controlled within a narrow range by negative feedback. Two hormones are involved in maintaining blood sugar levels at the set point (about 90 mg of glucose/100 mL of blood). When blood sugar levels rise above the set point, the hormone insulin signals the liver to absorb the excess sugar. When blood sugar levels drop below the set point, the hormone glucagon signals the liver to release its stored glucose to the bloodstream. In type 1 diabetes, the body doesn't produce enough insulin and insulin supplements are required.

1. Based on your understanding of homeostasis, for negative-feedback control of blood glucose levels to function properly
   (A) the control center for glucose must be somewhere in the digestive system.
   (B) there must be sensors that monitor blood glucose levels.
   (C) there must be several other hormones involved (in addition to insulin and glucagon).
   (D) the body must prevent glucose levels from changing even slightly.

2. If you hadn't eaten for several hours, which hormone would be responsible for returning your glucose levels to the set point?
   (A) insulin
   (B) glucagon
   (C) liver
   (D) pancreas

3. An animal's inputs of energy and materials would exceed its outputs
   (A) if the animal is an endotherm, which must always take in more energy because of its high metabolic rate.
   (B) if it is actively foraging for food.
   (C) if it is growing and increasing its mass.
   (D) never; homeostasis makes these energy and material budgets always balance.

4. Which statement best describes the difference in responses of effector B cells (plasma cells) and cytotoxic T cells?
   (A) B cells confer active immunity; cytotoxic T cells confer passive immunity.
   (B) B cells kill pathogens directly; cytotoxic T cells kill host cells.
   (C) B cells secrete antibodies against a pathogen; cytotoxic T cells kill pathogen-infected host cells.
   (D) B cells accomplish the cell-mediated response; cytotoxic T cells accomplish the humoral response.

5. Which of the following statements is NOT true?
   (A) An antibody has more than one antigen-binding site.
   (B) An antigen can have different epitopes.
   (C) A pathogen makes more than one antigen.
   (D) A lymphocyte has receptors for multiple different antigens.

6. Which part of this figure shows an active plasma B cell?

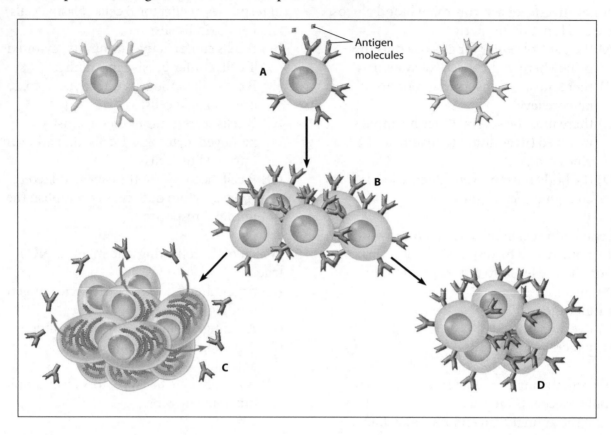

(A) part A
(B) part B
(C) part C
(D) part D

7. Where is the antigen-binding site of this antibody?

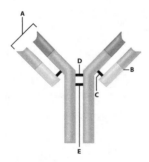

(A) site A
(B) site B
(C) site C
(D) site D
(E) site E

After reading the paragraph, answer the question(s) that follow.

To protect U.S. soldiers serving overseas, each soldier receives vaccinations against several diseases, including smallpox, before deployment. Following intelligence about an imminent smallpox threat, the Army wants to ensure that soldiers stationed in Iraq are fully protected from exposure to the disease, so all the soldiers in the threat zone are given a second vaccination against smallpox.

8. The first vaccination provides immunity because
   (A) a localized inflammatory response is initiated.
   (B) the vaccine contains manufactured antibodies against smallpox.
   (C) antigenic determinants in the vaccine activate B cells, which form plasma cells as well as memory cells.
   (D) the vaccine contains antibiotics and other drugs that kill the smallpox virus.

9. The second vaccination is beneficial because
   (A) it contains plasma cells that survive longer than 4–5 days.
   (B) it stimulates production of a higher concentration of antibodies in the bloodstream.
   (C) it requires two injections to stimulate antibody formation.
   (D) it keeps previously produced plasma cells circulating in the bloodstream.

10. Which molecule in this figure portraying water-soluble hormone action is the receptor protein?

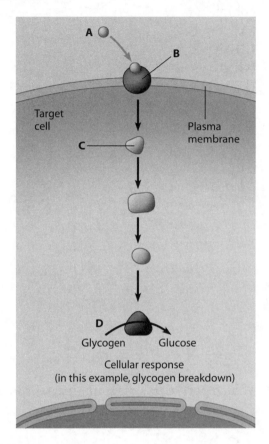

   (A) molecule A
   (B) molecule B
   (C) molecule C
   (D) molecule D

11. Which step in this figure portraying lipid-soluble hormone action shows transcription in response to the bound hormone-receptor complex?
    (A) step 1
    (B) step 2
    (C) step 3
    (D) step 4

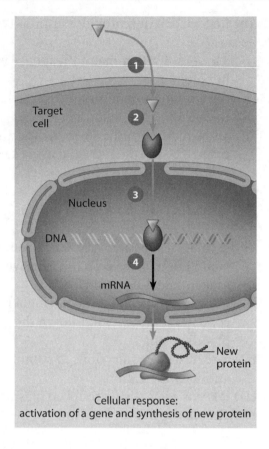

Target cell

Nucleus

DNA

mRNA

New protein

Cellular response:
activation of a gene and synthesis of new protein

12. Which part of this diagram of a neuron depicts the axon?

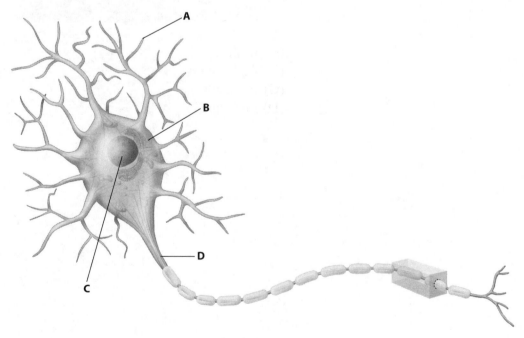

    (A) part A
    (B) part B
    (C) part C
    (D) part D

After reading the paragraphs, answer the question(s) that follow.

You recently sprayed your home with insecticide to remove an infestation of cockroaches. In your kitchen, you noticed some roaches lying on their backs twitching furiously before they died. This aroused your curiosity, so you decided to investigate exactly how the insecticide works on the nervous system.

In your research, you discover that the insecticide you used contains a permanent acetylcholinesterase inhibitor. Acetylcholine is a neurotransmitter that stimulates skeletal muscle to contract. Acetylcholinesterase removes acetylcholine from the synapse after the signal is received. Exposure to high pesticide concentrations has a similar effect on humans and can also be caused by exposure to the nerve gas Sarin and other chemical agents.

13. Why did the insecticide cause uncontrollable twitching in the roaches?
    (A) Acetylcholine was released, but the insecticide prevented it from diffusing across the synapse.
    (B) Acetylcholine was released, but the insecticide prevented it from binding to the receptor sites of the postsynaptic neurons.
    (C) The insecticide caused continuous stimulation of the muscles.
    (D) The insecticide prevented acetylcholinesterase from being removed from the synapse.

14. Since pesticides affect humans in a manner similar to that of roaches, it would be valid to conclude that
    (A) acetylcholinesterase affects the DNA of all animals.
    (B) the mechanism of stimulating skeletal muscle contraction must be similar in humans and roaches.
    (C) pesticides are harmful to roaches but not to humans.
    (D) the terminal end of the axon releases acetylcholine in roaches, but not in humans.

15. A common feature of action potentials is that they
    (A) cause the membrane to hyperpolarize and then depolarize.
    (B) can undergo temporal and spatial summation.
    (C) are triggered by a depolarization that reaches the threshold.
    (D) move at the same speed along all axons.

16. Where are neurotransmitter receptors located?
    (A) the nuclear membrane
    (B) the nodes of Ranvier
    (C) the postsynaptic membrane
    (D) the membranes of synaptic vesicles

17. Which of the following is a *direct* result of depolarizing the presynaptic membrane of an axon terminal?
    (A) Voltage-gated calcium channels in the membrane open.
    (B) Synaptic vesicles fuse with the membrane.
    (C) The postsynaptic cell produces an action potential.
    (D) Ligand-gated ion channels open, allowing neurotransmitters to enter the synaptic cleft.

18. Why are action potentials usually conducted in only one direction?
    (A) The nodes of Ranvier can conduct potentials in only one direction.
    (B) The brief refractory period prevents reopening of voltage-gated Na^+ channels.
    (C) The axon hillock has a higher membrane potential than the terminals of the axon.
    (D) Voltage-gated channels for both Na^+ and K^+ open in only one direction.

19. Which of the following conclusions is supported by this graph?

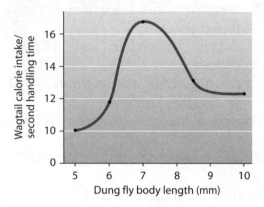

Graph from N. B. Davies. 1977. Prey selection and social behaviour in wagtails (Aves: Motacillidae). *Journal of Animal Ecology* 46: 37-57, fig. 9.

(A) Prey size does not affect the number of calories gained per second of handling time by wagtails.

(B) Wagtails get more calories per second of handling time with larger flies than with smaller ones.

(C) Wagtails get more calories per second of handling time with smaller flies than with larger ones.

(D) Wagtails get more calories per second of handling with 7-mm flies than with either larger or smaller ones.

After reading the paragraphs, answer the question(s) that follow.

A researcher is investigating the ability of salmon to migrate thousands of miles in the ocean yet return to the same location where they were hatched to spawn. Data from experiments suggest that more than one type of homing mechanism may be involved in this behavior. When salmon arrive at a river mouth from the open sea, they appear to use olfactory cues to find their home streams, but how do they find their way back to the correct spot along the coastline from the open ocean?

Several experiments are carried out to test the hypothesis that geomagnetic factors (the influence of Earth's magnetic field) play a key role in the ability of salmon to find the proper location along the coast. In one such experiment, salmon hatched in Ketchikan, Alaska, were subjected to the geomagnetic characteristics of a different location on the Alaska Peninsula, Cold Bay. The fish were then released to determine to which of the two locations they would return to spawn.

20. If the salmon return to spawn at Cold Bay, the behavior involved is primarily _____, but if they return to Ketchikan, the behavior is primarily _____.
(A) proximate . . . ultimate
(B) innate . . . learned
(C) learned . . . innate
(D) fixed . . . altruistic

21. What type of behavior would explain the ability of the salmon to return to their home streams?
(A) imprinting
(B) associative learning
(C) social learning
(D) habituation

22. Which of the following is NOT required for a behavioral trait to evolve by natural selection?
(A) In each individual, the form of the behavior is determined entirely by genes.
(B) An individual's reproductive success depends in part on how the behavior is performed.
(C) Some component of the behavior is genetically inherited.
(D) An individual's genotype influences its behavioral phenotype.

Free-Response Question

1. *Muscle cells are responsible for moving parts of the skeleton by contracting.*

   (a) **Describe** how skeletal muscle cells contract.
   (b) **Explain** what occurs in muscle cells when oxygen concentrations are low.

Ecology

Chapter 52: An Introduction to Ecology and the Biosphere

YOU MUST KNOW

- The role of abiotic factors in the formation of biomes.
- How biotic and abiotic factors affect the distribution of biomes.
- How changes in these factors may alter ecosystems.

Concept 52.1 *Earth's climate varies by latitude and season and is changing rapidly*

▍ **Ecology** is the scientific study of the interactions between organisms and the environment.

▍ **Climate** is the long-term prevailing weather conditions in a given area. The major components that make up the climate are temperature, precipitation, sunlight, and wind. Climate patterns can be described on two scales: *macroclimate* and *microclimate*.

- **Macroclimate patterns** work at the global, regional, or local level.
- The changing angle of the sun over the year, bodies of water, and mountains exert seasonal, regional, and local effects on **macroclimate**.
- **Microclimate** is determined by fine-scale variations, such as sunlight and temperature under a log.
- Increasing greenhouse gas concentrations in the air are warming Earth and altering the distributions of many species. Some species will not be able to shift their ranges quickly enough to survive.

Concept 52.2 *The structure and distribution of terrestrial biomes are controlled by climate and disturbance*

▍ **Biomes** are the major types of ecosystems that occupy very broad geographic regions.

▍ The importance of climate, especially precipitation and temperature, are reflected in the climograph for the major biomes of North America featured in Figure 10.1.

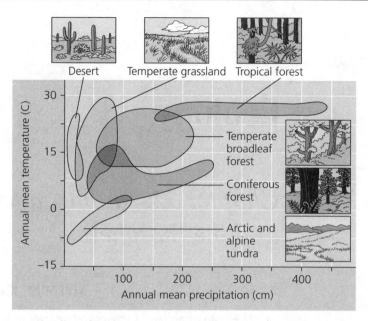

Figure 10.1   Climograph for some major biomes in North America

> **STUDY TIP** Why are some biomes rich in species abundance? Why is productivity higher in some ecosystems? You will not be expected to know the names of specific biomes, but you may be expected to predict and justify how alteration of a factor such as moisture, or light availability, or soil nutrients would impact the system. Always look for the big picture! Think *why? What if?*
> (Species abundance is related to primary productivity, which provides energy resources, and numerous niches. Productivity is higher in ecosystems that have abundant light energy, warmer temperatures, and longer growing seasons.)

▌ **Savannas** are characterized by grasses and also some trees. The dominant herbivores are insects, such as ants and termites. Fire is a dominant abiotic factor, and many plants are adapted for fire. Plant growth is quite substantial during the rainy season, but large grazing mammals must migrate during regular seasons of drought.

▌ **Desert** is marked by sparse rainfall, and desert plants and animals are adapted to conserve and store water. Deserts contain many CAM plants and plants with adaptations that prevent animals from consuming them, such as the spines on cacti. Temperature (either hot or cold) is usually extreme.

▌ **Chaparral** is dominated by dense, spiny, evergreen shrubs. These are coastal areas with mild rainy winters and long, hot, dry summers. Plants are adapted to fires.

▌ **Temperate grassland** is marked by seasonal drought with occasional fires and by large grazing mammals. All these factors prevent the significant growth of trees. Grassland soil is rich in nutrients, making these areas good for agriculture.

- **Temperate broadleaf forest** is marked by dense stands of deciduous trees that require sufficient moisture. These forests are more open than (and not as tall as) rain forests. They are stratified—the top layer contains one or two strata of trees; beneath that are shrubs; and under that is an herbaceous stratum. **Canopy** refers to the upper layers of trees in a forest. These trees drop their leaves in fall, and many mammals enter hibernation. Many birds migrate to warmer climates.
- **Coniferous forest** is dominated by cone-bearing trees such as pine, spruce, and fir. The conical shape of conifers prevents much snowfall from accumulating on—and breaking—these trees' branches.
- **Tundra** is marked by permafrost (permanently frozen layer of soil), very cold temperatures, high winds, and little rainfall. Tundra supports no trees or tall plants. It accounts for about 20% of Earth's terrestrial surface.
- **Tropical forest** has pronounced vertical stratification. The canopy is so dense that little light breaks through. These forests are marked by epiphytes, which are plants that grow on other plants instead of the soil. Rainfall is varied. Biodiversity is greatest of all the terrestrial biomes.

Concept 52.3 *Aquatic biomes are diverse and dynamic systems that cover most of Earth*

- **Aquatic biomes** make up the largest part of the biosphere, because water covers roughly 75% of Earth's surface. These biomes are classified into **freshwater biomes** and **marine biomes**.
- All aquatic biomes display vertical stratification, which forms the following ecologically unique areas:
  - The **photic zone**, in which there is enough light for photosynthesis to occur, and an **aphotic zone**, where very little light penetrates.
  - The **benthic zone** is located at the bottom of the biome, where it is made up of sand, inorganic matter, and organic sediments. Organic sediments also include **detritus**, which is dead organic matter.
  - **Thermoclines** are narrow layers of fast temperature change that separate a warm upper layer of water and cold deeper waters.
- The **two types of freshwater biomes** are standing bodies of water, such as lakes and wetlands, and moving bodies of water, such as streams and rivers.
  - In lakes, communities are distributed according to the water's depth. The **littoral zone** (well-lit shallow waters near the shore) contains rooted and floating aquatic plants, whereas the **limnetic zone** (well-lit open surface waters farther from shore) is occupied by phytoplankton.
  - **Oligotrophic lakes** are deep lakes that are nutrient-poor and oxygen-rich and contain sparse phytoplankton. **Eutrophic lakes** are shallower, and they have higher nutrient content and lower oxygen content with a high concentration of phytoplankton.
  - The prominent physical attribute of **streams and rivers** is current. Organisms are distributed in vertical zones and from the headwaters to the mouth.
  - **Estuaries** are areas where freshwater streams or rivers merge with the ocean.

▌ **Marine biomes** include the following:

- ■ The **intertidal zone**, where land meets the water, is periodically submerged and exposed by the twice-daily tides.
- ■ The **neritic zone**, beyond the intertidal zone, is the shallow water over the continental shelves.
- ■ The **pelagic zone** is a vast realm of open blue water found past the continental shelves.
- ■ A **coral reef** is a biome created by a group of cnidarians that secrete hard calcium carbonate shells, which vary in shape and support the growth of other corals, sponges, and algae. Coral reefs are among the most productive ecosystems on Earth.

Concept 52.4 Interactions between organisms and the environment limit the distribution of species

▌ The ecological study of species involves **biotic** (living) and **abiotic** (nonliving) influences.

- ■ **Biotic factors** may include behaviors as well as interactions with other species. Population and community ecology (Chapters 53 and 54) will explore many of the interactions between organisms.
- ■ The **abiotic components** of an environment are the nonliving, chemical, and physical components. Some important abiotic factors include temperature, water, salinity, sunlight, and soil.

Chapter 53: Population Ecology

YOU MUST KNOW

- How density, dispersion, and demographics can describe a population.
- The differences between exponential and logistic models of population growth.
- How density-dependent and density-independent factors can control population growth.

SCIENCE PRACTICES: CAN YOU . . .

- Pose scientific questions and apply mathematical routines to analyze interactions among community components?
- Predict and justify the effect of a change in one of the components on the interactions within the community and matter and energy flow?

Concept 53.1 ***Dynamic biological processes influence population density, dispersion,***
and demographics

▌ A **population** is a group of individuals of a single species living in the same
general area. **Population ecology** explores how biotic and abiotic factors influence the density, distribution, size, and age structure of populations.

▌ Three fundamental characteristics of the organisms in a population follow:

 ▪ **Density** is the number of individuals per unit area or volume. The density of a population increases by births or immigration and decreases
 by deaths or emigration.

 ▪ **Dispersion** is the pattern of spacing among individuals within the
 boundaries of the population.

   ▪ The most common pattern of dispersion is *clumped*, with individuals
   in patches, usually around a required resource. *Example*: cottonwood
   trees along a stream in the arid Southwest.

   ▪ A *uniform* dispersion pattern is often the result of antagonistic interactions. Animals that defend territories often show a uniform pattern.
   Example: red-winged blackbirds during mating season.

   ▪ *Random* dispersion shows unpredictable spacing. This is not a common spacing in nature, as there is usually a reason for a pattern of
   spacing.

 ▪ **Demography** is the study of vital statistics of a population, especially
 birth and death rates. A graphic way to show birth and death rates in a
 population is survivorship curves. Three types are shown in Figure 10.2.

   ▪ **Type I** shows low death rates during early and midlife; then the
   death rate increases sharply in older age groups.

   ▪ **Type II** shows a constant death rate over the organism's life span.

   ▪ **Type III** shows very high early death rates early, then a flat rate for
   the few surviving to older age groups.

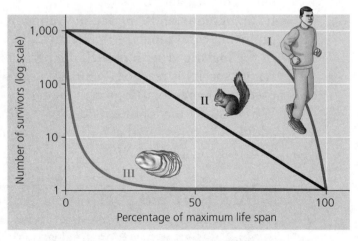

Figure 10.2   Survivorship curves: types I, II, and III

*Concept 53.2 The exponential model describes population growth in an idealized,
unlimited environment*

▍ **Exponential population** growth refers to population growth under ideal conditions. Figure 10.3 shows a graph of population growth as predicted by the exponential model. Any species, regardless of its life history, is capable of exponential growth if resources are abundant.

▍ Exponential population growth is shown by the equation $dN/dt = r_{max}N$.

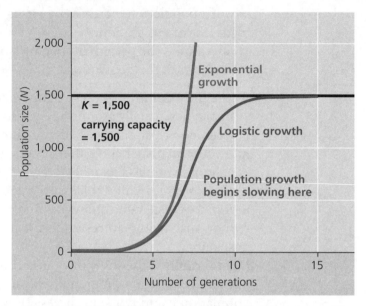

**Figure 10.3   Population growth predicted by the exponential
and logistic model**

*Concept 53.3 The logistic model describes how a population grows more slowly as
it nears its carrying capacity*

▍ The **carrying capacity** of a population is defined as the maximum population size that a certain environment can support at a particular time with no degradation of the habitat.

▍ If immigration and emigration are ignored, a population's growth rate (per capita increase) equals birth rate minus death rate: $dN/dt = B - D$.

▍ In the **logistic growth model**, the per capita rate of increase declines as carrying capacity is reached. Figure 10.3 shows a graph of population growth as predicted by the logistic growth model.

▍ We construct the logistic model by starting with the exponential model and adding an expression that reduces per capita rate of increase as N approaches K: $dN/dt = r_{max}N (K - N)/K$.

> **TIP FROM THE READERS**
> You will be given a formula sheet to use on the AP exam. The formulas for population growth, exponential growth, and logistic growth rate are included. You will need to practice using each of these! There is a practice problem at the end of this section.

Concept 53.4 Life history traits are products of natural selection

▌ Traits that affect an organism's schedule of reproduction and survival make up its **life history**. Life histories entail three variables: How early? How often? How many?

   ▪ How early in the life cycle does reproduction begin?
   ▪ How often does the organism reproduce? Some organisms save their resources for one big reproductive event (*big-bang reproduction*), whereas others produce offspring in *repeated reproduction*.
   ▪ How many offspring per reproductive event?

▌ Life history traits are evolutionary outcomes, *not* conscious decisions by organisms.

▌ Selection of life history traits that are sensitive to population density and carrying capacity are known as *K*-**selection**. *K*-selection operates in populations living close to the density imposed by the carrying capacity. By contrast, selection for life history traits that maximize reproductive success is called *r*-**selection**.

▌ The logistic growth model is sometimes associated with *K*-selection, whereas the exponential growth model is often associated with *r*-selection. Both *K*-selection and *r*-selection are two ends of a continuum of life history strategies.

Concept 53.5 Many factors that regulate population growth are density dependent

▌ A death rate that rises as population density rises and a birth rate that falls as population density rises are **density-dependent factors**. Examples of factors that reduce birth rates or increase death rates include the following:

   ▪ *Competition for resources*. As population density increases, competition for resources intensifies. This might include competition for food, space, or essential nutrients.
   ▪ *Territoriality*. Available space for territories or nesting may be limited, thus controlling the population.
   ▪ *Disease*. Increasing densities allow for easier transmission of diseases.
   ▪ *Predation*. As prey populations increase, predators may find the prey more easily.

▌ When a death rate does not change with increase in population density, it is said to be **density independent**. Natural disasters are examples of density-independent factors.

▌ All populations exhibit some size fluctuations. Many populations undergo regular boom-and-bust cycles that are influenced by complex interactions between biotic and abiotic factors.

Concept 53.6 The human population is no longer growing exponentially but is still increasing rapidly

▌ The exponential growth model in Figure 10.3 approximates the population explosion of humans over the last four centuries. However, since about 1970 the rate of growth has fallen by nearly 50%.

▌ One reason for falling human population growth is demographic transition. **Demographic transition** occurs when a population goes from high birth rates and high death rates to low birth rates and low death rates. Demographic transition may regularly take 150 years to complete. First, death rates fall, usually due to increased medical care and sanitation; however, falling birth rates take much longer, thus delaying transition.

■ **Age-structure pyramids** show the relative number of individuals of each age in a population, and can be used to predict and explain many demographic patterns. Study Figure 10.4. Why is Afghanistan poised for rapid growth? Why might economic growth in Italy be predicted to slow?

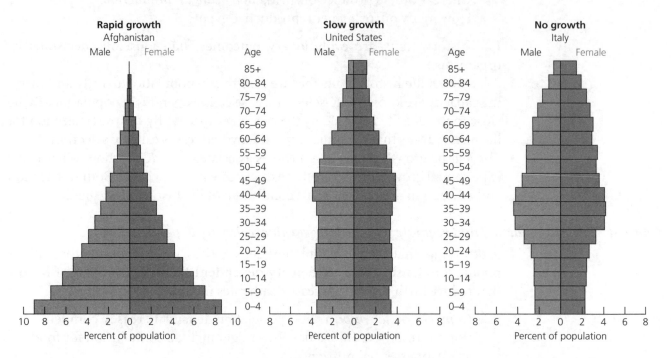

Figure 10.4 Age structure pyramids

■ Global carrying capacity for humans is not known. A concept termed the **ecological footprint** examines the total land and water area needed for all the resources a person consumes in a population. Currently, 1.7 hectares per person is considered sustainable. A typical person in the United States has a footprint of 10 hectares.

Practice problem: Can you use and apply a formula to calculate population growth rate? Use the table below to calculate the population growth rate of a hypothetical population where the carrying capacity (K) = 1,500 individuals and r_{max} is 1.0. See page 280 for the solution to the problem.

Population Size (N)	Maximum Rate of Increase (r_{max})	$\dfrac{K-N}{K}$	Per Capita Rate of Increase $r_{max}\left(\dfrac{K-N}{K}\right)$	Population Growth Rate $r_{max}N\left(\dfrac{K-N}{K}\right)$
1,600	1.0			
1,750	1.0			
2,000	1.0			

Chapter 54: Community Ecology

> **YOU MUST KNOW**
>
> - The difference between a fundamental niche and a realized niche.
> - The role of competitive exclusion in interspecific competition.
> - The symbiotic relationships of parasitism, mutualism, and commensalism.
> - The impact of keystone species on community structure.
> - The difference between primary and secondary succession.

Concept 54.1 Community interactions are classified by whether they help, harm, or have no effect on the species involved

▊ A **community** is a group of populations of different species living close enough to interact. **Interspecific interactions** may be positive for one species (+), negative (−), or neutral (0) and include *competition, predation,* and *symbioses.*

> ***STUDY TIP*** The prefix *inter-* means between different groups, whereas *intra-* means within the same group. *Intraspecific competition* is competition within the same species, like two males fighting over a territory. *Interspecific competition* is competition between two different species for resources, like food. Pay attention to the prefix! You could be asked to write about either type of competition in an essay.

▊ **Interspecific competition** for resources occur when resources are in short supply. Competition is a −/− interaction between the species involved. Central to the idea of competition and community structure are these two concepts:

- ▪ The **competitive exclusion principle** states that when two species are vying for a resource, eventually the one with the slight reproductive advantage will eliminate the other.
- ▪ An organism's **ecological niche** is the sum total of biotic and abiotic resources that the species uses in its environment. A species' **fundamental niche**, the niche potentially occupied by the species, is often different from the **realized niche**, the portion of the fundamental niche the species actually occupies.

▊ **Predation** is a +/− interaction between two species in which one species (the **predator**) eats the other species (the **prey**). Defenses for predators include the following:

- ▪ **Cryptic coloration**, in which the animal is camouflaged by its coloring.
- ▪ **Aposematic,** or **warning coloration**, in which a poisonous animal is brightly colored as a warning to other animals.
- ▪ **Batesian mimicry** refers to a situation in which a harmless species has evolved to mimic the coloration of an unpalatable or harmful species. In **Müllerian mimicry**, two bad-tasting species resemble each other, ostensibly so that predators will learn to avoid them equally.

- **Herbivory** is also a +/– interaction in which an herbivore eats part of a plant or alga. It is advantageous for an animal to be able to distinguish toxic from nontoxic plants. A plant's main protective devices are chemical toxins, spines, and thorns.
- **Symbiosis** occurs when individuals of two or more species live in direct contact with one another.

  - **Parasitism** is a +/– symbiotic interaction in which the parasite derives its nourishment from its host. Parasites may have a significant effect on the survival, reproduction, and density of their host population.
  - **Mutualism** is an interspecific interaction that benefits both species (+/+). Both pollinators and flowering plants benefit from their relationship.
  - **Commensalism** benefits one of the species but neither harms nor helps the other species. A fern growing in the shade of another plant could be a commensal relationship.

Concept 54.2 Diversity and trophic structure characterize biological communities

- **Species diversity** measures the number of different species in a community (species richness) *and* the relative abundance of each species. A community with an even species abundance is more diverse than one in which one or two species are abundant and the remainder are rare.
- The **trophic structure** of a community refers to the feeding relationships among the organisms. **Trophic levels** are the links in the trophic structure of a community.
- The transfer of food energy from plants through herbivores through carnivores through decomposers (from one trophic level to another) is referred to as a **food chain. Food webs** consist of two or more food chains linked together.
- **Dominant species** in a community have the highest **biomass** (the sum weight of all the members of a population) or are the most abundant.
- **Keystone species** exert control on community structure by their important ecological niches. Notice in Figure 10.5 the impact of the keystone predator *Pisaster* (a sea star) on the diversity of species present in a tidal pool.

Concept 54.3 Disturbance influences species diversity and composition

- A **disturbance**—storm, fire, flood, drought, or human activity—changes a community by removing organisms or changing resource availability.

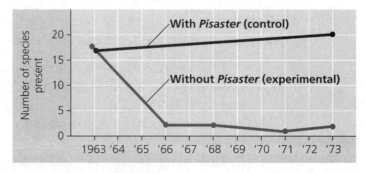

Figure 10.5   Impact of keystone predator on species diversity

Disturbance is not necessarily bad for a community. The **intermediate disturbance hypothesis** states that moderate levels of disturbance create conditions that foster greater species diversity than low or high levels of disturbance.

▌ **Ecological succession** refers to transitions in species composition in a certain area over ecological time.

▪ In **primary succession**, plants and animals gradually invade a region that was virtually lifeless where soil has not yet formed. The gradual colonization of a newly formed volcanic island would be an example.

▪ **Secondary succession** occurs when an existing community has been cleared by a disturbance that leaves the soil intact. An abandoned farm will show secondary succession as it starts with the soil intact.

Concept 54.4 Biogeographic factors affect community diversity

▌ Two biogeographic contributions are especially important in community diversity:

▪ *The latitude of the community.* Plant and animal life is generally more abundant and diverse in the tropics, becoming less so moving toward the poles.

▪ *The area of the community.* If all other factors are held equal, the larger the geographic area of a community is, the more species it has.

▌ Because of their isolation and limited size, islands are natural laboratories for studying biogeographic factors. In addition to actual islands, this idea also pertains to islands of land, like national parks surrounded by development.

▌ **Island biogeography** is primarily influenced by two factors:

▪ Rates of immigration and extinction are influenced primarily by the *size* of the island and the *distance* of the island from the mainland. The greater the sizes of the island, the higher the *immigration* rates and the lower the rates of *extinction*.

▪ As the distance from the mainland increases, the rate of *immigration* falls, whereas *extinction* rates increase.

Chapter 55: Ecosystems and Restoration Ecology

YOU MUST KNOW

- How energy flows through the ecosystem by understanding the terms in bold that relate to food chains and food webs.
- The difference between gross primary productivity and net primary productivity.
- The carbon and nitrogen biogeochemical cycles.
- Biogeochemical cycles such as the carbon and nitrogen cycles, and how they may impact individual organisms and/or populations and ecosystems.

Concept 55.1 Physical laws govern energy flow and chemical cycling in ecosystems

- An **ecosystem** is the sum of all the organisms living within its boundaries (biotic community) and all the abiotic factors with which they interact. Ecosystem ecology involves two unique processes: *energy flow* and *chemical cycling*.

- The flow of energy can be traced through the feeding or trophic levels in food chains and food webs. *Energy cannot be recycled*; therefore, energy must be constantly supplied to an ecosystem—in most cases by the sun.

  - **Primary producers** in an ecosystem are the **autotrophs** ("self-feeders"). They support all others organisms in the ecosystem.
  - Organisms that are in trophic levels above primary producers cannot make their own food and are therefore consumers or **heterotrophs** ("other-feeders").
  - *Herbivores* eat primary producers and are called **primary consumers**.
  - *Carnivores* that eat herbivores are called **secondary consumers**, whereas carnivores that eat secondary consumers are termed **tertiary consumers**.
  - **Detritivores**, or **decomposers**, are consumers that get their energy from *detritus*, which is nonliving organic material such as the remains of dead organisms, feces, dead leaves, and wood. Detritivores convert organic materials from all trophic levels to inorganic compounds that can be used by producers. In this way nutrients cycle through ecosystems.

- It is not uncommon for a species to feed at more than one trophic level. An animal's diet might consist of berries and fish or algae and insects. The feeding level may also change as the stage in a species' life cycle changes.

Concept 55.2 Energy and other limiting factors control primary production in ecosystems

- The amount of light energy converted to chemical energy by autotrophs is an ecosystem's **primary production**. The amount of all photosynthetic production sets the spending limit for the energy budget of the entire ecosystem.

  - Total primary production in an ecosystem is known as that system's **gross primary production (GPP)**.
  - GPP is not the amount of energy available to consumers, however. Some of the fuel molecules made by the producers must be used as fuel for their own cellular respiration. **Net primary production (NPP)** is equal to gross primary production minus the energy used by the primary producers for their "autotrophic respiration" (R_a):

$$NPP = GPP - R_a$$

- *Primary production* in aquatic ecosystems is affected primarily by light availability and nutrient availability. In the photic zone, light—and therefore photosynthesis—decreases with depth. The nutrient most often limiting marine production is either nitrogen or phosphorus. A lake that is nutrient-rich and that supports a vast array of algae is said to be **eutrophic**.

- Temperature and moisture are the key factors controlling primary production in terrestrial ecosystems. A measure of the amount of water transpired by plants and evaporated from the landscape, termed **evapotranspiration**, combines both key terrestrial factors.

Concept 55.3 Energy transfer between trophic levels is typically only 10% efficient

▌ Energy is lost at each level of transfer as heat, or for movement or reproduction or any of the many life processes that consume energy.

▌ If 10% of energy is transferred from primary producer to primary consumer to secondary consumer, only 1% of the net primary production (10% of 10%) is available to secondary consumers. The loss of energy from trophic level to trophic level is one of the factors that keeps food chains so short.

▌ *Ecological pyramids* can give insight into food chains. Try to sketch and explain each of these: a *biomass pyramid*, an *energy pyramid*, and a *pyramid of numbers*.

Concept 55.4 Biological and geochemical processes cycle nutrients and water in ecosystems

▌ **Biogeochemical cycles** are nutrient cycles that contain both biotic and abiotic components. Understanding these cycles allows scientists to trace how nutrients flow through ecosystems and how humans may have altered the flow.

▌ The **carbon cycle** is a balance between the amount of CO_2 removed from ecosystems by photosynthesis and added by cellular respiration. The burning of fossil fuels has added significant amounts of additional CO_2 to the atmosphere. Examine Figure 10.6 to see the generalized flow of carbon while also considering the effects of CO_2 on global warming.

> **TIP FROM THE READERS**
> Remember this:
> Matter cycles!
> Energy does not cycle!

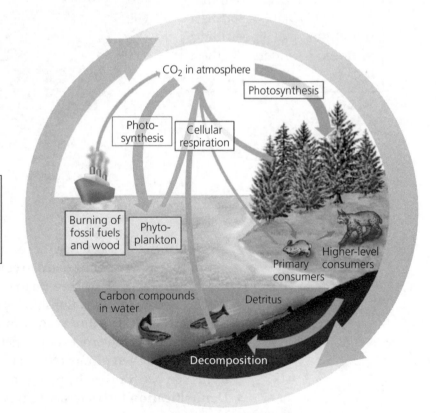

Figure 10.6   The carbon cycle

■ The **nitrogen cycle** moves nitrogen from the atmosphere through the living world. Nitrogen is a common limiting factor for plant growth, making its movement through ecosystems especially important. Note the important role of bacteria in the nitrogen cycle while tracing nitrogen flow through Figure 10.7.

> **TIP FROM THE READERS**
> Work through each figure verbally. You need to be able to explain how a change in the amount of nitrogen or carbon dioxide or another factor would impact an ecosystem. To understand the impact of acid rain, or fertilizer runoff, or global warming, you need to understand these cycles!

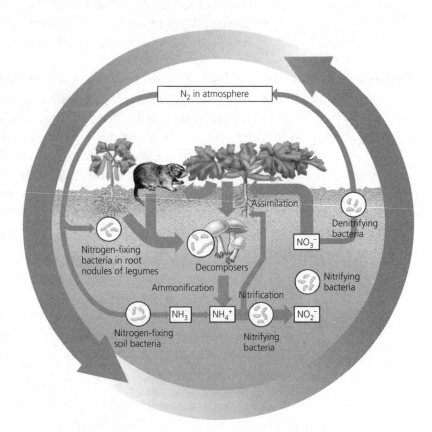

Figure 10.7   The terrestrial nitrogen cycle

■ Most of Earth's nitrogen is in the form of N_2, which is unusable by plants. The major pathway for nitrogen to enter an ecosystem is **nitrogen fixation**, the conversion of N_2 by bacteria to forms that can be used by plants. Earlier we noted this relationship between plants that are legumes and the bacterium *Rhizobium* as an example of mutualism.

■ **Nitrification** is the process by which ammonium (NH_4^+) is oxidized to nitrite and then nitrate (NO_3^-) by bacteria. Two inorganic nitrogen forms can be absorbed by plants: nitrates and ammonium.

■ **Denitrification** by bacteria releases nitrogen to the atmosphere.

■ Other important nutrient cycles involve water and phosphorus.

Concept 55.5 Restoration ecologists help return degraded ecosystems to a more natural state

▌ **Bioremediation** is the use of organisms, usually prokaryotes, fungi, or plants, to detoxify polluted ecosystems. It has been used to restore areas degraded by mining, or to remove oil or radioactive elements.

▌ **Bioaugmentation** is the introduction of desirable species such as nitrogen fixers to add essential nutrients.

Chapter 56: Conservation Biology and Global Change

YOU MUST KNOW

- The value of biodiversity, and the major human threats to it.
- How human activity is changing the Earth.

Concept 56.1 Human activities threaten Earth's biodiversity

▌ **Biodiversity**—short for biological diversity—can be considered at three main levels: genetic diversity, species diversity, and ecosystem diversity.

▌ Four major threats to biodiversity are habitat loss, introduced species, over-harvesting, and global change.

Concept 56.2 Population conservation focuses on population size, genetic diversity, and critical habitat

▌ When a population drops below a minimum viable population (MVP) size, its loss of genetic variation due to nonrandom mating and genetic drift can trap it in an *extinction vortex*. One key factor is the loss of genetic variation necessary to enable evolutionary responses to environmental change, such as the appearance of new strains of pathogens.

Concept 56.3 Landscape and regional conservation help sustain biodiversity

▌ The structure of a landscape can strongly influence biodiversity. As *habitat fragmentation* increases and edges become more extensive, biodiversity tends to decrease. *Movement corridors* can promote dispersal and help sustain populations.

▌ A **biodiversity hot spot** is a relatively small area with an exceptional concentration of endemic species and a large number of endangered and threatened species. Biodiversity hot spots are also hot spots of extinction and thus prime candidates for protection.

Concept 56.4 Earth is changing rapidly as a result of human actions

▌ Nutrient cycling is altered by human activities, particularly agriculture. For example, soil nitrogen is often depleted by crops. Excess nitrogen enters aquatic ecosystems as a result of livestock activities and can lead to eutrophication.

- **Acid precipitation** is defined as rain, snow, or fog with a pH less than 5.2. The burning of wood and fossil fuels releases sulfur oxides and nitrogen oxides into the atmosphere. These oxides react with water, forming sulfuric acid and nitric acid.
- In **biological magnification**, toxins become more concentrated in successive trophic levels of a food web. The toxins cannot be broken down biologically by normal chemical means, so they magnify in concentration as they move through the food chain.
- The **greenhouse effect** refers to the absorption of heat the Earth experiences due to certain atmospheric gases. Carbon dioxide and water vapor intercept and absorb much reflected infrared radiation, re-reflecting some back toward Earth.

  - Because of the burning of fossil fuels, CO_2 levels have been steadily increasing. One effect of this increase is that Earth is being warmed significantly (**global warming**).

- The **ozone layer** reduces the amount of UV radiation penetration from the sun through the atmosphere. Chlorine-containing compounds used by humans are eroding the ozone layer, allowing more DNA-damaging UV radiation to penetrate to the surface of the Earth.

Solution to practice problem on page 272:

Using a population size of 1,600 as an example,

$$\frac{dN}{dt} = r_{max} N \frac{(K-N)}{K} = \frac{1(1,600)(1,500-1,600)}{1,500}$$

and the population "growth" rate is −107 individuals per year. The population shrinks even faster when N is farther from the carrying capacity; when N equals 1,750 and 2,000 individuals, the population shrinks by 292 and 667 individuals per year, respectively.

Population Size (N)	Maximum Rate of Increase (r_{max})	$\dfrac{K-N}{K}$	Per Capita Rate of Increase $r_{max}\left(\dfrac{K-N}{K}\right)$	Population Growth Rate $r_{max}N\left(\dfrac{K-N}{K}\right)$
1,600	1.0	$(1{,}500 - 1{,}600)/1{,}500 = -0.067$	$1(-0.067) = (-0.067)$	$(-0.067)(1{,}600) = -107$
1,750	1.0			
2,000	1.0			

Level 1: Knowledge/Comprehension Questions

1. All of the following statements about Earth's ozone layer are *false* EXCEPT
   (A) it is composed of O_2.
   (B) it decreases the amount of ultraviolet radiation that reaches Earth.
   (C) it is thinning as a result of widespread use of certain chlorine-containing compounds.
   (D) it is a result of widespread burning of fossil fuels.
   (E) it allows green light in but screens out red light.

2. Which of the following is the major primary producer in a savanna ecosystem?
   (A) lion
   (B) gazelle
   (C) grass
   (D) snake
   (E) diatom

3. The carrying capacity of a population is defined as
   (A) the amount of time the parents in the population spend rearing and nurturing their offspring.
   (B) the maximum population size that a certain environment can support at a particular time.
   (C) the amount of vegetation that a certain geographic area can support.
   (D) the number of different types of species a biome can support.
   (E) the number of different genes a population can carry at a particular time.

4. Which of the following terms is used to describe major types of ecosystems that occupy broad geographic regions?
   (A) biome
   (B) community
   (C) chaparral
   (D) trophic level
   (E) photic zone

5. A lake that is nutrient-rich and that supports a vast array of algae is said to be
   (A) oligotrophic.
   (B) abyssal.
   (C) littoral.
   (D) eutrophic.
   (E) limnetic.

6. Which of the following best describes an estuary?
   (A) an area that is periodically flooded, causing its soil to be consistently damp
   (B) an area where a river changes course after being diverted from its original course by an obstacle
   (C) the area where a freshwater river merges with the ocean
   (D) the area where a mass of cold water and a mass of warm water meet in the pelagic zone
   (E) an outshoot of land that extends into the ocean

7. Which of the following is the term that refers to the layer of light penetration in aquatic ecosystems?
   (A) littoral zone
   (B) limnetic zone
   (C) photic zone
   (D) benthic zone
   (E) aphotic zone

Directions: The group of questions below consists of five lettered choices followed by a list of numbered phrases or sentences. For each numbered phrase or sentence, select the one choice that is most closely related to it. Each choice may be used once, more than once, or not at all.

Questions 8–12
   (A) Temperate grassland
   (B) Tropical forest
   (C) Temperate broadleaf forest
   (D) Tundra
   (E) Desert

8. Characterized by permafrost and few large plants

9. Characterized by epiphytes, a significant canopy, and abundant rainfall

10. Characterized by an understory of shrubs and trees that lose their leaves in the fall

11. Characterized by occasional fires, nutrient-rich soil, and large grazing animals

12. Characterized by sparse rainfall and extreme daily temperature fluctuations

13. A bacterial colony that exists in an environment displaying ideal conditions will undergo
    (A) logistic growth.
    (B) intrinsic growth.
    (C) hyperactive growth.
    (D) exponential growth.
    (E) unbounded growth.

14. A species' specific use of the biotic and abiotic factors in an environment is collectively called the species'
    (A) habitat.
    (B) trophic level.
    (C) ecological niche.
    (D) placement.
    (E) partitioning.

15. In which type of camouflaging does a non-toxic animal mimic the appearance of a toxic animal?
    (A) Müllerian mimicry
    (B) cryptic coloration
    (C) aposematic coloration
    (D) Batesian mimicry
    (E) parasitoidism

16. The dominant species in a community is the one that
    (A) has the greatest number of genes per individual.
    (B) is at the top of the food chain.
    (C) has the largest biomass.
    (D) eats all other members of the community.
    (E) bears the most offspring in each mating.

17. Which statement best describes energy transfer in a food web?
    (A) Energy is transferred to consumers, which convert it to nitrogen compounds and use it to synthesize amino acids.
    (B) Energy from producers is converted into oxygen and transferred to consumers.
    (C) Energy from the sun is stored in green plants and transferred to consumers.
    (D) Energy is transferred to consumers that use it to synthesize food.
    (E) Energy moves from autotrophs to heterotrophs to decomposers, which convert it to a form producers can use again.

18. A fire cleared a large area of forest in Yellowstone National Park in the 1980s. When the first plants pioneered this burned area, this was an example of
    (A) primary succession.
    (B) secondary succession.
    (C) biological evolution.
    (D) a keystone species.
    (E) the top-down model.

19. In the nitrogen cycle, the process by which nitrogen in the atmosphere is made available for use by plants is known as
    (A) ammonification.
    (B) denitrification.
    (C) nitrogen fixation.
    (D) nitrogen cycling.
    (E) nitrogenation.

20. The process in which CO_2 in the atmosphere intercepts and absorbs reflected infrared radiation and re-reflects it back to Earth is known as
    (A) global warming.
    (B) atmospheric insulation.
    (C) stratospheric insulation.
    (D) biological magnification.
    (E) the greenhouse effect.

21. A Type I survivorship curve is level at first, with a rapid increase in mortality in old age. This type of curve is
    (A) typical of many invertebrates that produce large numbers of offspring.
    (B) typical of human and other large mammals.
    (C) found most often in r-selected populations.
    (D) almost never found in nature.
    (E) typical of all species of birds.

22. Which of the following would not be a density-dependent factor limiting a population's growth?
    (A) increased predation by a predator
    (B) a limited number of available nesting sites
    (C) a stress syndrome that alters hormone levels
    (D) a very early fall frost
    (E) intraspecific competition

23. The human population is growing at such an alarmingly fast rate because
    (A) technology has increased our carrying capacity.
    (B) the death rate has greatly decreased since the Industrial Revolution.
    (C) the age structure of many countries is highly skewed toward younger ages.
    (D) fertility rates in many developing countries are above the 2.1 children per female replacement level.
    (E) all of the above are true.

24. When one species was removed from a tide pool, the species richness became significantly reduced. The removed species was probably
    (A) a strong competitor.
    (B) a potent parasite.
    (C) a resource partitioner.
    (D) a keystone species.
    (E) the species with the highest relative abundance.

25. Which of the following interspecific interactions is not an example of a +/− interaction?
    (A) ectoparasite and host
    (B) herbivore and plant
    (C) honeybee and flower
    (D) pathogen and host
    (E) carnivore and prey

Level 2: Application/Analysis/Synthesis Questions
After reading the paragraphs, answer the question(s) that follow.

The largest estuary in the United States is the Chesapeake Bay, which extends through six states, including Maryland, Virginia, and Pennsylvania. The bay is one of the most productive natural areas in the world. It is home to thousands of plants and animals, including many commercially important species.

The water of the bay is relatively shallow. Many areas are no more than 10 feet deep, with an average depth of 30 feet. Light penetrates the shallow water and supports the submerged plants that provide food and shelter for the many species living in the bay ecosystem. However, like many estuaries, the bay receives large amounts of fertilizer runoff from farms, lawns, and wastewater treatment facilities. This runoff introduces large amounts of nutrients.

1. Which of the following is the most probable sequence of events when fertilizer runoff reaches the bay?
   (A) submerged vegetation increases, more food for fish and shellfish, fish and shellfish populations increase
   (B) phytoplankton population increases, more food for fish and shellfish, fish and shellfish populations increase
   (C) phytoplankton population increases, sunlight blocked to submerged vegetation, submerged vegetation dies, fish and shellfish populations decrease
   (D) submerged vegetation decreases, fish and shellfish feed on decaying plants, phytoplankton feed on fish and shellfish, commercial fisheries decline

2. Which of the following pairs of nutrients would have the greatest effect on growth of phytoplankton?
   (A) carbon and hydrogen
   (B) oxygen and carbon dioxide
   (C) nitrogen and phosphorus
   (D) sulfur and magnesium

Use the graph below to answer questions 3 and 4.

3. According to this graph of the population growth of fur seals, in what year did the population first reach its carrying capacity?

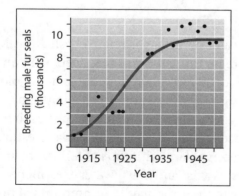

   (A) 1925
   (B) 1930
   (C) 1940
   (D) 1950

4. The formula $dN/dt = r_{max}N(K - N)/K$ describes the pattern of growth for the graph. Which of the following is true concerning the formula and the graph it describes in 1945?
   (A) $K = N$
   (B) $r_{max}N$
   (C) The variable K is constantly changing.
   (D) $(K - N)/K = 9,000$

After reading the paragraphs, answer the questions that follow.

Introduced species are a problem all over the world, and there are many examples in the United States. Several years ago, a fisherman caught a northern snakehead fish in a pond in Crofton, Maryland (a suburb of Washington, DC). Snakeheads are a favorite food of immigrants from China, and live fish can frequently be found in Asian markets. It's suspected that the fish in the Crofton pond were purchased locally and then intentionally released.

Snakeheads are top predators, and 90% of the northern snakeheads' diet consists of other fishes. The northern snakehead can breathe out of water and travel short distances (about 100 feet) across land. They also breed rapidly. Females can lay more than 100,000 eggs per year. Juveniles have also been identified in the Potomac River and other rivers in Pennsylvania.

5. When snakeheads enter aquatic ecosystems, biodiversity in these ecosystems would most likely
   (A) increase, since another species has been added to the environment.
   (B) decrease, since the snakehead will prey on native species.
   (C) remain the same, since local species will prey on the snakeheads and remove them.
   (D) remain the same, because the snakeheads will merge without problems into established communities.

6. Based on the characteristics of the snakehead described, which of the following is most likely to be a productive strategy to reduce the spread of this species?
   (A) extending the fishing season for prey fishes
   (B) introducing a natural predator to feed on juvenile snakeheads
   (C) introducing a fungus that prevents fish eggs from hatching
   (D) introducing algae and photosynthetic bacteria to reduce nutrient levels in the water

After reading the paragraph, answer the question(s) that follow.

You're a member of an influential African family that's been displaced from your home by civil war. You're trying to select a new country in which to settle to gain better economic opportunities. You know that Nigeria is a large country with rich natural resources and are considering it for your new home. You've learned some basic principles of population growth and did some research on the Internet. Among the data you found was the following diagram of the current age structure of the country.

7. The age-structure data for Nigeria shows that the country has many more individuals under the age of 15 than over the age of 40. What does this imply about the future population of Nigeria?
   (A) The population will probably remain stable.
   (B) The population will probably decrease.
   (C) The population will probably grow rapidly.
   (D) The number of older people will probably increase rapidly.

8. Based on the age structure of the country, which of the following situations would be most likely over the next 20 years?
   (A) strong economic gains stimulated by population growth
   (B) an increased demand for resources based on population growth
   (C) a decreased demand for medical services due to the small number of elderly citizens
   (D) a decline in housing prices based on lack of demand

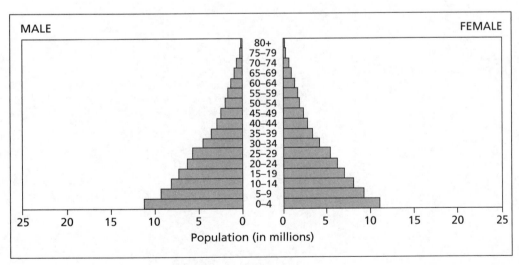

Populaion Pyramid for Nigeria, 2007

From U.S. Census Bureau International Data Base, 2007

1. *All of the organisms in a community are interrelated by the abiotic and biotic resources they use in the course of their lives.*

   (a) **Describe** the relationships that exist among a hawk, a mouse, a plant, and soil in a particular ecosystem.

   (b) **Discuss** two examples of the impact of human population growth on abiotic components of the environment. For each example, *explain* how a change in the abiotic component will impact the biotic component of the biosphere.

Part III

The Laboratory

AP Biology is designed to be equivalent to a two-semester college course, and laboratory experience must be included in all AP Biology courses. The College Board has published a collection of suggested investigations, and it is expected that all students will complete eight major investigations during the course. Of these, two should come from topics included in each Big Idea. Many of the investigations may be ongoing projects or require several days for completion. Since your teacher may select from many possible investigations, we present here only a small sample of labs. Your success in the course does not hinge on completion of a particular lab, but rather on your mastery of the underlying Science Practices.

Our students find that a review of the laboratory exercises is an excellent way to review many of the core concepts of their AP Biology course. A great resource for your review is LabBench, available at http://www.phschool.com/science/biology_place/labbench/. Although it was written to accompany the "old" lab manual, there is a wealth of information about some labs that your teacher may choose to use as a starting point for investigation. At LabBench you will find exercises that take you through 12 basic college labs, along with sample questions to test your understanding. Now, let's look at some selected investigations.

Inquiry-Based Labs

In laboratory, your teacher will engage you in inquiry labs in which you are not given step-by-step instruction, but rather are asked to select your own question, develop a hypothesis, and determine your own procedure for data collection. Although each lab is different, the following phrases indicate what you typically should understand and be able to do in order to demonstrate your lab expertise:

▌ Explain the purpose of each step of your procedure.
▌ Describe how you could determine whether . . .
▌ Summarize the pattern.
▌ Choose one of the variables that you identified, and design a controlled experiment to test . . .
▌ Explain how the concept is used to account for . . .
▌ Describe variables that were not controlled in the experiment, and describe how those variables might have affected the outcome of the experiment.
▌ Devise a method for data collection.
▌ Describe a method to determine . . .
▌ Indicate the results you expect for both the control and the experimental groups.
▌ Describe the results depicted in the graph.
▌ Describe the essential features of an experimental apparatus that could be used to measure . . .
▌ Graph the data . . . On the same axis, graph additional lines representing . . . (a prediction).

BIG IDEA 1: Investigation 2, Mathematical Modeling: Hardy-Weinberg

Overview of the Lab

▌ This investigation has you manipulate a computer spreadsheet to build a mathematical model to investigate the relationship between changing allelic frequencies in a population and evolution. You will develop an understanding of the Hardy-Weinberg equation, gain expertise with a spreadsheet program, and use your model to answer some question you pose.

▌ The skills you develop in creating the spreadsheet model and using it will be invaluable throughout many of your college courses, but not information that we can review with you in this guide. Instead, let's have a look at the basic content knowledge that should take away from your lab experience.

YOU MUST KNOW

- The Hardy-Weinberg equation and be able to use it to determine the frequency of alleles in a population.
- Conditions for maintaining Hardy-Weinberg equilibrium.
- How genetic drift, selection and the heterozygote advantage affect Hardy-Weinberg equilibrium.

SCIENCE PRACTICES: CAN YOU . . .

- Use a data set to reflect a change in the genetic makeup of a population over time and apply mathematical methods to investigate the cause(s) and effect(s) of this change?
- Apply mathematical methods to data from a real or simulated population to predict how the genetic composition of a population may change due to natural selection or gene flow or other factors?
- Evaluate data-based evidence that describes evolutionary changes in the genetic makeup of a population over time?
- Use data from mathematical models based on the Hardy-Weinberg equilibrium to analyze genetic drift and the effect of selection in the evolution of specific populations?
- Justify how data from mathematical models based on the Hardy-Weinberg equation can be used to analyze genetic drift and the effects of selection in the evolution of specific populations?
- Describe a model that represents evolution within a population?
- Evaluate data sets that illustrate evolution as an ongoing process?

▮ **The Hardy-Weinberg Law of Genetic Equilibrium** provides a mathematical model for studying evolutionary changes in allelic frequency within a population. It predicts that the frequency of alleles and genotypes in a population will remain constant from generation to generation if the population is stable and in genetic equilibrium.

> **Five conditions are required in order for a population to remain at Hardy-Weinberg equilibrium:**
>
> 1. No change in allelic frequency due to mutation
> 2. Random mating
> 3. No natural selection
> 4. Extremely large population size
> 5. No gene flow

▮ **No change in allelic frequency due to mutation**—Any mutation in a particular gene could change the balance of alleles in the gene pool. Mutations alone never change allelic frequency. Natural selection may make a mutation more common in a population over time. That is evolution.

▮ **Random mating**—In a population at equilibrium, mating must be random. In assortative mating, individuals tend to choose mates similar to themselves; for example, large blister beetles tend to choose mates of large size and small blister beetles tend to choose small mates. Though this does not alter allelic frequencies, it results in fewer heterozygote individuals than you would expect in a population where mating is random.

▮ **No natural selection**—No alleles are selected over other alleles. If selection occurs, those alleles that are selected for will become more common. For example, if resistance to a particular herbicide allows weeds to live in an environment that has been sprayed with that herbicide, the allele for resistance may become more frequent in the population.

▮ **Extremely large population size.** A large breeding population helps to ensure that chance alone does not disrupt genetic equilibrium. In a small population, only a few copies of a certain allele may exist. If for some chance reason the organisms with that allele do not reproduce successfully, the allelic frequency will change. This random, nonselective change is what happens in **genetic drift**.

▮ **No gene flow**—No new alleles can come into the population, and no alleles can be lost. Both immigration and emigration can alter allelic frequency. To estimate the frequency of alleles in a population, we can use the Hardy-Weinberg equation. According to this equation:

> p = the frequency of the dominant allele (represented here by A)
> q = the frequency of the recessive allele (represented here by a)
>
> For a population in genetic equilibrium:
>
> $p + q = 1.0$ (The sum of the frequencies of both alleles is 1.0.)
> $$(p + q)^2 = 1.0$$
>
> so
>
> $$p^2 + 2pq + q^2 = 1.0$$
> The three terms of this binomial expansion indicate the frequencies of the three genotypes:
>
> p^2 = frequency of AA (homozygous dominant)
> $2pq$ = frequency of Aa (heterozygous)
> q^2 = frequency of aa (homozygous recessive)

❚ This information in the box contains all the information you need to calculate allelic frequencies when there are two different alleles. Make sure that you can use it!

Sample Problem #1:

Consider a population of pigs where B = tan pigs and b = black pigs. We can use the Hardy-Weinberg to determine the percent of the pig population that is heterozygous for tan coat and so arrive at the frequency of the two coat-color alleles.

1. Calculate q^2.

Count the individuals that are homozygous recessive in the illustration above. Calculate the percent of the total population they represent. This is q^2.

Answer: Four of the sixteen individuals show the recessive phenotype, so the correct answer is 25% or 0.25.

2. Find q.

Take the square root of q^2 to obtain q, the frequency of the recessive allele.

Answer: $q = 0.5$

3. Find p.

The sum of the frequencies of both alleles = 100%, $p + q = 1$. You know q, so what is p, the frequency of the dominant allele?

Answer: $p = 1 - q$, so $p = 0.5$

4. Find $2pq$.

The frequency of the heterozygotes is represented by $2pq$. This gives you the percent of the population that is heterozygous.

Answer: $2pq = 2(0.5)(0.5) = 0.5$, so 50% of the population is heterozygous.

▌ You may want to go back to Topic 6, Chapter 23 where we also discuss Hardy-Weinberg.

TIP FROM THE READERS

There are a couple of ways you can get confused when doing these problems. Here's what to watch for:

1. If you are given the number of individuals showing the dominant trait, remember that this is *not* p^2 because it includes heterozygotes. However, you can use it to determine q^2 and from that get a value for q.

2. You also may be given a problem where you are told the frequency of an allele. You are being directly given p or q!

Sample Problem #2:

In a certain population of 1,000 fruit flies, 360 have red eyes, while the remainder have sepia eyes. The sepia eye trait is recessive to red eyes. How many individuals would you expect to be homozygous for red eye color?

Hint: The first step is always to calculate q^2! Start by determining the number of fruit flies that are homozygous recessive.

Answer: You should expect 40 to be homozygous dominant.

Calculations:

q^2 for this population is $640/1,000 = 0.64$

$q = \sqrt{0.64} = 0.8$

$p = 1 - q = 1 - 0.8 = 0.2$

The homozygous dominant frequency = $p^2 = (0.2)(0.2) = 0.04$

Therefore, you can expect 4% of 1,000, or 40 individuals, to be homozygous dominant.

Multiple-Choice Questions

1. If the frequency of two alleles in a gene pool is 90% A and 10% a, what is the frequency of individuals in the population with the genotype Aa?
   (A) 0.81
   (B) 0.09
   (C) 0.18
   (D) 0.01
   (E) 0.198

2. If a population experiences no migration, is very large, has no mutations, has random mating, and there is no selection, which of the following would you predict?
   (A) The population will evolve, but much more slowly than normal.
   (B) The makeup of the population's gene pool will remain virtually the same as long as these conditions hold.
   (C) The composition of the population's gene pool will change slowly in a predictable manner.
   (D) Dominant alleles in the population's gene pool will slowly increase in frequency, whereas recessive alleles will decrease.
   (E) The population probably has an equal frequency of A and a alleles.

3. In a population that is in Hardy-Weinberg equilibrium, the frequency of the homozygous recessive genotype is 0.09. What is the frequency of individuals that are homozygous for the dominant allele?
   (A) 0.7
   (B) 0.21
   (C) 0.42
   (D) 0.49
   (E) 0.91

BIG IDEA 2: Investigation 4, Diffusion and Osmosis

Overview of the Lab

In Part I you will create a cell model with agar cubes and use them to calculate surface area-to-volume ratios and make predictions about the rate of diffusion. In Part II, you will use dialysis tubing, which is selectively permeable, to create cell models that you can use to investigate questions about movement of molecules across the membrane. In Part III you will use a living tissue to observe and understand osmosis in cells.

YOU MUST KNOW

- Factors that affect diffusion across the membrane.
- How water potential is measured and its relationship to solute concentration and pressure potential of a solution.
- Water moves from a region where water potential is high to a region where water potential is low.
- The relationship of molarity to osmotic concentration.
- How to determine osmotic concentration of a solution from experimental data.

SCIENCE PRACTICES: CAN YOU . . .

- Design an experiment to measure the rate of osmosis in a model system?
- Analyze data and make predictions about molecular movement through cellular membranes?
- Connect the concepts of diffusion and osmosis to the structure of the membrane and molarity?
- Use the principles of water potential to predict and justify the movement of water into plant tissue?

Hints and Review

Get a firm fix on the terminology! This is one topic where you will not get any credit if you understand the concept but garble the vocabulary.

▌ **Osmosis** is the movement of water from a region of high concentration to a region of low concentration through a selectively permeable membrane.

▌ In **dynamic equilibrium** molecules are in motion, but there is no net change in concentration.

▌ Study Figure 4.1 to review some important terms.

▌ In Figure 4.1a the two solutions are equal in their solute concentrations. We say that they are **isotonic** to each other.

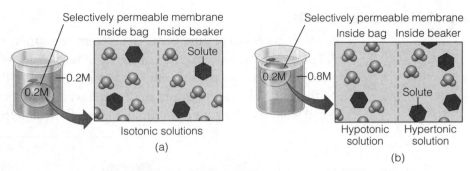

Figure 4.1   Diffusion across a selectively permeable membrane

❚ In Figure 4.1b, the solution in the bag contains less solute than the solution in the beaker. The solution in the bag is **hypotonic** (lower solute concentration) to the solution in the beaker. The solution in the beaker is **hypertonic** (higher solute concentration) to the one in the bag. Water will move from the hypotonic solution into the hypertonic solution.

Obviously, you must have a solid knowledge of the principles of diffusion to understand this lab, but what many students find most difficult in this laboratory is the concept of water potential. Here's a quick review.

❚ **Remember this:** *Water moves from a region where water potential is high to a region where water potential is low.*

❚ **Water potential** (Ψ) involves *two* components: solute potential (Ψ_S) and pressure potential (Ψ_P).

❚ An increase in solute concentration will cause solute potential (Ψ_S) to decrease.

❚ In plant cells, pressure increases as water flows in because the cell wall does not "give." This increases pressure potential (Ψ_P).

❚ A dehydrated potato slice does *not* have high water potential. Its water potential is low, and if placed in distilled water (which has high water potential), water will move into the potato cells.

Calculating Water Potential

Water potential is calculated using the following formula:

Water potential (Ψ) = pressure potential (Ψ_P) + solute potential (Ψ_S)	
Pressure potential (Ψ_P):	In a plant cell, pressure exerted by the rigid cell wall that limits further water uptake.
Solute potential (C_S):	The effect of solute concentration. Pure water at atmospheric pressure has a solute potential of zero. As solute is added, the value for solute potential becomes more negative. This causes water potential to decrease also.
	In sum, as solute is added, the water potential of a solution drops, and water will tend to move into the solution.
In this laboratory we use bars as the unit of measure for water potential; 1 bar = approximately 1 atmosphere.	

Factors That Affect Water Potential

The water potential of pure water in an open container is zero because there is no solute and the pressure in the container is zero. Adding solute lowers the water potential. When a solution is enclosed by a rigid cell wall, the movement of water into the cell will exert pressure on the cell wall. This increase in pressure within the cell will raise the water potential. Figure 4.2 will help you understand this.

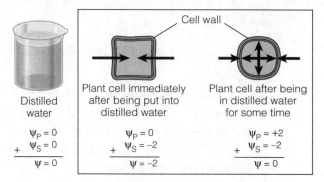

Figure 4.2   Water potential in a plant cell

Multiple-Choice Questions

1. In beaker b, shown in Figure 4.3, what is the water potential of the distilled water in the beaker, and of the beet core?
   (A) water potential in the beaker = 0; water potential in the beet core = 0
   (B) water potential in the beaker = 0; water potential in the beet core = −0.2
   (C) water potential in the beaker = 0; water potential in the beet core = 0.2
   (D) water potential in the beaker cannot be calculated; water potential in the beet core = 0.2
   (E) water potential in the beaker cannot be calculated; water potential in the beet core = −0.2

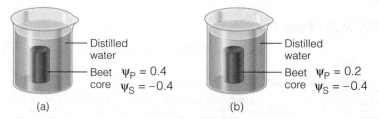

Figure 4.3   Water potential practice problem

2. Which of the following statements is true for the diagrams in Figure 4.3?
   (A) The beet core in beaker a is at equilibrium with the surrounding water.
   (B) The beet core in beaker b will lose water to the surrounding environment.
   (C) The beet core in beaker b would be more turgid than the beet core in beaker a.
   (D) The beet core in beaker a is likely to gain so much water that its cells will rupture.
   (E) The cells in beet core b are likely to undergo plasmolysis.

BIG IDEA 2: Investigation 5, Photosynthesis

Overview of the Lab

▍ What factors affect the rate of photosynthesis in living leaves? In the first part of this lab, you will learn a procedure to measure the rate of photosynthesis. There are several methods for this. Your teacher may elect to use the floating disk procedure to indirectly measure the rate of oxygen production, or a DPIP reduction method, or probes interfaced with computers.

▍ In the second part you will design and conduct your own investigation of one factor that affects the rate of photosynthesis.

YOU MUST KNOW

- The equation for photosynthesis and understand the process of photosynthesis.
- The relationship between light wavelength or intensity and photosynthetic rate.
- The anatomy of a typical leaf, and how the structures interact in photosynthesis.
- How to determine the rate of photosynthesis and then be able to design a controlled experiment to test the effect of some variable factor on photosynthesis.

SCIENCE PRACTICES: CAN YOU . . .

- Measure the rate of photosynthesis using a technique that gives consistent results?
- Apply mathematical routines to calculate the rate of photosynthesis?
- Apply the concepts you have learned in your investigation to describe relationships of cell structure and function?
- Describe strategies for capture, storage, and use of free energy by plants?

Hints and Review

▍ In photosynthesis, plant cells convert light energy into chemical energy that is stored in sugars and other organic compounds.

The equation for photosynthesis is
$$6\,H_2O + 6\,CO_2 \rightarrow C_6H_{12}O_6 + 6\,O_2$$

There are several ways you could measure the rate of photosynthesis. One would be to measure the rate of oxygen production. In the floating disk procedure, this is done by submerging a leaf and then measuring the accumulated gas over time. Another method uses the reduction of DPIP by electrons from chlorophyll that are excited when exposed to light.

You should be able to explain the chemical and physiological basis for the technique you use to measure the rate of photosynthesis.

1. In the floating disk procedure, a vacuum is created to remove accumulated gases from the air spaces in the leaf, which will cause the leaf disk to sink when placed in water. Since oxygen is produced in photosynthesis, most of these gas molecules are oxygen, and when the disk is exposed to bright light, it will float again as oxygen is generated.

2. In the DPIP reduction technique, you disrupt the chloroplast membrane by tossing the plant material in a blender. When exposed to light, the electrons in chlorophyll will be boosted to higher energy levels, but the electron acceptors embedded in the thylakoid membranes are not able to function. These high-energy electrons from chlorophyll will be picked up by DPIP, a chemical that is readily reduced. When oxidized, DPIP is deep blue. When oxidized, it will become colorless.

With a technique that will measure the rate of photosynthesis, such as DPIP reduction or oxygen accumulation, you can design an experiment to test a variable such as exposure to different light intensities (vary the distance from a bulb, or vary the wattage of the bulb) or exposure to different wavelengths of light (use colored cellophanes or filters).

Multiple-Choice Questions

1. If a student uses the DPIP technique described above, which of the following statements best describes the role of DPIP?
   (A) It mimics the action of chlorophyll by absorbing light energy.
   (B) It serves as an electron donor and blocks the formation of NADPH.
   (C) It is an electron acceptor and is reduced by electrons from chlorophyll.
   (D) It is bleached in the presence of light and can be used to measure light levels.

2. Some students were not able to get many data points when using the DPIP technique because the solution went from blue to colorless in only 5 minutes when they used chloroplasts exposed to light. What modification to the experiment do you think would be most likely to provide better results?
   (A) Double the volume of chloroplasts used.
   (B) Double the volume of DPIP so that the solution has a lower initial transmittance.
   (C) Boil the chloroplast in order to further disrupt the thylakoid membrane.
   (D) Select a different plant material, and blend it more thoroughly.

3. If a student performed this experiment using DPIP that was initially deep blue and got a flat curve, showing very little change in color over time, which of these would be a plausible explanation?
   (A) The rate of photosynthesis was very high.
   (B) The intensity of light may have been too great for a reaction to occur.
   (C) The chlorophyll was damaged and could not respond to light.
   (D) The DPIP was already reduced when the experiment began.

4. A student used the floating disk technique to measure the rate of photosynthesis. After 20 minutes under a bright light, none of the disks had floated. Based on your understanding of photosynthesis, which of the following might be a reasonable explanation for this?
   (A) Not all green plant material does photosynthesis when exposed to light.
   (B) A source of carbon dioxide was not provided.
   (C) A source of oxygen was not available.
   (D) There was no water available for photosynthesis.

BIG IDEA 2: Investigation 6, Cellular Respiration

Overview of the Lab

In this experiment you will learn how to use a respirometer to measure the rate of cellular respiration. You will select a question that interests you, and then design an experiment to test the effect of a single variable on the rate of respiration.

YOU MUST KNOW

- The equation for cellular respiration.
- The components of a respirometer and how they interact.
- The gas laws that affect volume changes within the respirometer.
- The relationship between movement of water in a respirometer and cellular respiration.
- The effect of temperature or increased metabolic activity on respiration.
- How to calculate the rate of respiration.

The equation for cellular respiration is

$$C_6H_{12}O_6 + 6\,O_2 \rightarrow 6\,H_2O + 6\,CO_2 + ATP$$

Hints and Review

▌ How can the rate of cellular respiration be measured? When you study the equation for cellular respiration, you will see that there are at least three ways:

1. Measure the amount of glucose consumed.
2. Measure the amount of oxygen consumed.
3. Measure the amount of carbon dioxide produced.

In this experiment, a *respirometer* is used to measure the amount of oxygen consumed.

▌ A **respirometer** is an air-tight chamber except for one opening for gases to enter or leave. Potassium hydroxide (KOH) soaks a cotton ball and will combine with the CO_2 produced by the organism you are using. A solid precipitate forms in the following reaction:

$$CO_2 + 2KOH \rightarrow K_2CO_3 + H_2O$$

Since CO_2 and O_2 are produced and consumed in equal amounts, any changes in the sealed container (assuming temperature and pressure remain constant) will be caused by a change in gas volume due to cellular respiration. With the CO_2 being removed as a solid precipitate, then it is O_2 that is consumed, lowering pressure within the respirometer, and allowing water to enter the pipette. Study Figure 6.1 to see the components of a respirometer.

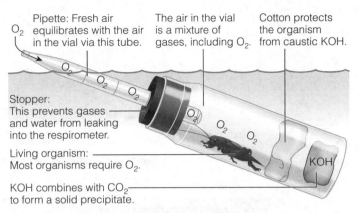

Pipette: Fresh air equilibrates with the air in the vial via this tube.

The air in the vial is a mixture of gases, including O_2.

Cotton protects the organism from caustic KOH.

Stopper: This prevents gases and water from leaking into the respirometer.

Living organism: Most organisms require O_2.

KOH combines with CO_2 to form a solid precipitate.

Figure 6.1   Components of a respirometer

▌ Although no single experimental organism is prescribed in this lab, many students will use some sort of seeds or small invertebrates. We will describe ideas related to seeds or invertebrates, though your experimental organism may differ.

▌ A seed contains an embryo plant and a food supply surrounded by a seed coat.

▌ *Germinating* (sprouting) seeds will show a higher rate of respiration than the *nongerminating* seeds because metabolic activity is increased. However, these dry, nongerminating seeds are not dead, but **dormant**. They can be stored for years and, when soaked in water, will germinate.

▌ As you investigate, consider the size and metabolic activity of the organism you have selected and its rate of cellular respiration. Your knowledge of metabolism should lead you to see that larger or more active or endothermic organisms require more energy.

▌ In most cases, chemical reactions occur more slowly at lower temperatures, so seeds or ectotherms that are chilled will show a lower rate of O_2 consumption. What comprises a valid control for your experiment? Why? A vial of glass beads alone might act as a *control* for this experiment. The control will compensate for any change in pressure or temperature.

▌ Students are often confused by the difference between a *control* (the glass beads in this experiment) and *factors that are held constant*. In this experiment, you will hold constant factors such the volumes within the containers, the number of peas or inverts, and the temperature of the water for each experimental group.

▌ Figure 6.2 shows a graph of typical results from an experiment. What is the question being tested?

▌ You need to be able to calculate the rate of these reactions. If you need to review how to do this, you will find a lesson on calculating rate in Investigation 13, Enzyme Activity.

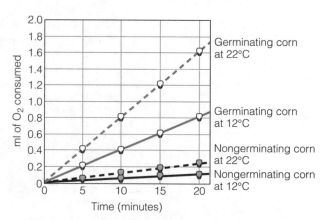

Figure 6.2   Effect of temperature on respiration rate

> **TIP FROM THE READERS**
> Students often become confused about the effect of cold on the rate of respiration. You may already know that there is more O_2 in colder water, but that is *not* the reason organisms respire more slowly when chilled! In general, within a normal physiological range for the organism, **the rate of metabolism is directly related to the temperature of the organism,** whether it be a pea seed, a cricket, or a goldfish.

Multiple-Choice Questions

1. What is the rate of oxygen consumption in germinating corn at 12°C as shown in Figure 6.2?
    (A) 0.08 ml/min
    (B) 0.04 ml/min
    (C) 0.8 ml/min
    (D) 0.6 ml/min
    (E) 1.00 ml/min

2. Using Figure 6.2, which of the following is a true statement based on the data?
    (A) The amount of oxygen consumed by germinating corn at 22°C is approximately twice the amount of oxygen consumed by germinating corn at 12°C.
    (B) The rate of oxygen consumption is the same in both germinating and nongerminating corn during the initial time period from 0 to 5 minutes.
    (C) The rate of oxygen consumption in the germinating corn at 12°C at 10 minutes is 0.4 ml O_2/minute.
    (D) The rate of oxygen consumption is higher for nongerminating corn at 12°C than at 22°C.
    (E) If the experiment were run for 30 minutes, the rate of oxygen consumption would decrease.

3. Which of the following conclusions is supported by the data shown on the graph in Figure 6.2?
   (A) The rate of respiration is higher in nongerminating seeds than in germinating seeds.
   (B) Nongerminating seeds are not alive and show no difference in rate of respiration at different temperatures.
   (C) The rate of respiration in the germinating seeds would have been higher if the experiment were conducted in sunlight.
   (D) The rate of respiration increases as the temperature increases in both germinating and nongerminating seeds.
   (E) The amount of oxygen consumed could be increased if pea seeds were substituted for corn seeds.

4. What is the role of KOH in a respirometer?
   (A) It serves as an electron donor to promote cellular respiration.
   (B) As KOH breaks down, the oxygen needed for cellular respiration is released.
   (C) It serves as a temporary energy source for the respiring organism.
   (D) It binds with carbon dioxide to form a solid, preventing CO_2 production from affecting gas volume.
   (E) Its attraction for water will cause water to enter the respirometer.

BIG IDEA 3: Investigation 7, Cell Division: Mitosis and Meiosis

This laboratory involves five parts:

1. **Modeling Mitosis**
   You will use beads or pipe cleaners or other materials to model the events of the cell cycle, including chromosome duplication and movement.

2. **Effects of Environment on Mitosis**
   You will set up and analyze an experiment using root squashes made from onion roots to investigate the effect of a protein (lectin) known to increase the rate of mitosis in the roots. After you collect data, you will use Chi-square analysis to statistically analyze the results.

3. **Loss of Cell Cycle Control in Cancer**
   For this part, you will consider HeLa cells and the Philadelphia chromosome and how the genetic changes they show lead to loss of cell cycle control and cancer.

4. **Modeling Meiosis**
   You will use the same materials as in Part I to model the events of meiosis, and show how meiosis and crossing-over events increase genetic variation. You will also demonstrate nondisjunction and explain its relationship to genetic disorders.

5. **Meiosis and Crossing over in *Sordaria***
   In this part, you will observe crossover events in meiosis of a fungus and calculate map distance.

YOU MUST KNOW

- The events of mitosis and meiosis in plant and animal cells.
- How mitosis and meiosis differ.
- How normal cells and cancer cells differ from each other.
- What may go wrong during the cell cycle in cancer cells.
- The roles of segregation, independent assortment, and crossing over in generating genetic variation.
- How to calculate map distance from experimental data.

SCIENCE PRACTICES: CAN YOU ...

- Make predictions about natural phenomena occurring in the cell cycle?
- Represent the connection between meiosis and increased genetic diversity?
- Use the mathematical routine of Chi-square analysis appropriately?

Hints and Review

Modeling Mitosis and Meiosis

Go back to Topic 2, Chapter 12 to review mitosis and the cell cycle. Go to Topic 4, Chapter 13 to review the important elements of meiosis.

Meiosis and Crossing Over in Sordaria

▌ When the growing filaments of two haploid strains of *Sordaria* that produce spores of different colors meet, fertilization occurs and zygotes form. Figure 7.1 shows spore formation in *Sordaria*.

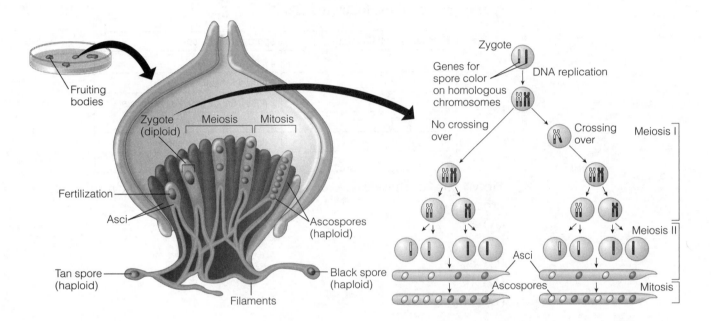

Figure 7.1   Crossing over in *Sordaria*

▌ Meiosis occurs within fruiting bodies to form four haploid *ascospores,* spores contained in *asci* (special sacs).

▌ Then one mitotic division doubles the number of ascospores to eight.

▌ The number of **map units** between two genes or between a gene and the centromere is calculated by determining the percentage of recombinants that result from crossing over. The greater the frequency of crossing over, the greater the map distance.

▌ Calculate the percent of crossovers by dividing the number of crossover asci (these are the ones with spores arranged 2:2:2:2 or 2:4:2) by the total number of asci × 100.

▌ To calculate the map distance, divide the percent of crossover asci by 2. The percent of crossover asci is divided by 2 because only half of the spores in each ascus are the result of crossing over.

Multiple-Choice Questions

1. Which of the following statements is correct?
   (A) Crossing over occurs in prophase I of meiosis and metaphase of mitosis.
   (B) DNA replication occurs once prior to mitosis and twice prior to meiosis.
   (C) Both mitosis and meiosis result in daughter cells identical to the parent cells.
   (D) Nuclear division occurs once in mitosis and twice in meiosis.
   (E) Synapsis occurs in prophase of mitosis.

2. A group of asci formed from crossing light-spored *Sordaria* with dark-spored produced the following results:

Number of Asci Counted	Spore Arrangement
7	4 light/4 dark spores
8	4 dark/4 light spores
3	2 light/2 dark/2 light/2 dark spores
4	2 dark/2 light/2 dark/2 light spores
1	2 dark/4 light/2 dark spores
2	2 light/4 dark/2 light spores

   How many of these asci contain a spore arrangement that resulted from crossing over?
   (A) 3
   (B) 7
   (C) 8
   (D) 10
   (E) 15

3. From this small sample, calculate the map distance between the gene and centromere.
   (A) 10 map units
   (B) 20 map units
   (C) 30 map units
   (D) 40 map units

BIG IDEA 3: Investigation 8, Biotechnology: Bacterial Transformation

Overview of the Lab

You will use antibiotic-resistance plasmids to transform *Escherichia coli*. A plasmid containing a gene for resistance to the antibiotic ampicillin is introduced into a strain of *E. coli* that is killed by ampicillin. If the susceptible bacteria incorporate the foreign DNA, they will become ampicillin resistant. You will apply mathematical routines to calculate transformation efficiency. Then, you may design and conduct an investigation to study transformation in more depth.

YOU MUST KNOW

- The principles of bacterial transformation, including how plasmids are engineered and taken up by cells.
- Factors that affect transformation efficiency.
- How to verify and screen for transformed cells.
- Bacterial transformation is a type of horizontal gene transfer, and increases genetic variation.

SCIENCE PRACTICES: CAN YOU . . .

- Calculate transformation efficiency and express the results in scientific notation?
- Predict and justify how a change in the basic protocol for bacterial transformation would affect transformation efficiency?

Key Concepts of Bacterial Transformation

▌ Genetic **transformation** occurs when a host organism takes in foreign DNA and expresses the foreign gene.

▌ Bacterial cells have a single main chromosome and circular DNA molecules called **plasmids**, which carry genetic information. All of the genes required for basic survival and reproduction are found in the single chromosome.

▌ **Plasmids** are circular pieces of DNA that exist outside the main bacterial chromosome and carry their own genes for specialized functions including resistance to specific drugs. In genetic engineering, plasmids are one means used to introduce foreign genes into a bacterial cell. To understand how this might work, consider the plasmid in Figure 8.1.

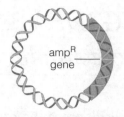

amp^R gene

Figure 8.1   Transformed plasmid

▌ Some plasmids have the *amp*^R gene, which confers resistance to the antibiotic ampicillin. *E. coli* cells containing this plasmid, termed "**+amp^R**" cells, can survive and form colonies on LB agar that has been supplemented with ampicillin.

▌ In contrast, cells lacking the amp^R plasmid, termed "**–amp^R**" cells, are sensitive to the antibiotic, which kills them. An ampicillin-sensitive cell (−amp^R) can be transformed to an ampicillin-resistant (+amp^R) cell by its uptake of a foreign plasmid containing the *amp*^R gene.

▌ **Competent cells** are cells that are most likely to take up extracellular DNA. Competent cells are in logarithmic growth, and chemical conditions are modified to induce the uptake of DNA. Study Figure 8.2 below to review the lab procedure used to prepare competent cells and get them to take up the amp^R plasmids.

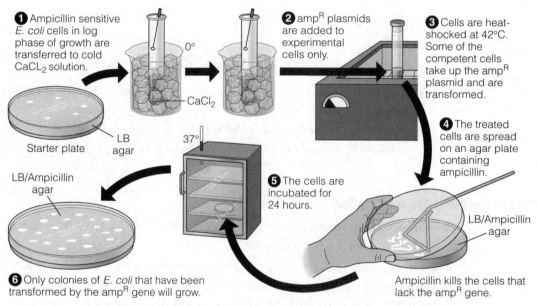

❶ Ampicillin sensitive *E. coli* cells in log phase of growth are transferred to cold CaCL₂ solution.

Starter plate

LB agar

0°

CaCl₂

❷ amp^R plasmids are added to experimental cells only.

❸ Cells are heat-shocked at 42°C. Some of the competent cells take up the amp^R plasmid and are transformed.

❹ The treated cells are spread on an agar plate containing ampicillin.

LB/Ampicillin agar

37°

❺ The cells are incubated for 24 hours.

LB/Ampicillin agar

❻ Only colonies of *E. coli* that have been transformed by the amp^R gene will grow.

Ampicillin kills the cells that lack the amp^R gene.

Figure 8.2   Procedure for bacterial transformation with amp plasmids

▌ Study Figure 8.3. It shows the expected results in this experiment.

▌ If there is no ampicillin in the agar, *E. coli* will cover the plate with so many cells it is called a "lawn" of cells (Plates A and C).

▌ Only transformed cells can grow on agar with ampicillin. Since only some of the cells exposed to the amp^R plasmids will actually take them in, only some

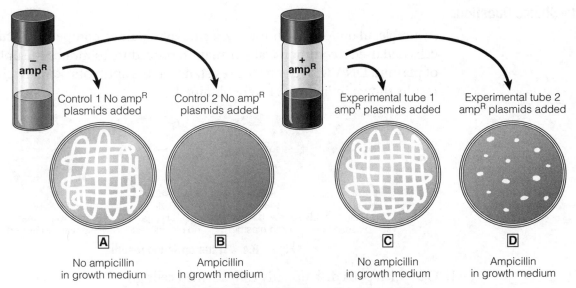

Figure 8.3 Sample transformation lab results

cells will be transformed. Thus, you will see only individual colonies on the plate (Plate D).

▌ If none of the sensitive *E. coli* cells have been transformed, nothing will grow on the agar with ampicillin (Plate B).

▌ **Restriction enzymes** or endonucleases are bacterial enzymes that will cut DNA at specific DNA sequences known as **recognition sites**. Often the enzymes cut the DNA so that the ends are single-stranded "sticky ends." A **gene of interest** (such as antibiotic resistance) can be introduced into a plasmid using restriction enzymes as described below.

▌ Here are the general steps used to introduce a gene of interest into bacteria:

1. Both the gene of interest and the plasmid are cut with the *same* restriction enzyme, so they have the same sticky ends.
2. DNA ligase is used to anneal and seal the sticky ends.
3. The recipient cells are transformed with the engineered plasmid.
4. Colonies carrying the plasmid are isolated.

▌ How do we know that transformation has been successful?

1. Use a **selection gene**, such as for antibiotic resistance. Only those cells that have incorporated the plasmid will have antibiotic resistance.
2. Use a **reporter gene** such as GFP (green fluorescent protein). Transformed cells will glow!

▌ **Transformation efficiency** is the number of transformed cells per microgram of the plasmid. High transformation efficiencies require cells that are in log phase of growth, suspended in ice-cold calcium chloride, have a rapid heat shock (this makes the membrane permeable to the plasmid), and plasmids that are not too large.

In a molecular biology laboratory, a student obtained competent *E. coli* cells and used a common transformation procedure to induce the uptake of plasmid DNA with a gene for resistance to the antibiotic kanamycin. The results shown in Figure 8.4 were obtained.

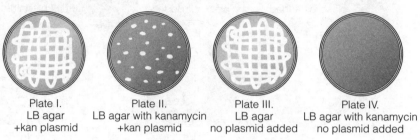

Plate I.
LB agar
+kan plasmid

Plate II.
LB agar with kanamycin
+kan plasmid

Plate III.
LB agar
no plasmid added

Plate IV.
LB agar with kanamycin
no plasmid added

Figure 8.4   Transformation results

1. On which petri dish do only transformed cells grow?
   (A) Plate I
   (B) Plate II
   (C) Plate III
   (D) Plate IV

2. Which of the plates is used as a control to show that nontransformed *E. coli* will not grow in the presence of kanamycin?
   (A) Plate I
   (B) Plate II
   (C) Plate III
   (D) Plate IV

3. If a student wants to verify that transformation has occurred, which of the following procedures should she use?
   (A) Spread cells from Plate I onto a plate with LB agar; incubate.
   (B) Spread cells from Plate II onto a plate with LB agar; incubate.
   (C) Repeat the initial spread of $-\text{kan}^R$ cells onto Plate IV to eliminate possible experimental error.
   (D) Spread cells from Plate II onto a plate with LB agar with kanamycin; incubate.
   (E) Spread cells from Plate III onto a plate with LB agar and also onto a plate with LB agar with kanamycin; incubate.

4. During the course of an *E. coli* transformation laboratory, a student forgot to mark the culture tube that received the kanamycin-resistant plasmids. The student proceeds with the laboratory because he thinks that he will be able to determine from his results which culture tube contained cells that may have undergone transformation. Which plate would be most likely to indicate transformed cells?
   (A) a plate with a lawn of cells growing on LB agar with kanamycin
   (B) a plate with a lawn of cells growing on LB agar without kanamycin
   (C) a plate with 100 colonies growing on LB agar with kanamycin
   (D) a plate with 100 colonies growing on LB agar without kanamycin

BIG IDEA 3: Investigation 9, Biotechnology: Restriction Enzyme Analysis of DNA

Overview of the Lab

You will use restriction endonucleases and gel electrophoresis to create and analyze genetic fingerprints. After electrophoresis, you will use your results to prepare a standard curve and estimate fragment sizes of an unknown sample.

YOU MUST KNOW

- The function of restriction enzymes and their role in genetic engineering.
- How gel electrophoresis separates DNA fragments.
- How to use a standard curve to determine the size of unknown DNA fragments.

SCIENCE PRACTICES: CAN YOU . . .

- Apply mathematical routines to construct a graph of DNA fragments of known size?
- Use this standard curve to determine the size of unknown DNA fragments?
- Use the results of gel electrophoresis to map the restriction sites of a bacterial plasmid?

Key Concepts of Restriction Enzyme Cleavage of DNA and Gel Electrophoresis

▍ **Gel electrophoresis** is a procedure that separates molecules on the basis of their rate of movement through a gel under the influence of an electrical field.

▍ The direction of movement is affected by the charge of the molecules, and the rate of movement is affected by their size and shape, the density of the gel, and the strength of the electrical field.

▍ DNA is a negatively charged molecule, so it will move toward the positive pole of the gel when a current is applied. When DNA has been cut by restriction enzymes, the different-sized fragments will migrate at different rates. Because the smallest fragments move the most quickly, they will migrate the farthest during the time the current is on. Keep in mind that the length of each fragment is measured in number of DNA base pairs.

▍ You may use restriction enzymes to create your own DNA fragments, or used fragments that are commercially prepared. In one lab we often purchase, students are given three samples of DNA obtained from a virus, the bacteriophage lambda. One sample is uncut DNA, one is incubated with the restriction enzyme *Hind*III, and one is incubated with *Eco*RI. The fragments of DNA are separated by electrophoresis, stained for visualization, and then analyzed.

■ After the DNA samples are loaded into wells in the gel, electricity is applied. The DNA fragments will migrate.

1. *DNA is negatively charged and will migrate toward the positive pole.*

2. *Smaller fragments of DNA will migrate faster than larger fragments.*

■ DNA is not visible to the naked eye. In order to visualize it, a dye must be added which will bind to the DNA, such as methylene blue.

■ Each fragment of DNA is a particular number of nucleotides, or base pairs, long. When researchers want to determine the size of DNA fragments produced with particular restriction enzymes, they run the unknown DNA alongside DNA with known fragment sizes. The known DNA acts as a **marker** and is used to help determine the unknown fragment sizes.

■ Figure 9.1 shows the results of electrophoresis. Semilog paper is used to plot the results of the *Hind*III digest. Since its fragments sizes are known, this is the *standard curve*. It can now be used to determine the other fragment sizes from DNA I and DNA II by interpolation.

■ In the commercial lab described above, *Hind*III is the marker and used to prepare the standard curve. The standard curve is used to determine the fragment sizes in the *Eco*RI digest.

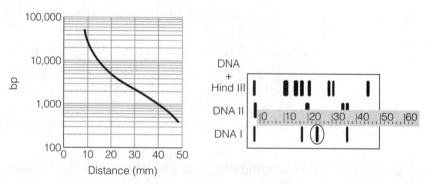

Figure 9.1   Using results of electrophoresis to determine fragment sizes for an unknown sample

Multiple-Choice Questions

1. How many base pairs is the fragment circled in Figure 9.1?
   (A) 350
   (B) 22
   (C) 2,200
   (D) 3,500

2. Which of the following statements is correct?
   (A) Longer DNA fragments migrate farther than shorter fragments.
   (B) Migration distance is inversely proportional to the fragment size.
   (C) Positively charged DNA migrates more rapidly than negatively charged DNA.
   (D) Uncut DNA migrates farther than DNA cut with restriction enzymes.

Here is a plasmid with restriction sites for *Bam*HI and *Eco*RI. Several restriction digests were done using these two enzymes either alone or in combination. Use Figure 9.2 to answer questions 3 and 4. **Hint:** Begin by determining the number and size of the fragments produced with each enzyme. "kb" stands for kilobases, or thousands of base pairs.

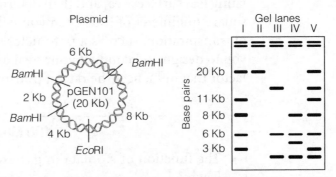

Figure 9.2 Plasmid with restriction sites for *Bam*HI and *Eco*RI

3. Which lane shows a digest with *Bam*HI only?
   (A) I
   (B) II
   (C) III
   (D) IV
   (E) V

4. Which lane shows a digest with both *Bam*HI and *Eco*RI?
   (A) I
   (B) II
   (C) III
   (D) IV
   (E) V

BIG IDEA 4: Laboratory 11, Transpiration

Overview of the Lab

Transpiration is the major mechanism that drives the movement of water through a plant. In the first section of this laboratory, you will begin by calculating leaf surface area, and then determine the average number of stomata per square millimeter of leaf. Then you will learn a technique to measure the rate of transpiration, such as a potometer or whole-plant method. This will allow you to design and conduct your own experiment to answer a question about a factor that influences the rate of transpiration.

YOU MUST KNOW

- The function of stomata in gas exchange in plants. What enters? What leaves?
- The role of water potential and transpiration in the movement of water from roots to leaves.
- The effects of various environmental conditions on the rate of transpiration.
- How to identify xylem and phloem cells and relate their structure to their function.

SCIENCE PRACTICES: CAN YOU . . .

- Predict and justify whether a plant cell will give or lose water based on water potential?
- Create and annotate a diagram to show what would happen to grass planted near a road that has been salted in winter? Include water potential in your representation.

Hints and Review

▌ Review **hydrogen bonding** (Topic 1, Chapter 3)! In water, a hydrogen bond is a weak bond between the hydrogen of one water molecule and the oxygen of another, and it accounts for the unique properties of water, including adhesion and cohesion.

▌ Water enters a plant through the root hairs, passes through the tissues of the root into the xylem, and travels up through the xylem vessels into the leaves.

▌ **Transpiration** is the evaporation of water from the leaves through the stomata. It is the major factor that pulls the water up through the plant.

▌ Study Figure 11.1 to see this process. When water enters the roots, hydrogen bonds link each water molecule to the next (*cohesion*) so the molecules of water are pulled up the thin xylem vessels like beads on a string. The water molecules also cling to the thin walls of the xylem cells (*adhesion*). The water moves up the plant, enters the leaves, moves into air spaces in the leaf, and then evaporates (transpires) through the *stomata* (singular, stoma).

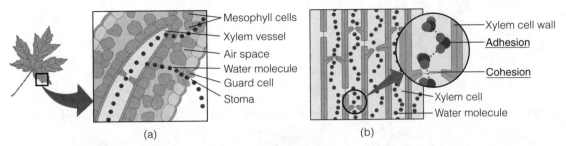

Figure 11.1a and b   Leaf anatomy showing movement of water molecules

- **Stomata** are the pores in the epidermis of a leaf. There are hundreds of stomata in the epidermis of a leaf. Most are located in the lower epidermis. This reduces water loss because the lower surface receives less solar radiation than the upper surface. Each stoma allows the carbon dioxide necessary for photosynthesis to enter, while water evaporates through each one in transpiration.

- **Guard cells** are cells surrounding each stoma. They help to regulate the rate of transpiration by opening and closing the stomata. To understand how they function, study the following figures. As you look at the figures, keep in mind that an increase in solute concentration lowers the water potential of the solution, and that water moves from a region with higher water potential to a region of lower water potential.

- Notice that in Figure 11.2a the guard cells are turgid, or swollen, and the stomatal opening is large. This turgidity is caused by the accumulation of K^+ (potassium ions) in the guard cells. As K^+ levels increase in the guard cells, the water potential of the guard cells drops, and water enters the guard cells.

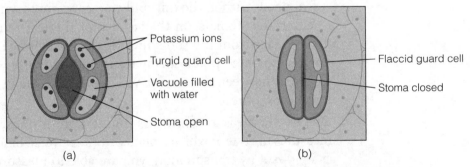

Figure 11.2a   Open stomata   **Figure 11.2b   Closed stomata**

- In Figure 11.2b, the guard cells have lost water, which causes the cells to become flaccid and the stomatal opening to close. This may occur when the plant has lost an excessive amount of water. In addition, it generally occurs daily as light levels drop and the use of CO_2 in photosynthesis decreases.

1 Assemble 4 potometers.

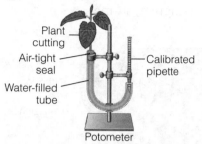

Plant cutting

Air-tight seal

Water-filled tube

Calibrated pipette

Potometer

2 Place each potometer in a different environment: room conditions, mist, wind, and bright light.

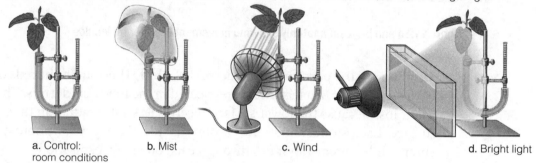

a. Control: room conditions

b. Mist

c. Wind

d. Bright light

3 Measure water loss in each potometer every 3 minutes for 30 minutes.

00:00

30:00

Figure 11.3   Procedure for transpiration lab (potometer method)

▌ A leaf needs carbon dioxide and water for photosynthesis. For carbon dioxide to enter, the stomata on the surface of the leaf must be open. Transpiration draws water from the roots into the leaf mesophyll. However, the plant must not lose so much water during transpiration that it wilts. The plant must strike a balance between conserving water and bringing in sufficient amounts of CO_2 for photosynthesis.

▌ One way to measure water loss from a plant is to use a *potometer*, a device that measures the rate at which a plant draws up water. Since the plant draws up water as it loses it by transpiration, you are able to measure the rate of transpiration. The basic elements of a potometer are shown above in Figure 11.3, step 1. They are

■ a plant cutting.
■ a calibrated pipette to measure water loss.
■ a length of clear plastic tubing.
■ an air-tight seal between the plant and the water-filled tubing.

■ Figure 11.3 shows a typical setup to investigate a variety of environmental effects on the rate of transpiration. Based on your knowledge of transpiration developed in this investigation, can you answer the following questions? (Answers will be found at the end of this investigation.)

1. What are some factors that you would need to hold constant for a valid *control*?

2. Predict the rate of transpiration in the plant that has been misted and placed in a plastic bag compared to the control, and justify your prediction.

3. Predict the rate of transpiration in the plant in front of the fan compared to the control, and justify your prediction.

4. Predict the rate of transpiration in bright light compared to the control, and justify your prediction.

5. Why is the bright light being shone through a tub filled with water?

■ Refer to Figure 11.4 as you read this section. Your teacher may expect you to be able to recognize and know the function of the following cell and tissue types.

- **Parenchyma** is the most abundant cell type, and cells are relatively unspecialized. The mesophyll of leaves (where most photosynthesis occurs), and root cortex (starch storage) are parenchyma.
- **Sclerenchyma** cells make up fibers, have thick secondary cell walls, and often serve a support function.
- **Collenchyma cells** have thickened cell walls and function in support.
- **Xylem** cells (tracheids and vessel elements) are dead at maturity. Their function is water transport.
- **Phloem** (sieve tubes and companion cells) transports solutes.
- **Epidermal cells** are outermost and serve a protective function.

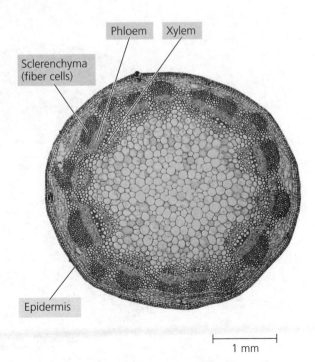

1 mm

Figure 11.4   Dicot stem

Multiple-Choice and Other Questions

1. All of the following enhance water transport in terrestrial plants EXCEPT
   (A) hydrogen bonds linking water molecules.
   (B) capillary action due to adhesion of water molecules to the walls of xylem.
   (C) evaporation of water from the leaves.
   (D) K^+ being transported out of the guard cells.

2. Under conditions of bright light, in which part of a transpiring plant would water potential be lowest?
   (A) xylem vessels in the leaves
   (B) xylem vessels in the roots
   (C) root hairs
   (D) spongy mesophyll of the leaves

Identify each of the structures in the micrograph of a monocot stem below (Figure 11.5) by answering the following questions using the appropriate choices from the list.
   Choices for tissue type: xylem, phloem, parenchyma, epidermis
   Choices for function: solute transport, water transport, food storage, protection

3. Name tissue type A and describe its function.

4. Name tissue type B and describe its function.

5. Name tissue type C and describe its function.

6. Name tissue type D and describe its function.

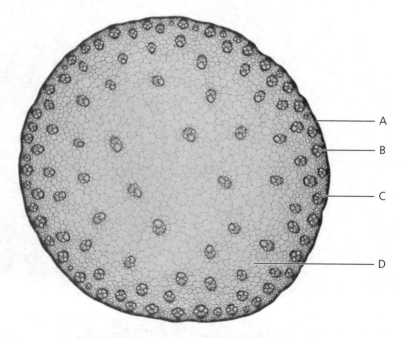

Figure 11.5 Cross section of a monocot stem

Answers for Questions about Figure 11.3

1. In order for the *control* to be valid, you should consider using the same type of plant, the same relative amount of leaf surface, have the water in all potometers at the same temperature, place the control and experimental in virtually identical conditions, with the exception of the single factor you are testing.

2. The plant cutting inside the plastic bag is in a situation of *high humidity*. Because of high water potential in the area surrounding the leaves, the rate of transpiration will be low.

3. The potometer in the *fan* shows increased water loss compared to the control. The reason is that the air movement results in greater evaporation, lowering water potential outside the plant surface, and more water is pulled from the roots.

4. The potometer in *bright light* generally shows a higher rate of water loss, indicating more photosynthesis than the control.

5. A bright light produces heat, which operates as another variable. The water-filled tub is a heat sink, and will minimize the effect of heat on the plant.

BIG IDEA 4: Investigation 12, Fruit Fly Behavior

▌ In this lab you will explore behaviors in an invertebrate, and design an experiment to answer a question about behavior. Animals exhibit a variety of behaviors, both learned and innate, that promote their survival and reproductive success in a variety of ways. In this investigation, you will make detailed observations of an organism's behavior, and design a controlled experiment to test a hypothesis about a specific case of animal behavior.

▌ If you use the 2012 "AP Investigations" manual, your experimental organisms may be fruit flies (as in the title of the investigation). It is equally possible that your teacher may select another species to use for similar studies, and in the 2001 Laboratory Manual, the organism used was the pill bug.

YOU MUST KNOW

- Some animal behaviors, such as orientation behavior, geotaxis, phototaxis, chemotaxis, and how they are adaptive.

SCIENCE PRACTICES: CAN YOU . . .

- Design a plan for collecting data to show how a particular species is affected by biotic or abiotic interactions?
- Analyze data collected to identify possible patterns and relationships between an organism or species and a biotic or abiotic factor?
- Apply mathematical routines, such as statistical analysis, to evaluate data?

▌ This lab is an opportunity to make detailed observations and learn about some interesting animal behaviors. Because the topic is so broad and there are so many local organisms and possibilities for your teacher to choose, we will only make a few comments about animal behavior here. This is a wonderful opportunity to teach experimental design, so this is where we will focus our discussion.

▌ Refer to the Introduction of this book, where we have given some hints about writing the lab essay. At the conclusion of this course, you should have conducted a number of investigations that you developed. You should now be able to design a controlled experiment to test a single variable, record data in a logical manner, and present your conclusions.

- There are hundreds of species of fruit flies—so how do members of the same species find each other and signal willingness to mate? Each species has evolved a complex series of behaviors that appear to be genetically programmed. Students who do Lab 11 in the 2001 AP Biology Lab Manual investigate this behavior in *Drosophila melanogaster*.

- How do fruit flies find their food sources? Orient to gravity? to light? The 2012 Investigation 12: Fruit Fly Behavior encourages a look at a number of behaviors in fruit flies, and directs you in the preparation of a choice chamber.

- In another behavior activity found in the 2001 Lab Manual, an experiment is done to observe how pill bugs respond to their environment. Pill bugs are placed in a choice chamber, half in the side lined with dry filter paper and the other half in the side lined with wet filter paper. Because pill bugs are crustaceans, they respire through gills and are generally found in a moist habitat. Because of this, most students hypothesize that more pill bugs will be found in the moist chamber. This is often what occurs, but not always!

- This brings us to an important consideration: If a student prepares a single-choice chamber, the exercise is not a controlled experiment. Could there be more light at one end of the choice chamber? More activity and vibration? A chemical residue on one side? Any of these conditions and more could possibly influence the organism's behavior. Without a control, it is very risky to state a conclusion.

- What's needed here is a **controlled experiment**. A controlled experiment begins with a *hypothesis*, a proposed solution for the problem being investigated. A hypothesis is often written as an IF, THEN statement that predicts the outcome we should expect if the hypothesis is correct. A hypothesis should not only predict results, but must be testable.

- In a controlled experiment, *all variables are held constant* except the one being tested or manipulated. For instance, if the goal is to test response to wet versus dry conditions, the light, temperature, chemicals in the filter paper or on the dish surface, and movement of the table must all remain constant. In addition, all the experimental organisms must be of the same approximate age, size, and state of health. It is not enough to say you will hold all variables constant; you must be explicit in your explanation of how you will do this.

- To be meaningful, the experiment must include a *large sample size* to be representative of a general condition.

- The *results must be measurable!* Are you going to count, measure, find the mass? Some way to quantify the results must be devised.

- Several *repetitions (or replicates)* of the experiment must be done. Like a large sample size, this lets you verify your result.

- Before you design an experiment, it may be useful to *search the literature* (including the World Wide Web) to see what has already been done and to help develop ideas for a reasonable study.

- Finally, *statistical analysis of your data* (such as the Chi-square test found at the end of this topic) should be done to validate experimental results.

Multiple-Choice Questions

1. A student wanted to study the effect of nitrogen fertilizer on plant growth, so she took two similar plants and set them on a window sill for a 2-week observation period. She watered each plant the same amount, but she gave one a small dose of fertilizer with each watering. She collected data by counting the total number of new leaves on each plant and also measured the height of each plant in centimeters. Which of the following is a significant flaw in this experimental setup?
   (A) There is no variable factor.
   (B) There is no control.
   (C) There is no repetition.
   (D) Measurable results cannot be expected.
   (E) It will require too many days of data collection.

2. Students placed five pill bugs on the dry side of a choice chamber and five pill bugs on the wet side. They collected data as to the number on each side every 30 seconds for 10 minutes. After 6 minutes, eight or nine pill bugs were continually on the wet side of the chamber, and several were under the filter paper. Which of the following is *not* a reasonable conclusion from these results?
   (A) It takes the pill bugs several minutes to explore their surroundings and select a preferred habitat.
   (B) Pill bugs prefer a moist environment.
   (C) Pill bugs prefer a dark environment.
   (D) Pill bugs may find chemicals in dry filter paper irritating.
   (E) Pill bugs demonstrate no significant habitat preference.

3. If a student wanted to determine whether pill bugs prefer a moist or a dry environment, what would be the best way to analyze data from the experiment?
   (A) Total the number of pill bugs on the dry side throughout the entire experiment and compare this with the number on the wet side throughout the experiment.
   (B) After waiting 5 minutes for the pill bugs to acclimate, count the number of pill bugs on the dry side every 30 seconds for 5 minutes. Total and average the results, and compare this with the number of pill bugs on the wet side during this same time interval.
   (C) Compare the number of pill bugs on the dry side at the end of 10 minutes with the number of pill bugs on the wet side at the end of 10 minutes.
   (D) Divide the number of pill bugs on the dry side throughout the experiment by the number on the wet side throughout the experiment.

4. Which of the following hypotheses is stated best?
   (A) If pill bugs are allowed free movement, then more will be found in a moist environment than in a dry environment.
   (B) If pill bugs like a moist environment, then they will move to the wet side of a choice chamber.
   (C) If an experiment with pill bugs is run for 10 minutes, then more pill bugs will be found in the most favorable environment.
   (D) Pill bugs are found in moist habitats, so I predict that more will be found where it is wet.

BIG IDEA 4: Investigation 13, Enzyme Activity

Overview of the Lab

▌ This experiment investigates enzymatic activity of peroxidase. In Procedure 1 you will learn how to measure the activity of peroxidase, and in Procedure 2 you will investigate the effect of varying pH on enzyme activity and then select your own question about factors that would affect enzyme activity and design an experiment to answer it.

▌ Your teacher might choose to use another enzyme-substrate system, and so the tips that follow will be generalized for any enzyme.

YOU MUST KNOW

- The factors that affect the rate of an enzyme reaction such as temperature, pH, enzyme concentration.
- How the structure of an enzyme can be altered, and how pH and temperature affect enzyme function.

SCIENCE PRACTICES: CAN YOU . . .

- Design a controlled experiment to measure the activity of a specific enzyme under varying conditions?
- Use mathematical routines to calculate the rate of a reaction from a graph or data chart?
- Predict and justify how changing an environmental factor such as temperature or pH would alter an enzyme's activity?

Hints and Review

▌ Enzymes are large globular proteins. Much of their three-dimensional shape is the result of interactions between the R (variable) groups of their amino acids. Anything that changes these interactions will change the shape of the enzyme and therefore alter the rate of reaction. The *active site* is the portion of the enzyme that will interact with the substrate.

▌ Remember, *change the shape, change the function!*

▌ Enzyme activity is affected by pH and temperature because these affect the 3-D shape. Extremes of pH and temperature result in *denaturation* when the 3-D shape is so altered the enzyme can no longer function.

▌ Enzymes are not denatured by cold, but the rate of reaction is decreased as temperature decreases.

▌ Be able to calculate rate from graphed data using Figure 13.1.

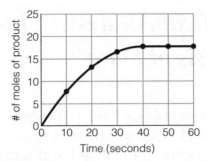

Figure 13.1 Enzyme activity over time

Enzyme Action over Time

We can calculate the rate of a reaction by measuring, over time, either the disappearance of substrate (as in our catalase example) or the appearance of product (as in Figure 13.1). For example, on the graph above, what is the rate, in moles/second, over the interval from 0 to 10 seconds?

$$\text{Rate} = \frac{\Delta y}{\Delta x}$$

so for this example, the rate would be

$$\frac{7 \text{ moles} - 0 \text{ moles}}{(10 \text{ seconds} - 0 \text{ seconds})} = \frac{7}{10}$$
$$= 0.7 \text{ moles/second}$$

▮ Note that the slope of the graph is steepest during the *initial* time period; this is when the rate of a reaction is greatest and occurs because the substrate is most abundant. The rate of the reaction decreases as substrate is consumed and the slope of the graph flattens.

▮ What is the rate for this same reaction between 40 and 60 seconds? (It is 0.) You should be able to explain this. It is because the substrate has been consumed.

Multiple-Choice Questions

1. In order to keep the rate of reaction constant over the entire time course, which of the following should be done?
   (A) Add more enzyme.
   (B) Gradually increase the temperature after 60 seconds.
   (C) Add more substrate.
   (D) Add H_2SO_4 after 60 seconds.
   (E) Remove the accumulating product.

2. What is the role of sulfuric acid (H_2SO_4) in this experiment?
   (A) It is the substrate on which catalase acts.
   (B) It binds with the remaining hydrogen peroxide during titration.
   (C) It accelerates the reaction between enzyme and substrate.
   (D) It blocks the active site of the enzyme.
   (E) It denatures the enzyme by altering the active site.

Statistical Analysis: Chi-Square Analysis of Data

Overview of Chi-Square

It is not sufficient to say, "My data looks really good!" or "The results were very close to what I expected." In science, we impose rigorous tests to support the validity of results. One of these is Chi-square analysis. Here is a short review of how to do this test, and at the same time a review of your knowledge of genetic ratios expected with different crosses.

YOU MUST KNOW

- What is meant by degrees of freedom, critical value, the null hypothesis, and how to do Chi-square analysis of data.

Chi-Square Analysis of Data

Assume you obtained the results shown in Figure 1 for the F_1 generation.

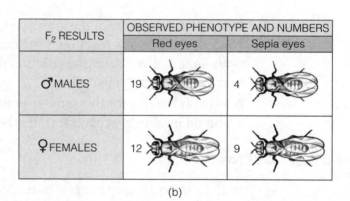

Figure 1   Data table of *Drosophila* (1)

▌ From the data presented, you can deduce that the F_1 cross was between individuals heterozygous for eye color: $+se \times +se$ (+ = red; *se* = sepia).

▌ The student kept data showing both males and females with the trait, because the unknown trait might be sex-linked. The data do not support a sex-linked trait, but do support an autosomal trait; thus the data are merged so that only the trait is considered in the cross.

▌ From this conclusion, you could write the following hypothesis: *If the parents are heterozygous for eye color, there will be a 3:1 ratio of red eyes to sepia eyes in the offspring.* Do your results support this hypothesis?

The formula for Chi-square is

$$\chi^2 = \text{the sum of } \frac{(o - e)^2}{e}$$

where:
o = observed number of individuals
e = expected number of individuals

■ The actual results of an experiment are unlikely to match the expected results precisely. But how great a variance is significant? One way to decide is to use the Chi-square (χ^2) test. This analytical tool tests the validity of a **null hypothesis**, which states that there is no statistically significant difference between the observed results of your experiment and the expected results. When there is little difference between the observed results and the expected results, you obtain a very low Chi-square value; your hypothesis is supported.

Using the Chi-Square Critical Values Table

The Chi-square critical values table provides two values that you need to calculate Chi-square:

■ **Degrees of freedom.** This number is one less than the total number of classes of offspring in a cross. In a monohybrid cross, such as our Case 1, there are two classes of offspring (red eyes and sepia eyes). Therefore, there is just one degree of freedom. In a dihybrid cross, there are four possible classes of offspring, so there are three degrees of freedom.

■ **Probability.** The probability value (p) is the probability that a deviation as great as or greater than each Chi-square value would occur simply by chance. Many biologists agree that deviations having a chance probability greater than 0.05 (5%) do not support the null hypothesis. Therefore, when you calculate Chi-square, you should consult the table for the p value in the 0.05 row.

Critical Values Table

Probability (p)	Degrees of Freedom (df)				
	1	**2**	**3**	**4**	**5**
0.05	3.84	5.99	7.82	9.49	11.1
0.01	6.64	9.21	11.3	13.2	15.1
0.001	10.8	13.8	16.3	18.5	20.5

Steps to Determining Chi-Square

1. **Set up a data chart as shown.** Since your hypothesis predicted a 3:1 ratio in the offspring, you would expect 3/4 of the total offspring (44) to have red eyes.

Phenotypes	Observed (o)	Expected (e)	($o - e$)	($o - e$)^2	$\dfrac{(o - e)^2}{e}$
Red Eyes	31	33	2	4	4/33 = 0.12
Sepia Eyes	13	11	2	4	4/11 = 0.36
					Total = χ^2 = 0.48

2. **Determine the degrees of freedom.** This is the number of categories (red eyes or sepia eyes) minus one. For this data, the number of degrees of freedom is 1.

3. **Find the probability (p) value for 1 degree of freedom in the 0.05 row.** This is the **critical value.** For this data, the critical value = 3.84.

4. **Accept or reject the null hypothesis.** The null hypothesis states that there is no statistically significant difference between the observed and expected data. Since the χ^2 value for this data is less than the critical value, you will accept the null hypothesis. This then supports your working hypothesis, *If the parents are heterozygous for eye color, there will be a 3:1 ratio of red eyes to sepia eyes in the offspring.*

▌ Don't forget this little nugget: **If the Chi-square value is greater than the critical value, the null hypothesis is rejected,** and you must consider reasons for this variation, such as errors in sample size or data collection.

> **TIP FROM THE READERS**
> There was a Chi-square problem on the AP Biology Exam a few years ago, and we found that a common error was that some students took the square root of the value they got for χ^2. Don't make this mistake! χ^2 is just the shorthand for the name of this mathematical technique, *Chi-square.*

Multiple-Choice Questions

1. You have been given a vial containing a red-eyed male with normal wings and a red-eyed female with normal wings. These are the F$_1$ generation. After 2 weeks, you collect the offspring from this pair and obtain the results shown in Figure 2. On the basis of the results shown in Figure 2, which statement is most likely true?

F$_2$ RESULTS	OBSERVED PHENOTYPE AND NUMBERS			
	Red eyes normal wings	Red eyes no wings	Sepia eyes normal wings	Sepia eyes no wings
♂ MALES	48	13	16	4
♀ FEMALES	50	9	10	10

Figure 2   Data table of *Drosophila* (2)

(A) The genes for red eyes and normal wings are linked.
(B) The gene for no wings is sex-linked.
(C) The gene for red eyes and the gene for no wings are both dominant.
(D) The gene for eye color is inherited independently of the gene for wings.
(E) The F$_1$ mates were both homozygous for both eye color and wings.

2. Based on the hypothesis that this is a dihybrid cross, with the two genes unlinked, calculate χ^2 using the data in the table of observed phenotypes.
    (A) 6.04
    (B) 7.81
    (C) 4.977
    (D) 24.0

3. Compare the Chi-square value obtained in Question 2 with the Critical Values Table on page 329 for $p = 0.05$. Which of the following statements would be true?
    (A) Since the calculated value for Chi-square is less than 7.82, the results support the hypothesis that the parents are heterozygous for two unlinked traits.
    (B) Since the calculated value for Chi-square is less than 7.82, the results support the hypothesis that eye color and wings are linked.
    (C) Since the calculated value for Chi-square is less than 7.82, the results are inconclusive. The experiment should be repeated.

AP Biology Equations and Formulas

The following is the formula list that you will receive as part of your testing materials. *Source:* AP Biolog—Course and Exam Description. © 2012. The College Board. www.collegeboard.org. Reproduced with permission.

Statistical Analysis and Probability

Standard Error

$$SE_{\bar{x}} = \frac{s}{\sqrt{n}}$$

Mean

$$\bar{x} = \frac{1}{n}\sum_{i=1}^{n} x_i$$

Standard Deviation

$$s = \sqrt{\frac{\sum (x_i - \bar{x})^2}{n-1}}$$

Chi-Square

$$\chi^2 = \sum \frac{(o-e)^2}{e}$$

s = sample standard deviation (i.e., the sample based estimate of the standard deviation of the population)

$\bar{x}$ = mean

n = size of the sample

o = observed individuals with observed genotype

e = expected individuals with observed genotype

Degrees of freedom equals the number of distinct possible outcomes minus one.

Chi-Square Table

p	Degrees of Freedom							
	1	2	3	4	5	6	7	8
0.05	3.84	5.99	7.82	9.49	11.07	12.59	14.07	15.51
0.01	6.64	9.32	11.34	13.28	15.09	16.81	18.48	20.09

Laws of Probability

If A and B are mutually exclusive, then P (A or B) = P(A) + P(B)

If A and B are independent, then P (A and B) = P(A) x P(B)

Hardy-Weinberg Equations

$$p^2 + 2pq + q^2 = 1$$

$$p + q = 1$$

p = frequency of the dominant allele in a population

q = frequency of the recessive allele in a population

Metric Prefixes

Factor	Prefix	Symbol
10^9	giga	G
10^6	mega	M
10^3	kilo	k
10^{-2}	centi	c
10^{-3}	milli	m
10^{-6}	micro	μ
10^{-9}	nano	n
10^{-12}	pico	p

Mode = value that occurs most frequently in a data set

Median = middle value that separates the greater and lesser halves of a data set

Mean = sum of all data points divided by number of data points

Range = value obtained by subtracting the smallest observation (sample minimum) from the greatest (sample maximum)

Rate and Growth		Water Potential (Ψ)
Rate dY/dt	dY = amount of change	$\Psi = \Psi p + \Psi s$
	t = time	Ψp = pressure potential
Population Growth dN/dt=B-D	B = birth rate	
	D = death rate	Ψs = solute potential
Exponential Growth $\dfrac{dN}{dt} = r_{max} N$	N = population size	The water potential will be equal to the solute potential of a solution in an open container, since the pressure potential of the solution in an open container is zero.
	K = carrying capacity	
Logistic Growth $\dfrac{dN}{dt} = r_{max} N\left(\dfrac{K - N}{K}\right)$	r_{max} = maximum per capita growth rate of population	

		The Solute Potential of the Solution
Temperature Coefficient Q$_{10}$ $Q_{10} = \left(\dfrac{k_2}{k_1}\right)^{\frac{10}{t_2 - t_1}}$	t_2 = higher temperature	$\Psi_s = -iCRT$
	t_1 = lower temperature	i = ionization constant (For sucrose this is 1.0 because sucrose does not ionize in water)
	k_2 = metabolic rate at t_2	
Primary Productivity Calculation	k_1 = metabolic rate at t_1	C= molar concentration
mg O$_2$/L x 0.698 = mL O$_2$/L		R= pressure constant (R = 0.0831 liter bars/mole K)
mL O$_2$/L x 0.536 = mg carbon fixed/L	Q_{10} = the *factor* by which the reaction rate increases when the temperature is raised by ten degrees	T = temperature in Kelvin (273 + °C)

Surface Area and Volume		Dilution - used to create a dilute solution from a concentrated stock solution
Volume of Sphere V = 4/3 π r^3	r = radius	$C_i V_i = C_f V_f$
	l = length	
Volume of a cube (or square column) V = l w h	h = height	i=initial (starting) C = concentration of solute f=final (desired) V = volume of solution
	w = width	
Volume of a column V = π r^2 h		**Gibbs Free Energy**
	A = surface area	$\Delta G = \Delta H - T\Delta S$
Surface area of a sphere A = 4 π r^2	V = volume	ΔG = change in Gibbs free energy
	Σ=Sum of all	ΔS = change in entropy
Surface area of a cube A = 6 a		ΔH = change in enthalpy
	a = surface area of one side of the cube	T = absolute temperature (in Kelvin)
Surface area of a rectangular solid A = Σ (surface area of each side)		**pH =** $- \log [H^+]$

Part IV

Sample Test

On the following pages is a sample examination that approximates the actual AP Biology Examination in format, types of questions, and content. Set aside three hours to take the test. To best prepare yourself for actual AP exam conditions, use only the allowed time for Section I and Section II.

Sample Test

Biology
Section 1

Time—1 hour and 30 minutes

Part A directions: Each of the questions or incomplete statements below is followed by four suggested answers or completions. Select the one that is best in each case, and then fill in the corresponding oval on the answer sheet. When you have completed Part A, you should continue on to Part B.

Questions 1 and 2 refer to the graph below.

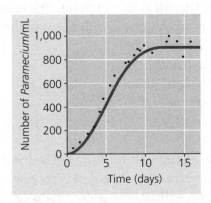

1. In the diagram above, what number of paramecia best represents the carrying capacity of the environment for the population shown?
   (A) 200
   (B) 500
   (C) 600
   (D) 900

2. The formula $dN/dt = r_{max}N(K - N)/K$ describes the pattern of growth for the *Paramecium* population graph. Which of the following is true concerning the formula and the graph it describes at carrying capacity?
   (A) $K = N$
   (B) $r_{max}N$
   (C) The variable K is constantly changing.
   (D) $(K - N)/K = 9,000$

Gene	Probability of Appearing in Gamete
P	1/4
Q	1/4
R	1/4

3. Three genes—P, Q, and R—are not linked. The probability of each gene appearing in a gamete is shown in the table above. Which of the equations below represents the probability that all three genes will appear in the same gamete?
   (A) $1/4 \times 1/4 \times 1/4$
   (B) $1/4 + 1/4 + 1/4$
   (C) $1/4 \div 1/4 \times 1/4$
   (D) $(1/4)1/3$

GO ON TO THE NEXT PAGE

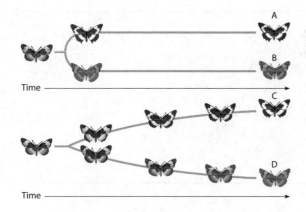

Time ——————————————▶

Time ——————————————▶

4. Which butterfly has changed gradually but significantly from its ancestor through microevolutionary events that were not part of a speciation event?
   (A) butterfly A
   (B) butterfly B
   (C) butterfly C
   (D) butterfly D

5. In deer, fur length is controlled by a single gene with two alleles. When a deer homozygous for long fur is crossed with a deer homozygous for short fur, the offspring all have fur of medium length. If these offspring with medium-length fur mate, what percentage of their offspring will have long fur?
   (A) 100%
   (B) 75%
   (C) 50%
   (D) 25%
   (E) 0%

6. Which of the following statements best supports the idea that certain cell organelles are evolutionarily derived from symbiotic prokaryotes living in host cells?
   (A) The process of cellular respiration in certain prokaryotes is similar to that occurring in mitochondria and chloroplasts.
   (B) Mitochondria and eukaryotes have similar cell wall structures.
   (C) Like prokaryotes, mitochondria have a double membrane.
   (D) Mitochondria and chloroplasts have DNA and ribosomes that are similar to those of prokaryotes.

7. Which of the following statements is NOT part of Darwin's theory of natural selection?
   (A) Individuals survive and reproduce with varying degrees of success.
   (B) Because there are more individuals than the environment can support, this leads to a struggle for existence in which only some of the offspring survive in each generation.
   (C) Individuals in a population vary in their characteristics, and no two individuals are exactly alike.
   (D) Members of the population that are physically weaker than others will be eliminated first by forces in the environment.

8. A farmer selects one green pepper plant that has all of the most desirable traits of the species. The farmer then produces a group of offspring plants using only genetic material from this ideal parent plant. The resulting plants are genetically identical to the parent and are said to be
   (A) hybrids.
   (B) recombinant.
   (C) clones.
   (D) transgenic.

9. In dogs, the trait for long tail is dominant (L), and the trait for short tail is recessive (l). The trait for yellow coat is dominant (Y), and the trait for white coat is recessive (y). Mating two dogs gives a litter of three long-tailed, yellow dogs and one long-tailed, white dog. Which of the following is most likely to be the genotype of the parent dogs?
   (A) $LLYY \times LLYY$
   (B) $LLyy \times LLYy$
   (C) $LlYy \times LlYy$
   (D) $LlYy \times LLYy$

10. In guinea pigs, black fur (*B*) is dominant to brown fur (*b*). No tail (*T*) is dominant over tail (*t*). What fraction of the progeny of the cross *BbTt* × *BbTt* will have black fur and tails?
    (A) 1/16
    (B) 3/16
    (C) 3/8
    (D) 9/16

11. A migrating flock of Canadian geese is nearly decimated by a severe storm. Only four members of the flock, which constitute the entire population of a specific region, survive and return north the next spring. These four start a new colony. This phenomenon is known as
    (A) gene flow.
    (B) natural selection.
    (C) genetic drift.
    (D) directional selection.

12. Allolactose stimulates *Escherichia coli* to produce mRNAs that code for the enzyme β-galactosidase, which breaks down lactose into glucose and galactose. In this case, the role of allolactose can best be described as that of
    (A) a DNA replication inhibitor.
    (B) a translation inhibitor.
    (C) an allosteric inhibitor of β-galactosidase.
    (D) a regulator of gene activity.

13. In a certain group of iguanas, the presence of brown skin is the result of a homozygous recessive condition in the biochemical pathway producing skin pigment. If the frequency of the genotype for this condition is 36%, which of the following is closest to the frequency of the heterozygote genotype in this population? (Assume that the population is in Hardy-Weinberg equilibrium.)
    (A) 16%
    (B) 24%
    (C) 48%
    (D) 64%

Use the pedigree below to answer the following two questions.

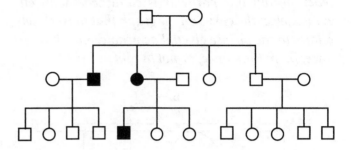

14. In the pedigree, squares represent males and circles represent females. Shaded figures represent individuals who possess a particular trait. Which of the following patterns of inheritance best explains how this trait is transmitted?
    (A) partially dominant
    (B) autosomal dominant
    (C) autosomal recessive
    (D) sex-linked recessive

15. What is the genotype of the shaded figure in the third generation if *A* is the dominant trait and *a* is the recessive trait?
    (A) *aa*
    (B) *Aa*
    (C) $X^a X^a$
    (D) $X^a y$

16. The movement of H^+ down their concentration gradient along the inner mitochondrial membrane, during chemiosmosis of cellular respiration, is an example of what type of movement across a membrane?
    (A) active transport
    (B) facilitated diffusion
    (C) osmosis
    (D) cotransport

GO ON TO THE NEXT PAGE

Questions 17–19 refer to the following graph. Each of the curves represents one pathway for the same reaction, but one pathway is catalyzed by an enzyme. Select the curve on the graph that most closely relates to the phrase given. Each choice may be used once, more than once, or not at all.

17. represents the activation energy of the uncatalyzed reaction

18. represents the activation energy of the catalyzed reaction

19. represents the transition state of the uncatalyzed reaction

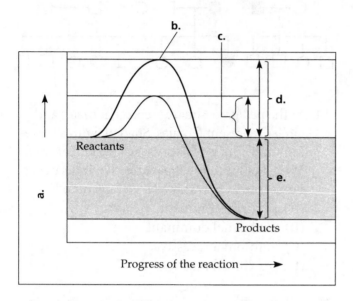

Use this figure for questions 20 and 21.

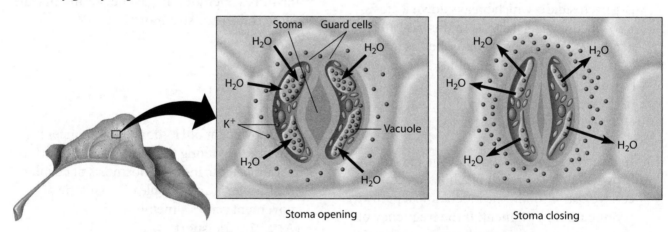

Stoma opening   Stoma closing

20. What is true of the guard cells shown in the right-hand panel of this figure?
    (A) Their turgor pressure is increasing.
    (B) Water is entering these cells.
    (C) These cells are allowing gas exchange for photosynthesis.
    (D) These cells are hypotonic to their immediate surrounding.

21. In the left-hand panel, the use of ATP is high in the guard cells. Why?
    (A) Energy is required to move water molecules into the guard cells.
    (B) Water is moved into the guard cell by active transport.
    (C) Energy is required to move oxygen into the leaf through the stoma.
    (D) Energy is required to move K^+ into the guard cells.

Questions 22–23 refer to the following chromosome map. Letters represent gene loci and numbers represent map units.

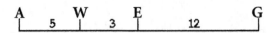

22. Considering the possibilities of recombination, which two genes on this chromosome are most likely to segregate together into a daughter cell?
    (A) A and W
    (B) A and E
    (C) A and G
    (D) W and E

23. If the section of the chromosome in the figure was involved in an inversion mutation, what would be the new order of genes?
    (A) AWEG
    (B) WEGA
    (C) GEWA
    (D) The genes are lost in this type of mutation and would not be present on the chromosome.

Questions 24–26 refer to the following figure, which shows a food web in a particular ecosystem. Each letter represents a species in this ecosystem, and the arrows show the flow of energy.

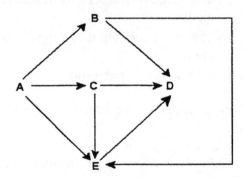

24. Which of the species in the food web is the primary producer?
    (A) A
    (B) B
    (C) D
    (D) E

25. Species B and C represent which of the following?
    (A) primary producers
    (B) primary consumers
    (C) secondary consumers
    (D) omnivores

26. Which of the following most accurately describes species E?
    (A) an herbivore, and a secondary consumer
    (B) an omnivore, and a secondary consumer
    (C) an omnivore, and both a primary and secondary consumer
    (D) an herbivore, and both a primary and secondary consumer

Questions 27–29 refer to the following gel, which was produced from four samples of a radioactively labeled strand of DNA that were cut with one type of restriction enzyme. The samples were separated by gel electrophoresis. Answer the questions on the basis of the bands you can visualize below.

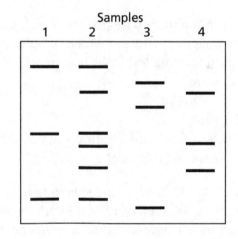

GO ON TO THE NEXT PAGE

27. The DNA fragments in the gel were separated when an electric field was applied across the gel and they migrated at different speeds. The differential migration speed of the different DNA fragments was due to the
    (A) amount of radioactivity in the samples.
    (B) degree to which the samples were negatively charged.
    (C) degree to which the samples were positively charged.
    (D) size of the fragments within the samples.

28. Which of the following is true about the DNA samples that were loaded onto the gel?
    (A) The DNA strand of sample 2 was originally the longest.
    (B) The DNA strand of sample 4 was originally the shortest.
    (C) Samples 2 and 4 are the same DNA sample.
    (D) Sample 2 was cut at more restriction sites than was sample 4.

29. Assuming that all fragments of each sample are imaged on the gel, which sample produced the smallest DNA fragment?
    (A) sample 1
    (B) sample 2
    (C) sample 3
    (D) samples 1 and 2 both produced the smallest fragment

Questions 30–32 refer to an experiment in which there is an initial setup of a U-tube with its two sides separated by a membrane that permits the passage of water and NaCl but not molecules of glucose. The U-tube is filled on one side with a solution of 0.4 M glucose and 0.5 M NaCl, and on the other, 0.8 M glucose and 0.4 M NaCl.

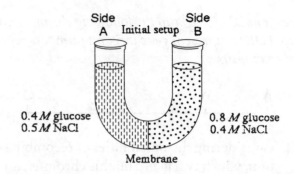

30. When this U-tube was set up, at time = 0 in the experiment, which of the following was true?
    (A) The solution on side A was hypertonic to the solution on side B.
    (B) The solution on side B was hypertonic to the solution on side A.
    (C) The two solutions were isotonic.
    (D) Active transport moved glucose from side A to side B.

31. After the experiment ran for one hour, the expected fluid levels in the U-tube would be
    (A) higher on side A that side B.
    (B) higher on side B than side A.
    (C) even.
    (D) equal, but the solutes would be reversed.

32. At the conclusion of the experiment, the NaCl would
    (A) show no movement.
    (B) show some movement from side A to side B.
    (C) show some movement from side B to side A.
    (D) make side A hypertonic.

Questions 33 and 34 are based on the following reading.

Recent studies have shown that the onset of puberty in American girls has decreased from an average of 12 to 13 years of age to as young as 8 to 10 years of age. Scientists who study premature puberty suggest that steroids in our food and in the environment may be contributing factors, since steroids are known to cross cell membranes and bind to receptors inside cells.

Why are hormones present in our foods? Synthetic testosterone compounds make young animals gain weight faster so they are ready for market sooner. Female animals receive synthetic estrogen to inhibit the reproductive cycle and divert all energy into weight gain. In the United States, up to two-thirds of meat animals are raised using hormones. In addition, hormones are used to increase milk production in dairy cattle.

33. How are steroid hormones able to cross cell membranes and enter cells?
    (A) Steroids and cell membranes both contain receptor proteins.
    (B) Steroids are polar, and the cell membrane is polar on the inside of the lipid bilayer.
    (C) Steroids are nonpolar lipids, and the cell membrane is lipid based.
    (D) Steroids can diffuse through open channel proteins in the membrane.

34. How do steroid hormones specifically alter the onset of puberty?
    (A) Steroid hormones bind to intracellular receptors and serve as transcription factors, turning on genes.
    (B) Steroid hormones bind to receptors on the cell membrane, turning on genes in the nucleus.
    (C) Steroid hormones act as receptors tyrosine kinase, complexing with other receptor molecules to alter genetic expression.
    (D) Steroid hormones bind directly to promoter areas, attracting RNA polymerase and starting transcription.

Use the pedigree for questions 35 and 36.

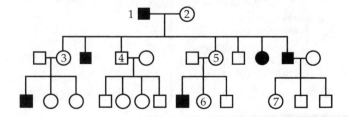

35. Which of the following patterns of inheritance best explains the transmission of the trait in the pedigree above?
    (A) sex-linked dominant
    (B) sex-linked recessive
    (C) autosomal dominant
    (D) autosomal recessive

36. In which of the numbered female individuals does the pedigree provide information that indicates that the female is heterozygous for the trait?
    (A) individual 2 only
    (B) individuals 2 and 3
    (C) individuals 2, 3, and 5
    (D) individuals 2, 3, and 6

37. A geneticist crosses two rabbits, both of which have brown fur. In rabbits, brown fur is dominant over white fur. Six of the eight offspring produced have brown fur, and the other two have white fur. The genotypes of the parents were most likely which of the following?
    (A) $BB \times bb$
    (B) $BB \times Bb$
    (C) $Bb \times bb$
    (D) $Bb \times Bb$

38. Which of the following is an example of simple diffusion across a membrane?
    (A) the movement of H^+ across the thylakoid membrane during photosynthesis
    (B) the uptake of neurotransmitters by the postsynaptic membrane during the transmission of a nerve impulse
    (C) the movement of oxygen in the alveoli across the epithelial membrane and into the bloodstream
    (D) the exchange of sodium and potassium across a cell membrane through the Na^+-K^+ pump

39. Which characteristic is NOT required of a population in Hardy-Weinberg equilibrium?
    (A) The population must be very large.
    (B) There must be no migration into or out of the population.
    (C) The members of the population must be mating randomly.
    (D) There must be only two alleles present for each characteristic in the population.

40. In humans, if red hair (R) is dominant to brown hair (r), and freckles (F) are dominant to no freckles (f), what fraction of the progeny of the cross $RrFf \times RRff$ will have red hair and no freckles?
    (A) 9/16
    (B) 1/2
    (C) 3/8
    (D) 3/16

41. When a species is split into two populations, separated by a geographic barrier that makes breeding between the populations impossible, this could eventually lead to
    (A) sympatric speciation.
    (B) allopatric speciation.
    (C) adaptive radiation.
    (D) polyploid speciation.

42. Which of the following best characterizes the reaction represented below?
    $A + B \rightarrow AB + energy$
    (A) exergonic reaction
    (B) endergonic reaction
    (C) oxidation-reduction reaction
    (D) hydrolysis

43. A scientist wanted to measure the metabolic rate of normal cells compared to the metabolic rate of cancerous cells. Which of the following would give the best data set for this question?
    (A) CO_2 is consumed by the cell.
    (B) O_2 is consumed by the cell.
    (C) Water is consumed by the cell.
    (D) Glucose is produced by the cell.

44. The graph above shows the rate of growth of a population of squirrels in a certain geographic area in Connecticut during the past several decades. This population is most closely exhibiting which of the following types of growth?
    (A) logistic growth
    (B) r-selected growth
    (C) K-selected growth
    (D) exponential growth

Use the figure showing the formation of complementary DNA (cDNA) from a eukaryotic gene for questions 45 and 46.

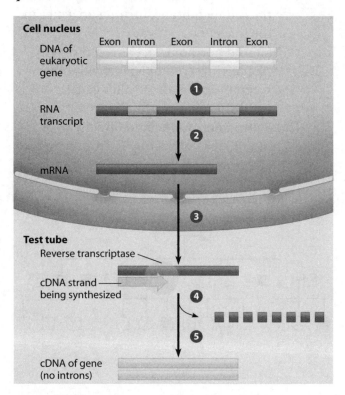

45. In the figure, what complex is responsible for reducing the RNA transcript in step 2?
    (A) DNA polymerase
    (B) RNA polymerase
    (C) spliceosomes
    (D) reverse transcriptase

46. What complex is responsible for the last step in the figure where the single strand of DNA, formed from RNA by reverse transcriptase, is converted to double-stranded DNA?
    (A) reverse transcriptase
    (B) RNA polymerase
    (C) DNA polymerase
    (D) small nuclear ribonucleoproteins

47. Two individuals who are carriers for cystic fibrosis (a recessively inherited disorder) have three children together. None of the children have cystic fibrosis. What is the probability that the couple's fourth child will be born with cystic fibrosis?
    (A) 0%
    (B) 25%
    (C) 50%
    (D) 75%

48. The statement that evolutionary changes are composed of rapid bursts of speciation that alternate with long periods in which species do not change significantly is known as
    (A) gradualism.
    (B) punctuated gradualism.
    (C) punctuated equilibrium.
    (D) uniformitarianism.

49. Which is thought to have been the first self-replicating genetic material?
    (A) DNA
    (B) RNA
    (C) cDNA
    (D) dsRNA

50. In photosynthesis, the functional product(s) of the light reactions
    (A) are ATP and NADPH.
    (B) are ATP and NADH.
    (C) is glyceraldehyde.
    (D) is glucose.

Questions 51–53 refer to the graphs below. The rate of reaction for three enzymes was calculated at different temperatures, and the rate of reaction for two additional enzymes was calculated at different pH levels. The results are shown in the following graphs. Assume that the y-axes share the same scale.

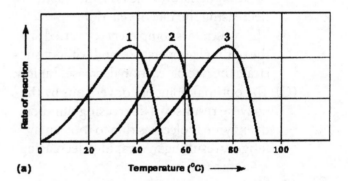

(a)

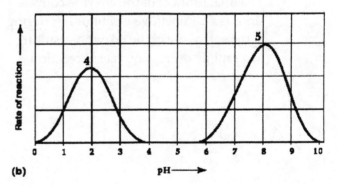

(b)

51. Which of these enzymes is most efficient—that is, has the highest rate of reaction?
    (A) 1
    (B) 3
    (C) 4
    (D) 5

GO ON TO THE NEXT PAGE

52. How many times more acidic is the optimal environment for enzyme 4 than the optimal environment for enzyme 5?
    (A) 6
    (B) 60
    (C) 10^6
    (D) 10^8

53. In enzymes 1, 2, and 3 why does the curve of the rate of reaction line for each enzyme drop sharply after the optimal temperature is reached?
    (A) The increasing thermal energy of the solution denatures the enzyme, rapidly decreasing the rate of reaction.
    (B) The substrate is completely converted at the optimal temperature, and without substrate the rate of reaction decreases rapidly.
    (C) The optimal point of conversions by the enzyme results in a decreasing number of enzyme molecules due to molecular fatigue, resulting in a rapidly decreasing rate of reaction.
    (D) The final product of the enzyme conversions is a competitive inhibitor interfering at the active site of the enzyme, resulting in a rapidly decreasing rate of reaction.

Question 54 refers to the graph shown below.

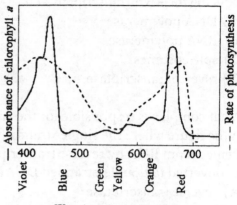

Wavelength of light (nm)

54. Which of the following is the best reason the curve for the absorbency of light by chlorophyll *a* does not perfectly match the rate of photosynthesis?
    (A) The rate of photosynthesis is always fractionally slower than the rate of absorbency by chlorophyll *a*.
    (B) The rate of photosynthesis is always fractionally faster than the rate of absorbency by chlorophyll *a*.
    (C) There are fewer chlorophyll *a* molecules in the cell than the other molecules involved in photosynthesis, so chlorophyll *a* is the rate-limiting reagent.
    (D) Chlorophyll *a* is not the only photosynthetically important pigment in chloroplasts.

Questions 55 and 56 refer to the following cladogram.

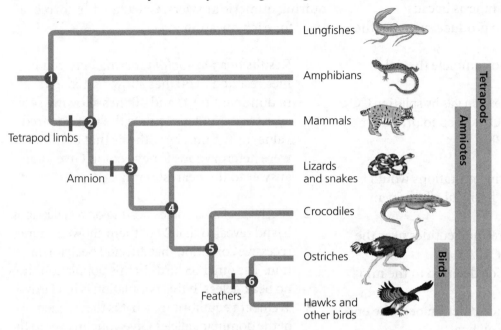

55. According to this figure, which pair of organisms shares the most recent common ancestor?
    (A) lungfish and amphibian
    (B) amphibian and mammal
    (C) lizard and ostrich
    (D) mammal and crocodile

56. Which branch point best supports the idea that birds are a member of the reptile clade?
    (A) 3
    (B) 4
    (C) 5
    (D) 6

After reading the paragraphs, answer questions 57 and 58 that follow.

You're conducting an experiment to determine the effect of different wavelengths of light on the absorption of carbon dioxide as an indicator of the rate of photosynthesis in aquatic ecosystems. If the rate of photosynthesis increases, the amount of carbon dioxide in the environment will decrease and vice versa. You've added an indicator to each solution. When the carbon dioxide concentration decreases, the color of the indicator solution also changes.

Small aquatic plants are placed into three containers of water mixed with carbon dioxide and indicator solution. Container A is placed under normal sunlight, B under green light, and C under red light. The containers are observed for a 24-hour period.

57. Based on your knowledge of the process of photosynthesis, the plant in the container placed under red light would probably
    (A) absorb no CO_2.
    (B) absorb the same amount of CO_2 as the plants under both the green light and normal sunlight.
    (C) absorb less CO_2 than the plants under green light.
    (D) absorb more CO_2 than the plants under the green light.

GO ON TO THE
NEXT PAGE

58. Carbon dioxide absorption is an appropriate indicator of photosynthesis because
    (A) CO_2 is needed to produce sugars in the Calvin cycle.
    (B) CO_2 is needed to complete the light reactions.
    (C) plants produce oxygen gas by splitting CO_2.
    (D) the energy in CO_2 is used to produce ATP and NADPH.

59. Which of the following mutations would be *most* likely to have a harmful effect on an organism?
    (A) a deletion of three nucleotides near the middle of a gene
    (B) a single nucleotide deletion in the middle of an intron
    (C) a single nucleotide deletion near the end of the coding sequence
    (D) a single nucleotide insertion downstream of, and close to, the start of the coding sequence

60. Which of the following is NOT true of RNA processing?
    (A) Exons are cut out before mRNA leaves the nucleus.
    (B) Nucleotides may be added at both ends of the RNA.
    (C) Ribozymes may function in RNA splicing.
    (D) RNA splicing can be catalyzed by spliceosomes.

Part B directions: Part B consists of questions requiring numeric answers. Calculate the correct answer for each question.

1. Results of a *Drosophila* mating between F_1 flies resulted in 60 flies showing red eyes (a dominant trait) and 40 flies showing sepia eyes (recessive). Calculate the Chi-squared value for the null hypothesis that both F_1 flies were heterozygous for eye color. Give your answer to the nearest tenth.

2. A census of albatrosses nesting on a Galápagos Island revealed that 24 of them showed a rare recessive condition that affected beak formation. The other 63 birds in this population show no beak defect. If this population is in Hardy-Weinberg equilibrium, what is the frequency of the dominant allele? Give your answer to the nearest hundredth.

3. Data taken to determine the effect of temperature on the rate of respiration in a goldfish is given in the table below. Calculate Q_{10} for this data.

Temperature	Respirations/Minute
16	16
21	22

Biology
Section II

Time—10 minutes to plan responses; 1 hour and 20 minutes for writing

Answer all questions. Number your answers as the questions are numbered below.

Answers must be in essay form. Outline form is NOT acceptable. Labeled diagrams may be used to supplement discussion, but in no case will a diagram alone suffice. It is important that you read each question completely before you begin to write.

1. The plasma membrane is one of the most important parts of a cell; it allows the selective passage of materials into and out of the cell, thereby maintaining a constant, desired internal composition. (10 points)
   (a) **Discuss** the components of a typical animal plasma membrane, as well as the roles each of these components plays in regulating the cell's internal environment.
   (b) **Discuss** the ways in which the following can enter an animal cell:
   - viral DNA.
   - hormones.
   - water molecules.

2. It is theorized that glycolysis was the first metabolic pathway for the production of ATP. Glycolysis begins the process of making ATP by breaking glucose into two molecules of pyruvate. **Justify** this claim with three pieces of evidence that support this point. (4 points)

3. Gene expression in a cell is influenced by a variety of factors. Not all genes on the eukaryotic chromosome are expressed, and in fact, only a small fraction of the genes are transcribed into working proteins. **Discuss** how chromatin packaging influences gene expression. (3 points)

4. A flowering plant in a ceramic pot is placed in a window that has light shining through most of the day, and it is given adequate water and soil nutrients. **Describe** the changes in the plant concerning light that would be induced by rotating the plant 180°. (3 points)

5. The pedigree on the next page traces the inheritance of alkaptonuria, a biochemical disorder. Affected individuals, indicated here by the darkened circles and squares, are unable to metabolize a substance called alkapton, which colors the urine and stains body tissues. Does alkaptonuria appear to be caused by a dominant allele or by a recessive allele? Fill in the genotypes of the individuals whose genotypes can be deduced. What genotypes are possible for each of the other individuals? (4 points)

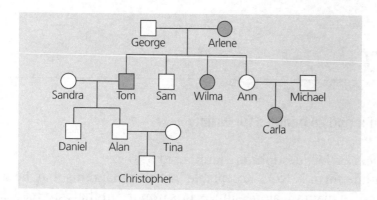

6. The template strand of a gene contains the sequence 3′-TTCAGTCGT-5′. Draw the nontemplate sequence and the mRNA sequence, indicating 5′ and 3′ ends of each. Compare the two sequences. (4 points)

7. Distinguish genetic drift from gene flow in terms of (a) how they occur and (b) their implications for future genetic variation in a population. (3 points)

8. Study this diagram, which represents different forest communities. (10 points)

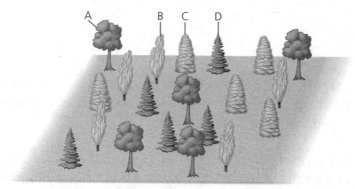

Community 1

Community 2

   (a) **Analyze** the factors involved in species diversity
   (b) **Calculate** the percent of each community represented by species A.
   (c) **Explain** how two communities that contain the same number of species can differ in species diversity.
   (d) **Predict** and **justify** the impact on community stability in each forest community of a disease that affects species A.

Part V

Answers and Explanations

Topic 1: The Chemistry of Life

ANSWERS AND EXPLANATIONS

Level 1: Knowledge/Comprehension Questions

▮ **1. (B) is correct.** RNA is made up of a phosphate group, a ribose sugar, and one of the following four nitrogenous bases: cytosine, guanine, uracil, and adenine. The phosphate group of RNA contains a phosphate atom and three atoms of oxygen, not two. DNA is similar to RNA in many ways but different in two important ones: It contains deoxyribose instead of ribose as its sugar and it contains the base thymine instead of uracil.

▮ **2. (C) is correct.** The answer is water, H_2O. Polar covalent bonds are those in which valence electrons are shared between atoms, but unequally. (The more electronegative atom will attract the electrons more strongly, and that end of the molecule will have a slightly negative charge, whereas the less electronegative atom will attract the electron less strongly and be slightly positive.) The two atoms involved in the bond must differ in electronegativity in order to form a polar covalent bond.

▮ **3. (E) is correct.** Use this question to review the unique properties of carbon. Since it has 4 valence electrons, carbon will form 4 covalent bonds. It does not form polar bonds.

▮ **4. (D) is correct.** The three terms you should keep in mind as you think of water traveling up through the xylem of a plant are transpiration (in which water evaporates from the plant's leaves); cohesion, in which the water molecules stick together due to the hydrogen bonds; and adhesion, whereby the water molecules stick to plant cell walls and resist the downward pull of gravity.

▮ **5. (D) is correct.** Unsaturated fatty acids contain one or more carbon-carbon double bonds, whereas saturated fatty acids contain no double bonds.

▮ **6. (A) is correct.** Lipids are the only one of the four major classes of biological molecules that are not polymers. They are grouped together because they are hydrophobic. Nucleic acids are polymers of nucleotide monomers, proteins are polymers of amino acid monomers, and carbohydrates are polymers of monosaccharide monomers.

▮ **7. (B) is correct.** The linkages between the amino acids of proteins are peptide bonds. Peptide bonds are covalent bonds formed in dehydration reactions. The carboxyl group of one amino acid is joined to the amino group of an adjacent amino acid, resulting in the loss of one molecule of water.

▮ **8. (C) is correct.** One common secondary structure of proteins is the alpha (α) helix; another is the beta (β) pleated sheet. The secondary structure of a protein refers to hydrogen bonding along the backbone (not the side chains) of the amino acid chain.

▮ **9. (E) is correct.** Cellulose is the polysaccharide that forms the strong cell walls of plant cells. It is a polymer of glucose.

10. (C) is correct. Denaturation is the process by which proteins lose their overall structure, or conformation, as a result of changes in pH, temperature, or salt concentration. Denatured proteins have reduced biological activity.

11. (B) is correct. The ratio of carbon, hydrogen, and oxygen atoms in carbohydrates is 1:2:1. For example, glucose is $C_6H_{12}O_6$.

12. (C) is correct. To get this correct, you should first add up all the C, H, and O in three fatty acid chains plus one glycerol. This would be 51 C, 74 H, and 9 O. Then, recall that to join to molecules by dehydration synthesis, one molecule of water must be removed. Since three fatty acid chains will be attached to one glycerol, three water molecules will be removed. Subtract 6 H and 3 O to arrive at the answer.

13. (B) is correct. Proteins have many functions, which encompass most of a cell's metabolic activity.

14. (B) is correct. Phospholipids are unique macromolecules. Their hydrophilic heads and hydrophobic tails contribute to the semipermeability of cell membranes.

15. (D) is correct. The negative charge comes from the electronegative oxygen of one water molecule attracted to the partial positive charge of hydrogen of another water molecule.

16. (C) is correct. The monomers in macromolecules are joined when a molecule of water is removed during dehydration, or condensation reactions.

17. (B) is correct. Since the pH scale is logarithmic, each unit change is by a factor of 10. A drop of pH means the solution is more acidic and has 10 times more H^+ ions.

18. (B) is correct. Transpiration refers to the evaporation of water from pores in leaves. Transpiration is possible because of cohesion and adhesion, but it is not an emergent property of water.

19. (C) is correct. An amino acid is composed of a central carbon, bonded to a hydrogen, with a variable (R) group, and with a carboxyl (the acid part) at one end, an amino group at the other end (the amino part). Aldehydes and ketones are found in sugars; the sulfhydryl group is found in one amino acid.

20. (E) is correct. Recall that hydrolysis means to use water to split a molecule, so look for a large molecule reduced to its monomers. Maltose is a disaccharide; glucose is a monosaccharide.

Level 2: Application/Analysis/Synthesis Questions

1. (D) is correct. Hydrogen bonds occur when a slightly positive hydrogen atom of a polar covalent bond in one molecule is attracted to a slightly negative atom of a polar covalent bond in another molecule. In living systems hydrogens are often attracted to the highly electronegative elements oxygen (as in this question with water) or nitrogen.

2. (D) is correct. Structural isomers have the same molecular formula, but differ in the covalent arrangement of their atoms. Answers A, B, and C would all change the molecular formula; thus, they would not be isomers. The location of the double-bonded oxygen is a change in covalent arrangement.

3. (B) is correct. Peptide bonds occur between the carboxyl group of one amino acid and the amino group of another amino acid. All that is required in this question is for you to carefully examine the information given in the diagram. Take a look at every question, even if you may think the content area is a hard one; often the answers are easier than you might expect.

▌ **4. (D) is correct.** Hydrolysis is a chemical process that splits molecules by the addition of water. Digestive enzymes work by hydrolysis. Water is removed to join the molecules; water is added to separate the molecules.

▌ **5. (D) is correct.** Recall that enzymes are proteins and proteins are made of amino acids.

▌ **6. (D) is correct.** Two environmental factors that affect the three-dimensional structure of enzymes are temperature and pH. As the structure of the enzyme is altered the enzyme will become less effective.

▌ **7. (C) is correct.** Ten glucose molecules would have a combined molecular formula of $C_{60}H_{120}O_{60}$. To form a polymer, a molecule of water would have to be removed as each glucose is added to the chain. Since ten glucose molecules are bonded together, nine H_2O must be removed, 18 hydrogen atoms and 9 oxygen atoms. This leaves a formula of $C_{60}H_{102}O_{51}$.

▌ **8. (D) is correct.** This conceptual question is based on your knowledge of the structure of DNA. Recall that the molecule is antiparallel, meaning one strand runs 5′ to 3′ while the opposite strand of the double helix runs 3′ to 5′. All of the answers are given with both strands running 3′ to 3′. To get the proper answer convert the second strand given in the answer to 3′ to 5′ and see which one has the proper, matching base-pair sequence.

Free-Response Questions

(a) A phospholipid molecule contains a hydrophilic "head" (containing a glycerol molecule and a phosphate group) and two hydrophobic fatty acid tails. In cell membrane surfaces, phospholipids are arranged in a bilayer, in which the hydrophilic heads are in contact with the cell's watery interior and exterior, while the tails are pointed away from water and toward each other in the interior of the membrane. The fatty acid chains of phospholipids can contain double bonds, which makes them unsaturated. Because of the kinks in the tails, phospholipids aren't packed together tightly, which contributes to the fluidity of the membrane. The fluidity of the cell membrane is very important in its function; the less fluid the membrane is, the more impermeable it is. There is an optimum permeability for the cell membrane, at which all the substances necessary for metabolism can pass into and out of the cell.

The fluidity of cell membranes enables hydrophobic molecules such as hydrocarbons, carbon dioxide, and oxygen to dissolve in the bilayer and easily cross the membrane. However, ions and polar molecules (including water, glucose, and other sugars) cannot pass through because of the hydrophobic interior. Protein channels and transport proteins allow these required substances to cross membranes.

(b) Proteins function as cell membrane transporters because they act as channels; substances that bind to them can help alter their conformation to permit the passage of molecules through them, and into the cell interior.

There are many different ways by which proteins can permit the passage of ionic and polar molecules through the lipid bilayer. Proteins associated with the membrane are either integral proteins, which actually penetrate the lipid bilayer (ones that completely go through the bilayer are called transmembrane proteins), or they are "peripheral proteins" that are associated with the outside of the membrane. Transmembrane proteins can form hydrophilic channels that permit the passage of certain hydrophilic substances that otherwise would not be able to cross the membrane. Other functions that membrane proteins serve are to attach the cell to the extracellular matrix, to stabilize it, and to function in cell-cell recognition. Membrane proteins are also important in cell-cell signaling; some have enzyme function and carry out important metabolic reactions, and they aid in joining adjacent cells.

This response shows thorough knowledge of the processes of the structure of phospholipids, cell membrane structure and components, and movement across membranes. A strong response to this item requires an understanding of topics from Units 1 and 2 of the textbook. Note that the response includes the following key terms in context, showing the writer's knowledge of their meanings and relatedness:

phospholipids	*metabolism*
hydrophilic head	*protein channels*
glycerol	*transporters*
phosphate	*conformation*
hydrophobic tails	*integral proteins*
lipid bilayer	*transmembrane proteins*
double bonds	*extracellular matrix*
unsaturated	*cell-cell signaling*
permeability	

The student's response would have been strengthened by a more explicit correlation of structure and function, especially by using protein functions other than transport. Specific examples of cell-cell signaling, enzymatic function, and others listed would have demonstrated a deeper knowledge of the topic.

Topic 2: The Cell

ANSWERS AND EXPLANATIONS

Level 1: Knowledge/Comprehension Questions

▌ **1. (C) is correct.** Light microscopes are good for viewing objects that are 0.2 μm or larger. With a light microscope, you can observe animal and plant cells, some bacterial cells, and some larger organelles such as nuclei mitochondria, and chloroplasts. To see the other organelles in the list of choices, you would need an electron microscope.

2. (C) is correct. The nucleoid region is the only cell structure on this list not found in both prokaryotes and eukaryotes. Eukaryotic cells have a true nucleus, which is surrounded by a double membrane called a nuclear envelope. The genetic material of prokaryotes is localized in a clump in one particular region of the cell that is not enclosed by a membrane.

3. (D) is correct. The endoplasmic reticulum (ER) is an organelle characterized by extensive, folded membranes, and it is often associated with ribosomes.

4. (B) is correct. The Golgi apparatus is the organelle that has a *cis* and *trans* face, and it acts as the packaging and secreting center of the cell. It consists of a series of flattened sacs of membranes called cisternae.

5. (E) is correct. Mitochondria are the powerhouses of the cell; cellular respiration takes place in the mitochondria, forming ATP, the cell's energy currency. Mitochondria are bound by double membranes, and the proteins involved in ATP production are embedded in the inner membranes of the mitochondria.

6. (A) is correct. Peroxisomes perform many metabolic functions in the cell, including the production of hydrogen peroxide in the process. The hydrogen peroxide, a poison, is immediately broken down in the peroxisome by the enzyme catalase. You may recall that catalase was the enzyme used in AP Lab 2.

7. (C) is correct. Lysosomes are characteristic of animal cells but not most plant cells. They are large membrane-bound structures that contain hydrolytic enzymes, and they are responsible for the breakdown of proteins, polysaccharides, fats, and nucleic acids. They function best at a low pH (around 5), so they pump hydrogen ions from the cytosol into their lumen to achieve this acidic pH.

8. (D) is correct. The only answer choice listed that names a molecule typically found in the plasma membranes of animal cells is D, carbohydrates. The major components of animal cell membranes are phospholipids, integral and peripheral proteins, and carbohydrates. One of the main functions of carbohydrates in the cell membrane is cell-cell recognition, which means carbohydrates are an important component of the immune system. Cell-surface carbohydrates are unique to each organism.

9. (D) is correct. Substances will move down their concentration gradient until their concentration is equal on either side of a membrane. For this reason, because the concentration of glucose on side B of the tube is 2.0 M, while the concentration of glucose on the A side of the tube is 1.0 M, glucose will move to side A.

10. (D) is correct. The only substance listed that can passively diffuse through the cell membrane unaided by proteins is carbon dioxide. Remember that passive diffusion occurs without the cell doing any work. Other choices need the processes of facilitated diffusion and transport proteins to cross the membrane. This is true of all of the answer choices listed except for D.

11. (A) is correct. This figure illustrates the process of cotransport. In cotransport, a pump that is powered by ATP transports a specific solute, protons in this case, out of the cell. The protons then travel down their concentration gradient back into the cell, passing through another transport protein and indirectly providing energy for the movement of another substance (sucrose in this case) against its concentration gradient.

12. (E) is correct. Large molecules are moved out of the cell by exocytosis. In exocytosis, vesicles that are to be exported from the cell (often coming from the Golgi apparatus) fuse with the plasma membrane, and their contents are expelled into the extracellular matrix. Pinocytosis, phagocytosis, and receptor-mediated endocytosis are types of endocytosis; cytokinesis is the division of the cytoplasm after mitosis.

13. (D) is correct. The shape of the curve in the art shown most closely depicts an exergonic reaction. The free energy of the products is lower than that of the reactants—meaning that in the course of the reaction, energy is given off. This is characteristic of exergonic reactions. Conversely, in an endergonic reaction, energy is taken in during the course of the reaction.

14. (C) is correct. The second law of thermodynamics states that every energy transfer that occurs increases the amount of entropy in the universe. The first law of thermodynamics states that the amount of energy in the universe is constant, and therefore energy can be neither created nor destroyed. Evolutionary theory refers to the myriad changes that have taken place to transform living organisms from the beginning of life on Earth until today.

15. (D) is correct. Catalysts speed up chemical reactions by providing an alternate reaction pathway that lowers the activation energy of the reaction. Less energy is required to start the reaction, so it runs more quickly.

16. (A) is correct. In allosteric regulation, the enzyme is usually composed of more than one polypeptide chain with more than one allosteric site remote from the active site, which is not part of the active site. When an allosteric activator binds to the allosteric site, the protein assumes a stable conformation with a functional active site, and the reaction can proceed. When an allosteric inhibitor binds, this stabilizes the inactive conformation of the protein.

17. (C) is correct. Competitive inhibitors compete for the active site of the enzyme. They are able to bind because they closely resemble the normal substrate. One way to overcome the effects of competitive inhibitors is to increase the amount of substrate so that chances are greater that a substrate molecule (rather than the competitive inhibitor) will bind.

18. (E) is correct. In cooperativity, the enzyme in question has more than one subunit with more than one active site, and it is able to bind more than one substrate—so multiple reactions can be taking place at once in the enzyme. The binding of one substrate molecule to the enzyme causes a conformation change that makes the binding of other substrate molecules, at the other active sites, more favorable.

19. (B) is correct. In feedback inhibition, the product of a metabolic pathway switches off the pathway by binding to and inhibiting an enzyme involved somewhere along the pathway.

20. (D) is correct. In noncompetitive inhibition, the inhibitor binds to a site other than the active site of the enzyme, and this causes the enzyme to change shape. The change in conformation makes the substrate unable to bind to the active site of the enzyme, and this prevents the action from taking place.

21. (A) is correct. Since the substance is being moved against the concentration gradient, energy is required. All the other choices are modes of passive transport and require no energy.

22. (D) is correct. The question requires you to know the structure and action of receptor tyrosine kinases. Recall that they exist as two proteins that come together when bound by ligands to form a dimer. The tyrosine kinases of each original monomer phosphorylate the tyrosine kinases of the other.

▮ **23. (A) is correct.** When a signal molecule binds to the receptor protein, the gate of the ion channel opens or closes, allowing or blocking the flow of specific ions.

▮ **24. (B) is correct.** Answer A is a reference to a G protein-coupled receptor, but is a quick pick if the question is not read carefully. The G protein is activated by the G protein-coupled receptor, which is a protein and eliminates E as a possible answer.

▮ **25. (D) is correct.** Kinase enzymes are involved with ATP. Protein kinase enzymes are used to amplify the signal during the transduction phase of cell signaling by activating cell proteins with a phosphate from ATP.

▮ **26. (C) is correct.** Many signaling pathways involve small, nonprotein water-soluble molecules or ions called second messengers. Calcium ions and cyclic AMP are two common second messengers. The second messengers, once activated (and always found on the inside of the membrane), can initiate a phosphorylation cascade resulting in a cellular response.

▮ **27. (D) is correct.** Intracellular receptors work with signal molecules that are hydrophobic compounds and are therefore able to cross the plasma membrane. Testosterone, as indicated in the question, is a steroid hormone and thus hydrophobic. Intracellular receptors often act as transcription factors.

▮ **28. (A) is correct.** In telophase, nuclear envelopes begin to form around the sets of chromosomes, which are now located at opposite ends of the cell. The chromatin becomes less condensed, and cytokinesis begins—the cytoplasm of the cell is divided.

▮ **29. (C) is correct.** During cytokinesis, the cytoplasm of the cell is divided approximately equally as the cell membrane pinches off (in animal cells), forming two daughter cells; a cell plate forms in plant cells.

▮ **30. (E) is correct.** In anaphase, the sister chromatids, which were lined up along the equator of the cell, begin to separate, pulled apart by the retracting microtubules. By the end of anaphase, the opposite ends of the cell contain complete and equal sets of chromosomes.

▮ **31. (B) is correct.** Interphase is not a part of mitosis; rather it is the part of the cell cycle when the cell gets ready to divide by replicating its DNA. There are three stages in interphase: G_1 phase, S phase, and G_2 phase. The genetic material is replicated in S phase.

▮ **32. (D) is correct.** Prometaphase is the phase of mitosis in which the nuclear envelope begins to fragment so that the microtubules can begin to attach to the kinetochores of the chromatids, which by this time are very condensed.

▮ **33. (B) is correct.** The depicted cell is in prometaphase. As you can see, the nuclear envelope is fragmenting, and the microtubules have already attached to some of the kinetochores at the centromeres of the chromosomes. The chromosomes are condensed and beginning to line up along the cell's equator.

▮ **34. (D) is correct.** The most crucial checkpoint of the cell cycle is the G_1 checkpoint. In the cell cycle, a checkpoint is a point at which there can be a signal to stop or to go ahead with division. If a cell receives the signal to go ahead at the G_1 checkpoint, it will usually complete the cycle and divide. If it does not receive the go-ahead signal, it will enter the (nondividing) G_0 phase for an indeterminate period of time.

Level 2: Application/Analysis/Synthesis Questions

▌ **1. (D) is correct.** The radioactive tracking starts with the formation of the protein, which occurs on the ER. The protein then moves to the Golgi and out of the cell via vesicles that will fuse with the membrane. Answer C is the proper pathway, but the radioactive amino acids will not be used in the nucleus, and would therefore not be tracked with this system.

▌ **2. (A) is correct.** During S phase of interphase, DNA is replicated. If this did not occur, the daughter cells would have half the genetic material found in the parent cell.

▌ **3. (A) is correct.** Water moves in a hypotonic to hypertonic direction. Cells A, B, and C are all animal cells (red blood cells) and lack a cell wall. Cell D is a plant cell with a cell wall. Cell A is in a hypotonic solution, cell B an isotonic solution, and cell C is in an hypertonic solution. When answering this type of question, pay close attention to whether the solution or the cell is referenced as being hypertonic or hypotonic. In A the solution is hypotonic, while the cell is hypertonic.

▌ **4. (C) is correct.** In plant cells the relatively inelastic cell wall exerts a back pressure on the cells, called turgor pressure. In cell D the plant cell is immersed in a hypotonic solution, causing the cell to uptake water, thus creating the highest levels of turgor pressure. The animal cell A lyses ("pops") because it has only a thin, flexible membrane.

▌ **5. (C) is correct.** The solution in Tank A started with more solutes, then as the water is purified its concentration of solutes increases through the process of reverse osmosis. Tank A becomes increasingly hypertonic over the course of the purification treatment.

▌ **6. (B) is correct.** Tank A with the tap water is hypertonic to the purified water. Since water flows from hypotonic to hypertonic, water would move from Tank B into Tank A.

▌ **7. (C) is correct.** Epinephrine is the ligand that activates the G protein-coupled receptor responsible for glycogen breakdown. Epinephrine does not enter the cell, suggesting a second messenger. Only in intact cells could the first messenger (epinephrine) be translated to a cellular response–glycogen breakdown.

▌ **8. (C) is correct.** G protein-coupled receptors are activated by their specific ligand, not by a phosphorylation event.

Free-Response Questions

(a) Some eukaryotic cell organelles might have evolved from free-living prokaryotic organisms. First of all, prokaryotic cells are much smaller than eukaryotic cells—they range from 100 nm to 10 μm, compared to the average size of eukaryotic cells: 10 to 100 μm. However, mitochondria (organelles unique to eukaryotic cells, and functioning in the creation of ATP in cellular respiration) and eukaryotic cell nuclei are comparable in size to prokaryotic cells, ranging from about 1 to 10 μm.

Another interesting characteristic of organelles that may tie them to prokaryotes is their structure and cell contents. To illustrate this, let's consider the structure of mitochondria. With few exceptions, mitochondria are found in all animal cells, plant cells, fungi, and protists. They can exist in great numbers in these cells, or cells can contain just one mitochondrion (depending on the metabolic activity of the cell). It has been observed that mitochondria can move around, alter their shape, and even divide in two—all of which are characteristic of living cells. Their structure consists of a double membrane exterior (the membrane is a typical combination of phospholipids and proteins, like the membrane of the cell itself); the outer membrane is relatively smooth, but the interior membrane has infoldings called cristae. This creates two different compartments in mitochondria: The inner compartment is the mitochondrial matrix, and the compartment in between the two membranes is called the intermembrane space. Mitochondria also contain mitochondrial DNA. Not very much DNA is contained in mitochondria, but the presence of DNA could be evidence that they were independent organisms at some time. Also similar to prokaryotes, mitochondria do not contain many interior structures other than their genetic material (which is not enclosed in a nucleus) and their cell membranes. All of the above indicates a close evolutionary relationship between prokaryotic cells and mitochondria.

(b) In order to trace the path of proteins in the cell from their creation to their expulsion, we must start in the nucleus. In the nucleus, mRNA is transcribed from DNA, and mRNA travels out of the nucleus through a nuclear pore to the cytoplasm, ending up at ribosomes, some of which are associated with the endoplasmic reticulum (called rough endoplasmic reticulum because of this association). Within the lumen of the ER, the mRNA is translated into protein, which then undergoes folding to assume its final shape, or conformation.

Secretory proteins travel from the endoplasmic reticulum to the series of flattened membranous sacs known as the Golgi apparatus. They enter at the *cis* face and eventually bud from the *trans* face, after undergoing a series of modifications to prepare them for secretion. The vesicles may then fuse with the cell membrane, and the contents are released from the cell in a process called exocytosis.

This response shows that the writer used the following key terms in context, showing the writer's knowledge of their meanings and relatedness:

organelles	*mRNA*
prokaryote	*ribosomes*
eukaryote	*endoplasmic reticulum*
mitochondria	*conformation*
phospholipids	*enzymes*
proteins	*secretory proteins*
crista	*Golgi apparatus*
mitochondrial matrix	*cis/trans face*
DNA	*vesicle*

The student response in (a) would be improved by including the chloroplast with the discussion on mitochondria. Specific sizes of cells and organelles is not expected. An additional argument for endosymbiosis could have been the presence of ribosomes in mitochondria and chloroplasts that translate genes unique to the organelles.

Additional points would probably be awarded in (b) if the student included a discussion of signal peptides and signal recognition particles. (This is covered in Chapter 17.)

Topic 3: Respiration and Photosynthesis

ANSWERS AND EXPLANATIONS

Level 1: Knowledge/Comprehension Questions

▌**1. (C) is correct.** The purpose of cellular respiration in eukaryotes is to produce energy for cellular work in the form of ATP. Respiration is an aerobic process, meaning that it requires oxygen. Answer choices A and B are incorrect because respiration involves the breakdown (not the synthesis) of carbohydrates, fats, and proteins. Choice D is wrong because ADP is the product of the dephosphorylation of ATP—it is left over after the energy from ATP has been released. Choice E is wrong because oxygen is required for cellular respiration.

▌**2. (B) is correct.** In the course of the reaction shown, potassium (K) is oxidized. Oxidation involves the loss of an electron, whereas reduction is the gain of an electron by an atom or molecule. In this reaction, potassium is oxidized and bromine is reduced. Both cellular respiration and photosynthesis involve numerous oxidation-reduction (redox) reactions.

▌**3. (D) is correct.** The net energy result of glycolysis is the production of two molecules of ATP and two molecules of NADH. Glycolysis is the first of the three stages of respiration—the second being the citric acid cycle and the third being oxidative phosphorylation. During glycolysis, glucose is oxidized to form two molecules of pyruvate. Glycolysis occurs in the cytosol, and the pyruvate it produces travels to the mitochondria, where it is used in the citric acid cycle.

▌**4. (B) is correct.** Note that this question is framed per glucose, not per single turn of the citric acid cycle. In the breakdown of glucose in the citric acid cycle, 2 ATP are produced. The citric acid cycle takes in a molecule called acetyl CoA (pyruvate is converted into acetyl CoA before it enters the citric acid cycle), and this is joined to a four-carbon molecule of oxaloacetate to form a six-carbon compound citrate that is then broken down again to produce oxaloacetate; the oxaloacetate reenters the cycle. In the course of the citric acid cycle, the following are produced *per glucose*: 4 CO_2, 2 ATP, 6 NADH, and 2 $FADH_2$.

▌**5. (D) is correct.** The process that produces the most ATP during cellular respiration is oxidative phosphorylation. $FADH_2$ and NADH donate electrons to the electron transport chain, which is coupled to ATP synthesis by chemiosmosis. The movement of electrons down the electron transport chain creates an H^+ gradient across the mitochondrial membrane, which drives the synthesis of ATP from ADP. A maximum of 26 to 28 ATP can be produced in oxidative phosphorylation per glucose molecule.

▌**6. (D) is correct.** In glycolysis, glucose is oxidized to two molecules of pyruvate. This is the first step in cellular respiration, showing a net production of 2 ATP and 2 NADH.

7. (A) is correct. In chemiosmosis, the hydrogen ion gradient created by the transfer of electrons in the electron transport chain provides the power to synthesize ATP from ADP.

8. (E) is correct. Fermentation is a way of harvesting chemical energy without using either oxygen or an electron transport chain. It consists of glycolysis and several reactions that serve to regenerate NAD^+. Electrons are transferred from NADH to pyruvate or its derivatives; then NAD^+ can return to glycolysis to once again accept electrons, continuing the production of small amounts of ATP. There are two main types of fermentation: alcohol fermentation (which creates ethanol as a product) and lactic acid fermentation (which creates lactate).

9. (B) is correct. The electron transport chain is a series of inner mitochondrial matrix membrane-embedded molecules that are capable of being oxidized and reduced as they pass along electrons. The energy produced from the passage of these electrons down the chain is used to create an H^+ gradient across the membrane, and the flow of H^+ down the gradient and back across the membrane powers the phosphorylation reaction of ADP to form ATP.

10. (C) is correct. The citric acid cycle includes the final reactions for the breakdown of glucose that began in glycolysis. The pyruvate from glycolysis is converted into acetyl CoA, which enters the cycle and is joined to oxaloacetate to create citrate, which is then converted to oxaloacetate again and reused. This cycle gives off CO_2 and forms 1 ATP, 3 NADH, and 1 $FADH_2$. The cycle goes through one rotation to break down each of the molecules of pyruvate produced in glycolysis (which of course is first converted to acetyl CoA), so the net result of the breakdown of one glucose molecule is 2 ATP, 6 NADH, and 2 $FADH_2$.

11. (A) is correct. Groups of photosynthetic pigment molecules in the thylakoid membrane are called photosystems. The two photosystems involved in photosynthesis are photosystem I and photosystem II. Both contain chlorophyll molecules and many proteins and other organic molecules, and both have a light-harvesting complex that harnesses incoming light. Each of these photosystems contains a reaction center, where chlorophyll *a* and the primary electron acceptor are located.

12. (B) is correct. The main products of the light reactions of photosynthesis are NADPH and ATP. NADPH and ATP are used to convert CO_2 to sugar in the Calvin cycle. The enzyme rubisco combines CO_2 with ribulose bisphosphate (RuBP), and electrons from NADPH and energy from ATP to synthesize a three-carbon molecule called glyceraldehyde 3-phosphate.

13. (B) is correct. The process in photosynthesis that bears the closest resemblance to chemiosmosis and oxidative phosphorylation in cellular respiration is linear electron flow. In this process, energy from the transfer of electrons down the electron transport chain is used to create a hydrogen ion gradient used in the making of ATP. Later, the energy stored in this ATP is used during the formation of carbohydrates in the Calvin cycle.

14. (E) is correct. The organic product of the Calvin cycle, which may be used later to build large carbohydrates in the cell, is glyceraldehyde 3-phosphate, or G3P. This molecule is created as a result of the fixation of three molecules of

CO_2, which costs the cell ATP and NADPH that were created in the light reactions of photosynthesis.

15. (E) is correct. C_4 and CAM plants both grow better than do C_3 plants under conditions of increased median air temperature and decreased relative humidity. Both C_4 and CAM plants use an alternative method of carbon fixation that enables them to fix carbon into an acid intermediate for later deposit into the Calvin cycle.

16. (D) is correct. The electron transport chains pump protons across membranes from regions of low H^+ concentrations to regions of high H^+ concentrations. This proton pumping occurs in both mitochondria and chloroplasts, and the protons then diffuse (with the concentration gradient) back across the membrane through ATP synthases. This drives the synthesis of ATP.

17. (D) is correct. ATP synthase is located in the thylakoid membrane. Notice the direction of flow of protons through ATP synthase in photosynthesis: The protons flow from the thylakoid space to the stroma. ATP is produced in the stroma, where it will be used by the Calvin cycle.

18. (C) is correct. When water is split, three products are formed: two protons, an oxygen atom that immediately bonds with another oxygen to form O_2, and two electrons. The electrons immediately feed the P680 chlorophyll *a* in the reaction center of photosystem II. The ultimate electron donor in photosynthesis is water.

19. (A) is correct. CAM plants separate the two stages of photosynthesis temporally to reduce photorespiration. This is accomplished by fixing CO_2 at night using PEP carboxylase and storing the carbon in organic acids. During the day when CAM plants have their stomata closed to conserve water, the carbon from the organic acids is chemically released and used in the Calvin cycle. Choice C would be correct for C_4 plants, but not for CAM plants.

20. (D) is correct. Each turn of the Calvin cycle involves the enzyme rubisco fixing one atom of carbon. It follows that it would take six turns to produce the six-carbon sugar glucose. Students sometimes miss this question by confusing the Calvin cycle with the Krebs or citric acid cycle of cellular respiration. Read these questions carefully, being disciplined enough to carefully identify what the question is asking.

21. (E) is correct. The light reactions of photosynthesis move electrons from their low-energy state in water to a higher energy level when the electrons are donated to $NADP^+$ to make NADPH. In cellular respiration electrons pass down the electron transport chain from high to low potential energy, ultimately combining with O_2 and hydrogen ions to form water.

Level 2: Application/Analysis/Synthesis Questions

1. (D) is correct. Plant leaves are green because they reflect and refract green light, which is not utilized in photosynthesis. Red light is used in photosynthesis, meaning the plant would absorb CO_2 for photosynthesis. Check the action spectrum for photosynthesis in your text and be prepared to explain the peaks and valleys shown in the graph.

2. (A) is correct. The light reactions convert solar energy to the chemical energy of ATP and NADPH, which are utilized in the Calvin cycle to reduce CO_2 to sugar.

3. (A) is correct. The Calvin cycle occurs in the stroma.

■ **4. (C) is correct.** In step 3 the five-carbon α-ketoglutarate is converted to the lower energy four-carbon compound succinate. The drop in energy allows for the production of 1 ATP and the reduction of NAD^+ to NADH. No other step in the citric acid cycle accomplishes this.

■ **5. (D) is correct.** Normally the potential energy of the H^+ gradient across the inner mitochondrial membrane is coupled with ATP synthase in the production of ATP. The drug allows for the leaking of H^+ across the membrane, effectively uncoupling ATP synthesis from the H^+ gradient.

■ **6. (A) is correct.** With the uncoupling of ATP synthase from the H^+ gradient, the inner membrane no longer has a sufficient electrochemical gradient to generate normal ATP production.

Free-Response Questions

(a) Although plants have two photosystems, they both work in the same way. Both photosystems have two components: the light-harvesting complexes and the reaction center. The light-harvesting complex is made up of many chlorophyll and accessory pigment molecules. When one of the pigment molecules absorbs light energy in the form of photons, one of the molecule's electrons is raised to an orbital of higher potential energy. The pigment molecule is then said to be in an "excited" state. The increase in potential energy is transferred to the reaction center of the photosystem. The reaction center consists of two chlorophyll *a* molecules, which use the increased potential energy passed to them by the photosynthetic pigments to donate electrons to the primary electron acceptor. The solar-powered transfer of an electron from the reaction-center chlorophyll *a* pair to the primary electron acceptor is the first step of the light reactions. This is the conversion of light energy to chemical energy.

(b) Glycolysis is the first stage of cellular respiration and occurs in the cytoplasm. Glycolysis involves the breakdown of glucose to two pyruvate molecules. To accomplish this, 2 ATP molecules are invested, which helps to destabilize glucose, making it more reactive and allowing glucose to break into two three-carbon molecules. By the time the pathway has produced pyruvate, 4 ATP molecules have been produced along with 2 NADH molecules. This gives a net energy gain of 2 ATP and 2 NADH. Thus, one important role of glucose is to produce energy molecules for the cell to use in its life processes. The second role is to produce pyruvate, which can feed into the citric acid cycle in the mitochondria and ultimately into the electron transport chain, where most of the ATP in cellular respiration is produced.

(c) In cellular respiration water is a product of the reaction, whereas in photosynthesis water is a reactant. In cellular respiration water is formed when the electrons at the end of the electron transport chain in the cristae membrane of the mitochondria combine with hydrogen ions and an atom of oxygen to form water. Water is the ultimate electron acceptor in cellular respiration. In photosynthesis an enzyme splits a water molecule into two electrons, two hydrogen ions, and an oxygen atom. The electrons are supplied as needed directly to the chlorophyll molecules in the reaction center of photosystem II. In photosynthesis water is the ultimate electron donor.

This response shows that the writer used the following key terms in context, showing the writer's knowledge of their meanings and relatedness:

photosystems	light-harvesting complexes
reaction center	chlorophyll
glycolysis	net energy production
electron donor	electron acceptor
accessory pigments	primary electron donor
ATP	NADPH

This response also contains an explanation of the following subjects and processes:

—role of photosystems in photosynthesis
—how light energy becomes transformed to chemical energy
—function of glycolysis in overall scheme of cellular respiration
—ATP and NADPH as important energy molecules in the cell
—role of water as an electron sink or an electron donor

This student has written a particularly clear essay. Notice how carefully sequenced the responses are. Also note the clarity of each sentence and the absence of third-person pronouns. Avoiding pronouns may require a little more time and patience, but it pays off in added clarity and higher scores.

Topic 4: Mendelian Genetics

ANSWERS AND EXPLANATIONS

Level 1: Knowledge/Comprehension Questions

▌ **1. (E) is correct.** The probability that the woman will have a seventh child who is a daughter is 1/2. Since the probability that a sperm carrying an X chromosome and the probability that a sperm carrying a Y chromosome will fertilize an egg is equal—both 50%—fertilization is considered an independent event. The outcome of independent events is unaffected by what events occurred before or will occur after. Therefore, the probability that this woman's next child will be a girl is 1/2. Likewise, the probability that she will have a child that is a boy is also 1/2.

▌ **2. (A) is correct.** If the probability of allele R segregating into a gamete is 1/4, and that of S segregating is 1/2, you can calculate the probability of two independent events occurring in a specific combination, order, or sequence by multiplying their probabilities. So in this case, you need to multiply 1/4 by 1/2.

▌ **3. (D) is correct.** Note that this cross involves incomplete dominance because there are phenotypes in the offspring. Let's say that the yellow coat parent is $CYCY$, and the homozygous brown coat parent is $CBCB$. Because the yellow coat parent can produce only gametes CY, and the brown coat parent can produce only gametes CB, the F_1 generation will all have genotype $CYCB$. Crossing two members of this generation would give you a ratio of 1 yellow coat: 2 gray coats: 1 brown coat. This means that 25% of the offspring would have brown coats, 25% would have yellow coats, and 50% would have gray coats.

▪ **4. (A) is correct.** All of the statements about meiosis are true except A. The spindle fibers attach during prophase, not metaphase.

▪ **5. (D) is correct.** To find the answer to this problem, first look at the ratio of the offspring. It's 6:2, which can be reduced to 3:1. Next, you can quickly work through the crosses listed. You can immediately rule out answers A and B, because A would give you only offspring that exhibited the dominant traits—short hair and green eyes—and B would give you all offspring that had the recessive traits—long hair and blue eyes. If you look carefully at the remaining answers, you will want to choose the one that will give you all short-haired offspring, so you will need the dominant allele to be present in both parents. This rules out answer E. If you still cannot choose between C and D, write out what gametes the parents could produce, and then use a Punnett square to determine their offspring. By doing this, you can see that D is correct: the ratio of offspring is 12:4, or 3:1, which matches the ratio in the original question.

▪ **6. (A) is correct.** If the boy is afflicted with hemophilia, then he must have inherited the recessive hemophilia gene from his mother. Sex-linked genes are usually located on the X chromosome. In order for the child to be a boy, he must have inherited a Y chromosome from his father. Because the gene causing hemophilia is located on the X chromosome, you can rule out answers B and C (it would not matter if the father possessed the allele for hemophilia because he can't pass on his X to a son). Therefore, you need to look for an answer choice that shows that the source of his X chromosome was a carrier of the allele—afflicted or not. This is answer A.

▪ **7. (E) is correct.** While genes that are on the same chromosome tend to be inherited together, the process of crossing over enables "linked" genes to sort independently. Those that are linked but located farther apart on the chromosome will undergo crossing over more frequently than those located very close together on a chromosome simply because there are more sites between the two genes at which crossing over can take place.

▪ **8. (D) is correct.** Autosomal dominant traits appear with equal frequency in both sexes, and they do not skip generations. These qualities are all exhibited by the trait that is illustrated in the pedigree. All three generations are affected with the trait, and both sexes are affected.

▪ **9. (A) is correct.** The father either has type A, B, or O blood. The mother, who has the phenotype for type A blood, has the genotype $I^A i$. In order to produce a son with genotype ii (the genotype of people with type O blood), she would need to reproduce with a man who had genotype $I^A i$, genotype $I^B I$, or genotype ii. Try writing out the Punnett square if you aren't confident of this.

▪ **10. (C) is correct.** If the individual with type O blood were to mate with an individual who has type AB blood, since I^A and I^B are both dominant over i, the genotype would be 1 I^A:1 I^B, and the phenotype would be a ratio of 1:1 offspring with type A blood or type B blood.

▪ **11. (B) is correct.** The most likely reason for this 2:1 ratio in the offspring is that Y is lethal in homozygous form, and this caused the death of all of the YY

individuals in the litter. The expected ratio of this cross would be 1 *YY*:2 *Yy*: 1 *yy*. If you remove the *YY*, you get a ratio of 2 *Yy* (yellow mice, since the gene for yellow, *Y*, is dominant) to 1 *yy* (nonyellow mouse).

■ **12. (D) is correct.** The only process listed that does not lead directly to genetic recombination, or the recombining (scrambling) of genes in the offspring, is gene linkage. If genes are linked, they are located on the same chromosome and are more likely to segregate together into the same cell, reducing genetic recombination

■ **13. (C) is correct.** Recall that a dihybrid cross between two heterozygotes produces a 9:3:3:1 offspring ratio. The question is asking for one of the heterozygotes (which would be one of the 3s in the above ratio). To come up with the answer, you could complete a Punnett square and see that the ratio of offspring produced is 9 warty green, 3 warty orange, 3 dull green, and 1 dull orange. The total number of off-spring produced is 144, and 3/16 of 144 is 27; the closest answer choice to 27 is 28.

■ **14. (B) is correct.** Fertilization restores the diploid number in a sexually reproducing organism. The two major events in the life cycle of sexually repro-ducing organisms are meiosis and fertilization.

■ **15. (E) is correct.** All of the responses are correct. Homologous chromosomes are a key conceptual point in Mendelian genetics. If you missed this question, review more thoroughly the concept of homologous chromosomes.

■ **16. (B) is correct.** The number of possible gametes can be determined by using the formula $2n$ where *n* equals the haploid number. With a diploid number of 6, thus a haploid number of 3, the answer is $2 \times 2 \times 2 \times 8$.

■ **17. (A) is correct.** Recall that in G_1 the chromosomes have not replicated. After meiosis I the chromosomes have replicated, which doubles the amount of DNA. Next, the homologs separate in meiosis I. This separation reduces the amount of DNA by one-half and back to the original amount.

■ **18. (A) is correct.** The synaptonemal complex forms during prophase I and is the platform from which crossing over occurs.

■ **19. (A) is correct.** In meiosis II the most important event is the separation of sister chromatids. Mitosis involves only the separation of sister chromatids; homologs do not form in mitosis.

■ **20. (B) is correct.** Synapsis is the process of homologs moving to the tetrad position. This occurs in prophase I, not S phase as indicated in the question.

■ **21. (E) is correct.** Begin by drawing a straight line to represent a chromosome, and then place genes A and B 8 units apart. Note that A and C are 28 units apart with B and C 20 units apart. This means the genes are found in order A, B, C. Next, consider the map distances between A and D and B and D. Since map distance represents the frequency of crossing over, it will only work if D precedes A on the chromosome. The correct gene sequence must be D-A-B-C.

■ **22. (B) is correct.** Since there are regions of the X chromosome that have no cor-responding region on the Y, alleles that are found on this unmatched area of the X in males may be expressed, even though the male has a single copy of the gene.

Level 2: Application/Analysis/Synthesis Questions

■ **1.** Albino (*b*) is a recessive trait; black (*B*) is dominant. First cross: parents *BB* × *bb*; gametes *B* and *b*; offspring all *Bb* (black coat). Second cross: parents *Bb* × *bb*; gametes 1/2 *B* and 1/2 *b* (heterozygous parent) and *b*; offspring 1/2 *Bb* and 1/2 *bb*.

■ 2.

Parents
$GgIi$ × $GgIi$

Sperm

	GI	Gi	gI	gi
GI	$GGII$	$GGIi$	$GgII$	$GgIi$
Gi	$GGIi$	$GGii$	$GgIi$	$Ggii$
gI	$GgII$	$GgIi$	$ggII$	$ggIi$
gi	$GgIi$	$Ggii$	$ggIi$	$ggii$

Eggs

9 green-inflated : 3 green-constricted :
3 yellow-inflated : 1 yellow-constricted

■ **3.** Parental cross is $AAC^RC^R \times aaC^WC^W$. F_1 genotype is AaC^RC^W, phenotype is all axial-pink. F_2 phenotypes are 3 axial-red : 6 axial-pink : 3 axial-white : 1 terminal-red : 2 terminal-pink : 1 terminal-white.

■ **4. (D) is correct.** Given that the diploid ($2n$) number is four and the cells in the diagram with an (n) or haploid number show two chromosomes, this must be meiosis. Nondisjunction in meiosis I would involve homologs not separating, whereas nondisjunction in meiosis II would involve sister chromatids not separating. The diagram shows sister chromatids not separating, thus leaving one cell $n + 1$ and the other cell $n - 1$.

■ **5. (B) is correct.** Type O blood is recessive. This means that the woman could be heterozygous for blood type A (I^Ai) and the father heterozygous for type B blood (I^Bi). If each parent donates their recessive allele, the baby would be ii, type O blood.

■ **6. (C) is correct.** It appears the mother has inherited the color-blind gene from her father and passed it to her son. This is a typical inheritance pattern seen in sex-linked genes.

■ **7. (C) is correct.** The question and correct answer describe independent assortment as it occurs in meiosis. Independent assortment increases variability, or as phrased in the question, increases the possible combinations of characteristics.

Answer from page 100: The karyotype in Figure 4.1 is that of a male. Note the unpaired X and Y in the bottom right corner.

Topic 5: Molecular Genetics

ANSWERS AND EXPLANATIONS

Level 1: Knowledge/Comprehension Questions

■ **1. (C) is correct.** Operons are gene expression mechanisms of bacteria. They are not found in eukaryotic cells.

■ **2. (B) is correct.** Viruses are made up of nucleic acid surrounded by a protein coat. The answer cannot be A because retroviruses consist of RNA. Viruses reproduce by injecting their genetic material into a host cell and using the cell's replicative machinery to replicate their DNA and proteins. The new viruses leave the cell to infect more cells, sometimes killing the host cell in the process.

■ **3. (D) is correct.** The process of genetic engineering is possible because the processes of transcription and translation are so similar in all eukaryotic cells. Once the spider gene or genes that were responsible for coding for the silk proteins were isolated and then inserted into a bacterial plasmid (which would serve as the vector), the cloning vector would be taken up by the goat's cells, and the goat's cells' transcription/translation machinery would begin the process of producing the spider protein, along with its own proteins.

■ **4. (D) is correct.** Restriction enzymes can be used to cut DNA at specific locations, and this enables researchers to perform recombinant DNA techniques. When specific restriction enzymes are added to the DNA, they produce cuts in the sugar-phosphate backbone and create "sticky ends," which can bind to DNA fragments from a different source to produce recombinant DNA. DNA ligase is then added to seal the strands together permanently.

■ **5. (C) is correct.** Transposons are also called transposable genetic elements, and they are pieces of DNA that can move from location to location in a chromosome—or a genome. Transposons are also called "jumping genes," and most of them are capable of moving to many different target sites in the genome.

■ **6. (D) is correct.** One of the two important ways that the cell has of controlling gene expression is through DNA methylation. In DNA methylation, methyl groups are attached to certain DNA bases after DNA is synthesized. This appears to be responsible for the long-term inactivation of genes.

■ **7. (B) is correct.** The process by which genetic information flows from mRNA to protein is called translation. Translation occurs in the cytoplasm of the cell, at ribosomes. A molecule of mRNA is moved through the ribosome, and codons are translated into amino acids one at a time. Transfer RNAs add their associated amino acids onto a growing polypeptide as its anticodon pairs with a codon on the mRNA and then departs from the ribosome to bind more free amino acids.

■ **8. (A) is correct.** In transcription, RNA is synthesized using the genetic information encoded by DNA. Transcription occurs in the nucleus of the cell. The double-stranded DNA helix unwinds, allowing enzymes and proteins to synthesize a new complementary single-stranded mRNA molecule from the template strand of DNA.

■ **9. (E) is correct.** In histone acetylation, acetyl groups are attached to certain amino acids of histones. Deacetylation is the process by which they are removed. Acetylation makes the histones change shape so they are less tightly bound to DNA, and this allows the proteins involved in transcription to move in and begin the process. Therefore, acetylation is one way for the cell to initiate transcription and to control the expression of its genes.

■ **10. (A) is correct.** This art portrays the lytic cycle of phage reproduction. In the lytic cycle, the phage first attaches to the cell surface and injects its DNA into the cell. It then hydrolyzes the host cell's DNA and uses the cell's machinery to produce phage proteins and to replicate its genome. The phage proteins are then assembled in the cell until the host cell lyses (breaks open), and the new phages are released to infect other cells. In the lysogenic cycle, the phage genome becomes incorporated into the host cell's DNA without destroying the host cell.

11. (A) is correct. In eukaryotic cells the enzyme telomerase catalyzes the lengthening of telomeres in selected cells, such as the germ cells that produce gametes.

12. (D) is correct. PCR, the polymerase chain reaction, is a technique by which any piece of DNA can be copied many times without the use of cells. The DNA is heated to separate its strands and then cooled to allow primers to attach to the single strands. DNA polymerase is added, which begins to add nucleotides to the 3′ end of each primer on the two strands. With each turn of the cycle, the amount of DNA is multiplied by two.

13. (C) is correct. The fragments of DNA separated out from one another along the gel once the electric field was applied because they differ in size. For DNA in gel electrophoresis, how far a molecule travels through a gel (while the current is applied) is inversely proportional to its size. The larger a fragment is, the more slowly it will migrate.

14. (D) is correct. The restriction enzyme used to cut the DNA that was placed into the first well of the gel must have cut the DNA at eight sites, because it produced nine DNA fragments. The number of fragments produced is always one more than the number of restriction sites cut.

15. (D) is correct. You should recall that a T4 phage injects its DNA, which is used to assemble new protein coats and viral DNA. The protein coat of the infecting T4 phage is left outside the bacterium.

16. (B) is correct. Reverse transcriptase is an example of an enzyme that is required to produce DNA from the RNA of a retrovirus.

17. (C) is correct. In genetic engineering (the manipulation of genes for practical purposes), DNA ligase is an enzyme that is used to seal the strands of newly recombinant DNA (DNA that is spliced together from two different sources) by catalyzing the formation of phosphodiester bonds.

18. (C) is correct. The addition of a poly-A tail after transcription is one example of posttranscriptional modifications that the mRNA undergoes. This poly-A tail inhibits the degradation of the newly synthesized mRNA strand and is thought also to help ribosomes attach to it. Another important modification that mRNA undergoes is the addition of a 5′ cap. The 5′ cap helps protect mRNA from degradation and also acts as the point of attachment for the ribosomes, just prior to translation.

19. (D) is correct. RNA polymerase is the most prominent enzyme involved in the transcription of DNA to make mRNA. It is responsible for binding to the promoter sequence on the template DNA, prying the two DNA strands apart, and hooking the RNA nucleotides together as they base-pair along the DNA template. RNA polymerases add nucleotides to the 3′ end of the growing chain until a terminator sequence is reached—it transcribes entire transcription units.

20. (E) is correct. Ribosomal RNA (rRNA), together with proteins, makes up ribosomes. Ribosomes, the sites of protein synthesis, are composed of two subunits, the large and the small subunit. The large subunit of the ribosome contains the A, P, and E sites, which shuttle through the tRNA and mRNA during translation.

▎ **21. (A) is correct.** tRNA, or transfer RNA, interprets the genetic message coded in mRNA. It transfers amino acids taken from the cytoplasmic pool to a ribosome, which adds the specific amino acid brought to it by tRNA to the end of a growing polypeptide chain. Each type of tRNA binds to a specific amino acid at one end; its other end contains an anticodon, which base-pairs with a complementary codon on the mRNA strand.

▎ **22. (B) is correct.** mRNA, also known as messenger RNA, is a type of RNA that is synthesized from DNA and attaches to ribosomes in the cytoplasm to specify the primary structure of a protein. Since mRNA is the product of transcription, which occurs in the nucleus, it must travel out of the nucleus and into the cytoplasm in order to participate in translation.

▎ **23. (E) is correct.** Acetylation occurs as a mechanism of gene regulation when histones are acetylated, not DNA. Histone acetylation causes the histone tails to no longer bind to neighboring nucleosomes, allowing the transcription proteins to have easier access to genes in the acetylated area.

▎ **24. (B) is correct.** In the modification of mRNA that occurs after transcription, a process called RNA splicing occurs. In this process, noncoding regions of nucleic acid that are situated between coding regions are cut out. These noncoding regions are called introns. The remaining regions are called exons, and these are spliced together to form the final mRNA product. When you think of exons, think expressed—because they are actually translated into proteins, whereas introns are not.

▎ **25. (E) is correct.** Wobble refers to the relaxing of base pairing between the third base of a codon and the corresponding base of a tRNA anticodon. For example, the base U in the third position of the anticodon can pair with either A or G in the third position of the mRNA codon. This is why only 45 tRNAs are needed to translate the 61 codons that specify an amino acid. (Recall that 3 of the 64 codons specify stop, instead of an amino acid.)

▎ **26. (E) is correct.** A missense mutation is a base-pair substitution (the replacement of a nucleotide and its partner in the cDNA strand with a different pair of nucleotides) that still enables the codon to code for an amino acid. The amino acid may or may not ultimately contribute to a functional protein, but a missense mutation is one where an amino acid is still chosen to be added to the polypeptide chain, and translation will continue.

▎ **27. (B) is correct.** The amount of adenine in DNA will equal the amount of thymine; guanine will equal cytosine, which leads to the mathematical equality A + G = C + T. Try this: If a certain species has 15% adenine in its DNA, what would be the percentage of guanine? (It would be 35%. A + T = 30%, so G + C = 70%, and G = 35%.)

▎ **28. (D) is correct.** DNA replication is semiconservative. Each new daughter molecule contains one newly synthesized strand, and one strand that used to belong to the parent double helix DNA.

▌ **29. (B) is correct.** In gel electrophoresis the smallest fragments travel farthest; the largest fragments are closest to the well. Fragment b is shortest, followed by a and c. We would expect a gel with c closest to the well, then a, with b at the far end.

▌ **30. (A) is correct.** Several of the other possible answers have errors that would be instructive to note. Answer B is false because the genetic code is near universal; answer C is false because introns are not part of the inserted gene, as they have been removed; answer D is false because ribosomes are not governed by the length of the gene; answer E is false because prokaryotes do not have splicing enzymes to take out introns. Answer A is correct because the signals that control gene expressions and promoter regions are different between prokaryotes and eukaryotes.

▌ **31. (A) is correct.** Retrotransposons move by means of an RNA intermediate that is a transcript of the retrotransposon DNA. They always leave a copy at the original site during transposition, since they are initially transcribed into an RNA intermediate. Sometimes this mechanism is called "copy and paste." Choice C refers to the general action of transposons, "cut and paste," in which the original segment is moved, but not copied.

▌ **32. (B) is correct.** In order to be a probe, it would have to contain complementary nucleotides, not amino acids. It is a common error to confuse the matching of nucleotides with nucleic acids (DNA or RNA) and amino acids with proteins.

▌ **33. (B) is correct.** The number of genes in the human genome has been revised sharply downward since the Human Genome Project. Less than a decade ago the number of human genes was at 100,000; now it is about 20,000! How can only 20,000 genes be enough for a human? The answer is alternative gene splicing yielding multiple proteins per gene.

▌ **34. (E) is correct.** An example of a multigene family is seen with the various forms of hemoglobin; the α-globin gene family is on chromosome 16, whereas the β-globin gene family is on chromosome 11.

▌ **35. (A) is correct.** Homeotic genes are "master control genes" and when activated will in turn activate numerous other genes important in development.

Level 2: Application/Analysis/Synthesis Questions

▌ **1. (D) is correct.** In HIV (the virus that causes AIDS) the envelope glycoprotein enables the virus to bind to specific receptors on certain white blood cells. After binding, the virus fuses with the cell's plasma membrane, entering the cell. The mutated gene CCR5 prevents binding and subsequent entering of the cell by HIV.

▌ **2. (D) is correct.** In HIV infections, recall that the infectious agent is a retrovirus with RNA as the genetic material. In the cytoplasm of the host cell the viral RNA is converted to DNA, which then enters the nucleus and incorporates as a provirus into the cell's DNA. The provirus can direct the production of new viral particles.

▌ **3. (D) is correct.** Bacteria do not perform mitosis, meiosis, or sexual reproduction. However, the processes of transformation, transduction, and conjugation are processes that bacteria use to increase their genetic diversity. Be sure to review all three processes.

4. (B) is correct. These profiles were done using short tandem repeats as genetic markers. All 13 sites must match as this shows the highest degree of certainty between the pilot's DNA and the DNA from the three families.

5. (D) is correct. Family 3 matches in all 13 sites. This is a good time to review the principles of gel electrophoresis, making sure you understand which poles on the gel are positive and negative, how the bands are arranged by size on the gel, and the role of restriction enzymes in preparing the DNA.

6. (D) is correct. A nucleotide insertion downstream and close to the start of the coding sequence creates a frameshift mutation, causing the regrouping of codons and a completely new list of amino acids, leading to a nonfunctional protein. If the amino acids sequence is changed after the insertion, a *missense mutation* occurs. If the regrouping leads to the formation of a stop codon, a *nonsense mutation* occurs, which terminates the forming protein. Types of mutations are common questions on the exam, but you should give emphasis to the *effect* of a mutation, and be able to predict which type in a particular location would cause the greatest change in nucleotide sequence.

7. (A) is correct. Introns, not exons, are removed before mRNA leaves the nucleus. The rest of the choices in this question are true and worthy of your perusal and understanding. RNA editing is a fundamental concept in molecular genetics.

8. (C) is correct. The polymerase enzymes, both DNA and RNA, can only add nucleotides to the 3' end of a growing strand. This is the underlying reason for the leading strand, where DNA polymerase is adding a continuous, unbroken new strand of nucleotides and the lagging strand where DNA polymerase is moving away from the unwinding replication fork forming new nucleotides in Okazaki fragments. Knowing all the enzymes of replication is not required for the exam, but understanding how DNA polymerase works and the formation of leading and lagging strands is fundamental and worth your time investment.

Free-Response Questions

(a) There are many ways by which chromosomes can be altered to cause problems for the cell. Among these is nondisjunction—when during mitosis or meiosis the chromosomes fail to separate properly, and one cell ends up with two copies of a chromosome while the other gets no copies. This results in a condition called aneuploidy. Smaller chromosomal mutations are deletions (in which part of the chromosome breaks off and is lost), inversions (in which a chromosome segment is reversed within a chromosome), duplications (in which a chromosome segment is repeated in a chromosome), and translocation (in which part of a chromosome is moved from one chromosome to another).

The reason that it is detrimental to an organism to have an abnormal chromosome number is that genes, which are located on chromosomes, code for proteins, which have specific functions in the cell. If an organism has extra copies of a particular gene, then this gene will be transcribed in excess, creating more of the usual gene product. This alters the relative amounts, or doses, of interacting products in the cell, and this can cause serious developmental problems. Likewise, if a gene is missing from a chromosome, it will not be transcribed, and its corresponding protein will not be produced. If that protein has an important cellular function, the organism will be seriously affected.

(b) Colorblindness and hemophilia are more common in males than females because males are hemizygous for these traits. These traits are found on the X chromosome, and because males have only one copy of the X chromosome, they will show the phenotype for whatever allele they receive. Females, on the other hand, have two copies of the X chromosome and, if one X chromosome has an allele for colorblindness, for example, and the other X chromosome carries an allele for normal vision, the female will have normal vision. She is a "carrier" for colorblindness because she can pass it on to her sons.

(c) When fertilization occurs and a sperm carrying a Y chromosome penetrates the egg first, a male zygote with one X and one Y chromosome is produced. If a sperm carrying an X chromosome penetrates the egg first, a female zygote with XX is produced. Although it seems as though the female zygote would have twice the cell product as the male, due to its double dose of genes located on the two X chromosomes, this is not the case. The reason for this is that, in every cell of the female human body, one of the X chromosomes is inactivated. The mechanisms for this are not fully understood, but the X chromosome that is inactivated condenses into a structure called a Barr body, which then associates with the nuclear envelope. As a Barr body, most of the X chromosome's genes are not expressed. As a result of this, females are a mosaic consisting of cells with the X chromosome from their mother activated and other cells with the X chromosome from their father activated in about a 50:50 ratio. This is also the reason sex-linked disorders are usually not expressed in females. Though one of the X chromosomes may be incapable of producing a crucial gene product, this mosaic effect ensures that the other half of the somatic cells produce sufficient amounts of the protein in question.

This response demonstrates knowledge of the following terms and processes:

chromosome	*inversion*
gene	*duplication*
gene product	*translocation*
doses	*X and Y chromosomes*
transcription	*zygote*
nondisjunction	*X chromosome inactivation*
aneuploidy	*Barr body*
deletion	*somatic cell*

Topic 6: Mechanisms of Evolution

ANSWERS AND EXPLANATIONS

Level 1: Knowledge/Comprehension Questions

1. (B) is correct. When two members of the same species are prevented from breeding by a geographic barrier such as a mountain range or river, the fact that they live in different ponds, or any other physical obstruction, the two individuals are said to be geographically isolated.

2. (C) is correct. The concept of punctuated equilibrium was put forth recently by Niles Eldredge and Stephen Jay Gould. It explains periods of apparently little charge (stasis) "punctuated" by sudden changes observed in the fossil record.

3. (B) is correct. Since selection favors the average-sized wings and reduces the frequency of two extremes, this is an example of stabilizing selection. This would be a good time to also review directional and disruptive selection, shown in Figure 6.1 of this guide.

4. (A) is correct. Darwin's observations can be summarized as follows: Organisms produce more offspring than the environment can support. Further, the offspring vary in their heritable characteristics. Those individuals well suited to their environment leave more offspring than other individuals (differential reproductive success). Over time, favorable traits accumulate in the population.

5. (D) is correct. If the frequency of the recessive allele is 0.10, then we know that the frequency of the other allele is 1 – 0.10, which is equal to 0.90. The Hardy-Weinberg equation states that, if a population contains just two alleles for a given trait, and we know the frequency of one of the alleles, we can calculate the frequency of the other using the equation $p + q = 1$. If we designate the frequency of the occurrence of the recessive allele as q, and use a value of 0.1, we can rearrange the equation to read $p + 0.10 = 1$. Then, $1 – 0.10 = 0.9$, which is equal to p, or the frequency of the other allele—in this case the dominant one.

6. (A) is correct. Artificial selection is the selective breeding of domesticated plants and animals in order to modify them to better suit the needs of humans. Humans have been practicing artificial selection for thousands of generations, and many of the common foods we eat are a result of this process.

7. (D) is correct. The founder effect occurs when a few individuals from a population colonize a new, isolated habitat. The smaller the number of individuals who start this new population, the more limited will be the starting gene pool for the population—and the less the new gene pool will resemble that of the parent population.

8. (B) is correct. Homology is the result of descent from a common ancestor. It can be described as the underlying structural or molecular similarities (even in structures that are no longer used for the same function) that exist in organisms as a result of common ancestry.

9. (E) is correct. Bottleneck effects are often the result of a natural disaster such as a flood, drought, fire, or anything that destroys most members of a population. The gene pool of the surviving members of the population may not resemble the gene pool of that of the parent population—some genes will be overrepresented, and some will be underrepresented. Bottlenecking reduces the genetic variability in a population because of the loss of alleles.

▌ **10. (C) is correct.** The gene pool is the collection of all the alleles that exist in a population—a population is defined as a group of individuals living in a certain geographic location that are capable of interbreeding.

▌ **11. (E) is correct.** Prezygotic barriers are those that prevent or hinder the mating of two species, or they prevent fertilization even if two species mate. All the answers are examples of prezygotic barriers except the last one. Hybrid breakdown is an example of a postzygotic barrier (postzygotic barriers are those that prevent hybrid zygotes from developing into viable adults) in which the second generation of offspring from two species is either weak or sterile.

▌ **12. (B) is correct.** Species that are found in only one geographic location are said to be endemic. Some examples of endemic species are kangaroos (endemic to Australia) and blue-footed boobies (endemic to the Galápagos Islands).

▌ **13. (A) is correct.** The allele that causes sickle-cell disease also prevents severe malarial infections. Individuals who have two copies of the sickle-cell allele will come down with the disease and most likely die. Those who are heterozygous, however, have an advantage. Because they have one normal allele, they will not contract full-blown sickle-cell disease; however they will have an increased protection against malaria. Individuals with two normal alleles will not have this same protection and are more likely to die from malaria.

▌ **14. (C) is correct.** Did you select B? You were too hasty and forgot a very important concept: The 64% of the population that show the dominant trait includes heterozygotes, so you cannot simply take the square root of 0.64. You must subtract 0.64 from 1.00 to obtain $0.36 = q^2$. Taking the square root of q^2 yields $q = 0.6$, so $p = 0.4$.

▌ **15. (E) is correct.** Recall that mutations are not "created" in response to some change in the environment or need. All the other responses are factors that contribute to allopatric speciation. Understand them!

▌ **16. (C) is correct.** In convergent evolution, species from different evolutionary branches appear alike as a result of undergoing evolution in very similar ecological roles and environments. Similarity between species that have undergone convergent evolution is known as analogy, and structures they share are analogous (not homologous) structures. (Remember, similar problem, similar solution. However, convergence does *not* indicate a common ancestor.) Homologous structures are those that are shared in two species as a result of those species' having a common ancestor.

▌ **17. (D) is correct.** One mechanism that can lead to sympatric speciation, which is the formation of a small new population within the parent population in plants, is the formation of autopolyploids through nondisjunction in meiosis. These plants have $4n$ chromosomes, instead of the normal $2n$ number, and they are unable to breed with members of the parent population—though they are still able to breed with other tetraploids.

▌ **18. (A) is correct.** For this question recall the endosymbiotic hypothesis proposes that mitochondria and plastids were engulfed by other cells and have their own genomes. It is most likely that over time some of their genes were incorporated into the host cell's genome. (*Hint:* Review the endosymbiotic hypothesis.)

▌ **19. (C) is correct.** It is important to recall that the early atmosphere did not have oxygen, and that the evolution of linear electron flow in photosynthesis by cyanobacteria resulted in the accumulation of oxygen in the atmosphere.

▌ **20. (C) is correct.** The smallest unit capable of evolution is the population. Individuals cannot undergo evolution because they exist for only one generation, and evolution is the changing and refinement of a group's gene pool to best fit the group's environment.

Level 2: Application/Analysis/Synthesis Questions

▌ **1. (C) is correct.** The key to this selection is the phrase "natural selection can only edit existing variations." The other answers deal mostly with misconceptions. Evolution does not produce perfect organisms and it does not operate by a preconceived plan. Read and note the misconceptions in the wrong answers in this question.

▌ **2. (D) is correct.** The lizard and ostrich share the most recent branch point, which is feathers. Each branch point represents the common ancestor; in this case, the most recent common ancestor for reptiles with feathers.

▌ **3. (D) is correct.** Of the groups listed, hawks and other birds share the most recent common ancestor with snakes and lizards, branch point 4. Do not be confused by which group of animal is closest on the figure. Mammals are shown next to snakes and lizards, but they share branch point 3, a more distant branch point than 4.

▌ **4. (C) is correct.** Directional selection is a form of natural selection where individuals at one end of the phenotypic range (darker fur in this question) survive or reproduce more successfully than do other individuals. Use this question to review directional, stabilizing, and disruptive selection.

▌ **5. (C) is correct.** Polyploidy is a common form of sympatric speciation in plants. Many plant species originated from accidents of cell division that resulted in extra *sets* of chromosomes being inherited. Note that sympatric speciation takes place without geographic isolation. Of the choices offered only *T. turgidum* is a polyploid. You may have noticed that *T. aestivum* is also a polyploid, but was not offered as an answer.

▌ **6. (D) is correct.** The art in this question shows two models for the tempo of speciation in evolution. The top model is punctuated equilibrium, whereas the bottom model is gradualism. The butterfly marked D, in the gradualism model, shows gradual change over time, much as Darwin discussed. Both models are common questions.

▌ **7. (B) is correct.** Allopatric speciation is the formation of new species in populations that are geographically isolated. The stem of the question gives us information indicating that *H. erectus* became isolated on an island, out of genetic contact with the rest of its species.

▌ **8. (D) is correct.** Reproductive barriers impede members of two closely related species from interbreeding and producing viable, fertile offspring. Barriers are divided into prezygotic and postzygotic barriers and have been used as the basis for essay questions in the past.

Free-Response Questions

(a) Microevolution is the change in the allele frequencies of a population that occurs from one generation to the next, and three factors that contribute to microevolution are genetic drift (including both the bottleneck effect and the founder effect), natural selection, and gene flow.

The term *genetic drift* refers to any change in a population's allele frequencies due to chance. Two examples of what this "chance" can consist of are the bottleneck effect and the founder effect. The bottleneck effect occurs after a natural disaster such as a violent storm or fire causes a drastic reduction in the size of the population. Such an event leaves only a few individuals to continue to produce offspring, so bottlenecking usually reduces the genetic variability in a population because some alleles are lost from the gene pool and others are overrepresented. Similarly, the founder effect occurs when a few members of a population colonize an isolated location that isn't accessed by members of the parent population. The smaller the number of founders, the more limited the variability of the genes in the new population.

Gene flow refers to genetic exchange due to the migration of individuals or gametes between populations. In this way, gene flow can introduce new alleles into a population. This can take place in the course of one generation to the next, so this is another valid contributor to microevolution.

Natural selection is another route by which microevolution can take place. This term refers to the differing reproductive success of all the individuals in a population. Those who are best suited to their environment will survive to pass on their alleles to the next generation, causing the allele frequencies to change over time. Natural selection is the only one of these factors that is adaptive.

(b) Macroevolution is the process of the evolution of new taxonomic groups above species level through evolution. It differs from microevolution in that microevolution refers only to changes that occur in populations from generation to generation—no new species or other taxonomic groups need arise in the course of microevolution. Microevolution occurs on a small scale, whereas macroevolution concerns the "bigger picture."

This is a good free-response answer because it shows knowledge of the following key terms that you will be expected to know for the exam:

microevolution	*founder effect*
gene pool	*natural selection*
population	*gene flow*
genetic drift	*macroevolution*
bottleneck effect	*taxonomic groups*

Essay questions on evolution are common occurrences on the AP Biology exam. When answering essay questions on evolution, avoid clichés like "survival of the fittest," as the phrase does not convey any specific information about evolution. Be descriptive in your discussions of evolution and remember to define key terms as they are used.

Topic 7: The Evolutionary History of Biological Diversity

ANSWERS AND EXPLANATIONS

Level 1: Knowledge/Comprehension Questions

1. (B) is correct. Animals are heterotrophic—they are not capable of fixing carbon and must obtain it from phototropic organisms. They belong to the Eukarya and are multicellular. The fungi are also eukaryotic, multicellular heterotrophs but have a cell wall made of chitin.

2. (B) is correct. The three domains into which all the living organisms are placed by systematists are Bacteria, Archaea, and Eukarya. Domains are one taxonomic level above kingdoms. Prokaryotes make up Archaea and Bacteria, whereas eukaryotes make up Eukarya.

3. (C) is correct. In mutualistic symbiosis, both organisms benefit from the association; in commensalistic symbiosis, one organism benefits, while the other is neither helped nor harmed; and in parasitic symbiosis, one organism benefits, while the other is harmed.

4. (A) is correct. Endosymbiosis—the theory of serial endosymbiosis—proposes that mitochondria and chloroplasts evolved from small prokaryotes that were engulfed by larger host cells. Eventually they became permanent functional parts of the cell.

5. (E) is correct. Photosynthesis evolved long before terrestrial life. The other options have evolved as plants adapted to a terrestrial existence.

6. (C) is correct. Lichens are symbiotic associations of millions of photosynthetic organisms in a network of fungal hyphae. They are very hardy and frequently colonize newly broken rock faces.

7. (B) is correct. Mycorrhizae are mutualistic associations of fungi and plant roots. They exchange minerals extracted from the soil and nutrients produced by the plants.

8. (B) is correct. Liverworts are bryophytes and so lack vascular tissue, ferns have vascular tissue and reproduce with spores, and gymnosperms reproduce with cones. Angiosperms (flowering plants) were the last to appear.

9. (E) is correct. The first four answer choices represent characteristics common to almost all animals; the last one does not. The characteristic of being diurnal (the opposite of nocturnal—being active during the day) is not a requirement for belonging to the kingdom Animalia.

10. (A) is correct. The amniotic egg is composed of extraembryonic membranes that function in gas exchange, waste storage, and the delivery of nutrients to the embryo. The amniotic egg was a key evolutionary adaptation for terrestrial life, allowing the embryo to have its own "pond" in which to develop.

11. (D) is correct. You are looking for the one false statement here, so be sure you understand why each of the other answers is correct. Remember that each node or branch point represents a common ancestor of the two branches that come from the node.

- **12. (E) is correct.** You should be able to explain the three ways in which bacteria receive new genes (transduction, transformation, and conjugation) as well as know that mutation is the ultimate source of genetic variation. Mitosis and meiosis do not occur in bacteria.

- **13. (A) is correct.** This is the definition of a shared derived character. An example that may help you to remember is that hair is a shared derived characteristic for all mammals. All mammals have hair, and no organism that is not a mammal has hair.

Level 2: Application/Analysis/Synthesis Questions

- **1. (A) is correct.** This question hinges on knowing the difference between shared ancestral characters and shared derived characteristics. Shared derived characters are evolutionary novelties unique to a specific clade, whereas shared ancestral characters originate in the ancestor of the taxon. Having four limbs is not unique to birds and mammals but is an ancestral trait shared by other clades.

- **2. (C) is correct.** An outgroup is a species from an evolutionary lineage that is known to have diverged before the lineage under study. Outgroups are used as points of comparison to help determine shared ancestral and shared derived characters.

- **3. (A) is correct.** Conjugation is one of the ways bacteria increase genetic diversity. During the process of an Hfr cell transferring an integrated F plasmid, host chromosome genes may also be transmitted. In this case the host bacterium uses enzymes much like crossing over enzymes in eukaryotes to remove its host genes and incorporated donor genes.

- **4. (C) is correct.** Transformation is the uptake of external DNA by a cell. Transformation is one of the ways in which bacteria are able to increase genetic diversity. When bacteria take up DNA from a different species, transformation results in horizontal gene transfer.

Free-Response Question

(a) Three methods or types of evidence that scientists use to classify organisms and study their degree of evolutionary relatedness are fossil evidence, the structure and development of organisms, and molecular evidence.

Systematists can also study patterns of homologous structures to determine evolutionary relationships. Species that were derived from the same ancestor should have similarities, called homologies. Here, scientists would look for homologous structures—structures that are similar in different species—and use these to tie organisms together.

The third way that systematists can study evolutionary relationships is on a molecular level. Because DNA is heritable, related species should share common genes, and the more recently the species branched off from a common ancestor, the more similar their DNA should be. Studying the DNA of organisms makes it possible for systematists to determine the degree of evolutionary difference between two species that are nearly identical in appearance, and it also allows systematists to judge the relatedness of two species that they might not guess would be related at all, based on external appearance. It is a much more precise and quantitative method for appraising evolutionary relatedness.

This response uses the following key terms in context, showing the writer's knowledge of their meanings and relatedness:

systematists
homologous structures
DNA

This response clearly describes three methods for studying the evolutionary relatedness of species, thus answering the question in a complete and organized way. Always check your final response to make sure it addresses the original question.

Topic 8: Plant Form and Function

ANSWERS AND EXPLANATIONS

Level 1: Knowledge/Comprehension Questions

1. **(E) is correct.** When the light source on a plant is uneven, the plant will grow toward the light source. In some plants this is the result of auxin moving from the apex down to the cells that are less exposed to light, and causing them to elongate faster than the cells on the side that is illuminated.

2. **(C) is correct.** The driving force behind the movement of sap in xylem (in the direction from the roots to the leaves) is the transpiration of water through the stomata on the leaves. The mechanism responsible for movement up through the xylem is the transpiration-cohesion-tension mechanism, and it occurs through bulk flow, in which fluid moves because of a pressure difference at opposite ends of a tube. The pressure is created by transpiration from the leaves, and contributing to the movement of water and minerals up the plant are gradients of water potential from cell to cell within the plant.

3. **(D) is correct.** In the transpiration-cohesion-tension mechanism, water is lost through transpiration from the leaves of the plant due to the lower water potential of the air. The cohesion of water due to hydrogen bonding plus the adhesion of water to the plant cell walls by hydrogen bonding enables the water to form a water column. Water is drawn up through the xylem as water evaporates from the leaves, each evaporating water molecule pulling on the one beneath it through the attraction of hydrogen bonds.

4. **(A) is correct.** In plants, phloem is responsible for carrying sugar made in the leaves to other locations that are incapable of photosynthesis. This process is called translocation. In angiosperms phloem is made up of sieve-tube members that are arranged end to end in long sieve tubes. In phloem, sugar travels from a sugar source—any site in the plant involved in photosynthesis,

for example—to a sugar sink. A sugar sink is any tissue that is a net consumer or depository or sugar.

■ 5. **(A) is correct.** The production of ATP in plant cells occurs in the mitochondria, as it does in the cells of animals, but it also occurs in the chloroplasts during photosynthesis. The light reactions of photosynthesis convert solar energy to the chemical energy of ATP and NADPH, and these light reactions take place in the chloroplasts, in the mesophyll cells of the plant leaf.

■ 6. **(E) is correct.** All of the factors listed aid in the uptake of water and minerals by the roots of a plant except the last choice, gravity. Water and minerals flow from the soil into the epidermis of the plant, then through the root cortex and into the xylem of the plant. The xylem then transports water and minerals throughout the plant body.

■ 7. **(C) is correct.** The Casparian strip is a belt made of a waxy material that runs through all of the endodermal cells, creating a ring that protects the vascular tissue from unwanted materials. The water and mineral solution must pass through the endodermis to enter the xylem in the stele of the root. The Casparian strip forces the solution through the plasma membrane of an endodermal cell, allowing the screening of the solution.

■ 8. **(D) is correct.** All of the answers listed, with the exception of excessive rainfall, are factors that would cause the stomata of a leaf to close during the day. All of the factors with the exception of answer D could lead to dehydration.

■ 9. **(A) is correct.** This question requires you to know that the sperm of angiosperms do not swim! This is the adaptive value of pollen: It allows fertilization without water. Sperm nuclei enter a pollen tube that grows down the style, discharging the two sperm nuclei for double fertilization.

■ 10. **(C) is correct.** The ovum is another name for an egg, the pollen grain contains the sperm nucleus, and an ovary becomes the fruit. Be sure to sort out all this vocabulary!

■ 11. **(A) is correct.** A fruit forms from the ovary, and contains the seeds.

■ 12. **(D) is correct.** This is another vocabulary question—the male reproductive part of an angiosperm is the anther.

■ 13. **(B) is correct.** The phloem transports the food made in mature leaves to the roots and other nonphotosynthetic parts of the plant, such as roots. The xylem is responsible for carrying water and minerals up through the plant from the roots. Those are the two types of vascular tissue in plants.

Level 2: Application/Analysis/Synthesis Questions

■ 1. **(B) is correct.** Short-day plants require a period of continuous darkness longer than a critical period in order to flower. These plants flower in early spring or fall. Short-day plants are actually long-night plants; that is, what the plant measures is the length of the night.

■ 2. **(B) is correct.** Long-day plants require a night period that is shorter than a critical period. The critical length for flowering is 9 hours, but the question asks which choice prevents flowering. Only one choice has a 24-hour period with more than 9 consecutive hours of darkness, thus preventing flowering.

■ 3. **(A) is correct.** Understanding signal transduction pathways is an important concept for this unit. Keep its role in allowing plants (and other organisms) to adapt to their environment in your mind as you study. A hormone or environmental stimulus activates the receptor protein in a signal transduction pathway, not a relay molecule.

▌**4. (B) is correct.** Salicylic acid activates signal transduction pathways throughout the plant to start production of pathogen-related proteins. These proteins include enzymes that hydrolyze the cell walls of a variety of pathogens. Note the importance of signal transduction pathways.

▌**5. (C) is correct.** This experiment by Went centered on phototropism—the growth of a shoot in response to light. Light causes auxins to move to the dark side of the stem, where the increase in auxins results in cell elongation. Cell elongation causes the dark side of the stem to expand more rapidly, causing the plant to grow toward the light. Consider this question in the context of the environment causing changes in the behavior of an organism.

▌**6. (C) is correct.** The term photoperiod refers to seasonal changes in the relative lengths of night and day. This is a second question that examines how a behavior, plant dormancy as a preparation for winter, results from an environmental queue.

▌**7. (A) is correct.** Seed germination is a continuation of a life cycle, not a beginning. Think of this question in the context of changes in the environment resulting in changes in gene expression as the seed germinates. Again, changes in the environment result in changes in the organism. The other three options in this question are correct and worth a quick review.

Free-Response Questions

(a) Photoperiodism is defined as any physiological response to a photoperiod (meaning a specific length of daylight or darkness). One example of a photoperiodism is flowering. It has been described in plants that are termed short-day or long day plants. For example, short-day plants will not set flower buds unless they receive a critical period of darkness, and any interruption in this will keep them from flowering. These plants will therefore, in nature, flower in the earliest days of spring for example, whereas long-day plants will not flower until the longer days of summer. In each case, the time of flowering will enhance reproductive success.

(b) It was originally thought that plant flowering depended on the length of the daylight, but then scientists concluded that it is actually night length that determines when a plant will flower. Plants monitor the length of the night by the molecular switching of two forms of the phytochrome pigment. Phytochromes respond to a shift in red (r) light to far-red (fr) light. This is a photoreversible response, and the threshold to trigger flowering is called the critical dark period. The accumulation of specific phytochrome isomers combined with the biological clock of plants allows for the specific monitoring of the length of the night. Long-day plants and short-day plants actually monitor the length of the night, not the day. Short-day plants must have a period of darkness longer than a specific critical period to flower, whereas long-day plants must have a period of darkness shorter than a critical period. Since in many plants, flowering is related to day length, this serves as an adaptation to ensure flowering resources are not committed until growing conditions are most favorable for survival of flowers, fruits, and seeds.

This response uses the following key terms in context, showing the writer's knowledge of their meanings and relatedness:

flower

phytochromes

photoperiodism

short-day and long-day plants

It also shows an understanding of the following processes: plant growth via stem elongation, gravitropism, and photoperiodism.

Topic 9: Animal Form and Function

ANSWERS AND EXPLANATIONS

Level 1: Knowledge/Comprehension Questions

1. (C) is correct. The only condition listed that is necessary in all organisms that breathe is the presence of moist membranes. The movement of O_2 and CO_2 across the membranes between the environment and the respiratory surface occurs by diffusion. Respiratory surfaces are generally thin and, since living animal cells must be wet in order to maintain their plasma membranes, these respiratory surfaces must be moist.

2. (C) is correct. The three main stages of development in animals are cleavage, in which a multicellular embryo forms from the zygote through a series of mitotic cell divisions; gastrulation, in which cells migrate and rearrange to form three germ layers; and organogenesis, in which rudimentary organs are formed from the germ layers. It is important for you to know these key terms.

3. (D) is correct. Epinephrine is a hormone that is secreted by the adrenal glands—specifically the adrenal medulla. It functions in raising the blood glucose level, increasing metabolic activities, and constricting blood vessels; all of this prepares the animal for the fight-or-flight response that is elicited in the body during stressful times.

4. (A) is correct. Erythrocytes are red blood cells, and they transport oxygen around the body. They are the most numerous blood cells, and are small and disk-shaped. In mammals, erythrocytes have no nuclei. Instead, they contain millions of molecules of hemoglobin, which is the iron-containing protein that transports oxygen. One molecule of hemoglobin can bind four oxygen molecules.

5. (D) is correct. The proximal tubule is the site of secretion and reabsorption that substantially changes the content and volume of the filtrate. It secretes hydrogen ions and ammonia to regulate the pH of the filtrate and also reabsorbs about 90% of the bicarbonate, which is an important buffer.

6. (A) is correct. Salivary amylase is contained in saliva; it is an enzyme that hydrolyzes starch, a glucose polymer found in plants, and glycogen, a glucose polymer found in animals. After hydrolysis, smaller polysaccharides and maltose remain.

7. (E) is correct. The anterior pituitary produces FSH, LH, TSH, ACTH, along with several other hormones. It is regulated by releasing hormones from the hypothalamus.

8. (C) is correct. The posterior pituitary gland releases two main hormones: oxytocin, which stimulates the contraction of the uterus and mammary gland cells, and antidiuretic hormone (ADH), which promotes the retention of water by the kidney. The actions of the posterior pituitary are regulated by the nervous system and the water/salt balance in the body.

9. (A) is correct. The ovaries secrete hormones called estrogens, which stimulate the growth of the uterine lining and promote the development of secondary sex characteristics in females. They are regulated by two other hormones, FSH and LH, released from the anterior pituitary.

10. (B) is correct. The thyroid gland releases the hormone thyroxine, which stimulates and maintains metabolic processes, and calcitonin, which lowers the blood calcium levels. The thyroid gland secretions are regulated by TSH and by the level of calcium in the blood.

11. (D) is correct. The traditional example of a negative feedback system is the thermostat example. In the body, one very prominent example of negative feedback is the regulation of our body temperature at about 37°C. A section of the brain is responsible for keeping track of the temperature of the blood, and if the blood is too warm, for example, it tells the sweat glands to increase production. Remember, more gets you less!

12. (E) is correct. Both HCl and pepsin can convert pepsinogen to its active form, pepsin. As more pepsin is converted, it can convert even more pepsinogen. This is an example of positive feedback. (More gets you more!) Other proteolytic enzymes such as trypsin and chymotrypsin are also secreted as zymogens and require cleavage of a portion of the molecule in order to be activated.

13. (B) is correct. Pepsin is an enzyme in the gastric juice of the stomach. It begins the hydrolysis of proteins by breaking peptide bonds between amino acids, thereby cleaving proteins into smaller polypeptides. The digestion of proteins continues in the small intestine.

14. (D) is correct. The large intestine, also known as the colon, is responsible for recovering water from the alimentary canal. It is also responsible for compacting the wastes into feces, which are stored in the rectum and then eliminated.

15. (D) is correct. The role of the sinoatrial (SA) node, or pacemaker, is to control the rate and timing of the contraction of heart muscles. It generates electric impulses that spread rapidly through the walls of the atria, making them contract in unison.

16. (C) is correct. The lymphatic system collects fluid and some proteins lost during regular circulation and returns them to the blood. This system is composed of a network of lymph vessels throughout the body, with lymph nodes, which are the sites at which lymph is filtered and viruses and bacteria are encountered by the immune system.

▌ **17. (D) is correct.** Carbon dioxide is most commonly transported in the blood in the form of bicarbonate—it reacts with water in the presence of the enzyme carbonic anhydrase to form carbonic acid, and a hydrogen dissociates from carbonic acid to produce the bicarbonate ion (HCO_3^-). Less commonly, carbon dioxide is transported by hemoglobin, or transported in solution in the blood.

▌ **18. (E) is correct.** All of the answers listed—except phagocytes—are examples of barrier defenses. Phagocytosis is one of the body's cellular innate defenses against infectious agents; it is the process by which invading organisms are ingested and destroyed by white blood cells.

▌ **19. (B) is correct.** When a lymphocyte is activated by an antigen, it is stimulated to divide and differentiate, and it forms two clones. One clone is of effector cells that combat the antigen. One clone is of memory cells that stay in circulation, recognize the antigen if it infects the body in the future, and launch an attack against it.

▌ **20. (D) is correct.** The nephron, which is the functional unit of the kidney, is composed of a long tubule and the glomerulus, which is a ball of capillaries. The Bowman's capsule surrounds the glomerulus. Filtration occurs as blood pressure forces fluid from the blood in the glomerulus into the lumen of the Bowman's capsule.

▌ **21. (D) is correct.** The sliding-filament model of muscle contraction states that the thin and thick filaments do not shrink during muscle contraction. Instead, the filaments slide past each other so that the degree of their overlap increases; this sliding is based on the interactions of actin and myosin molecules that make up the filaments.

▌ **22. (B) is correct.** Three successive stages of development follow fertilization. The first is cleavage, which is rapid cell division that produces a mass of new cells that share the cytoplasm of the original cell. The new cells all have their own nuclei and are called blastomeres. The second stage is gastrulation, and the third is organogenesis.

▌ **23. (B) is correct.** The reflex arc is the simplest type of nerve circuit (automatic response). A sensory neuron receives information from a receptor; it is passed to interneurons in the spinal cord and then to a motor neuron, which signals an effector cell (such as a muscle fiber) to respond to the stimulus.

▌ **24. (C) is correct.** Neurotransmitters are excreted by the synaptic vesicles and act as intercellular messengers, transmitting the nerve impulse from one neuron to the next neuron or another cell. A single postsynaptic neuron can receive signals from many neurons that secrete different neurotransmitters.

▌ **25. (C) is correct.** Homeostasis is the ability of many animals to regulate their internal environment. They do this through thermoregulation, which is the maintenance of internal temperature in a certain range, and osmoregulation, which is the regulation of solute balance within certain parameters.

▌ **26. (B) is correct.** Fertilization is the fusion of the egg cell (ovum) and the sperm cell, and it results in the formation of a zygote. The zygote is diploid, whereas the egg cell and the sperm cell, both the products of meiosis, are haploid.

27. (D) is correct. Fixed action patterns are a sequence of behavioral acts that are virtually unchangeable and usually carried to completion once they are initiated. Sign stimuli trigger fixed action patterns. These stimuli may be a feature of another animal, such as an aspect of its appearance, or some other event.

28. (C) is correct. Habituation is one type of learning. Learning is defined as the ability of an animal to modify its behavior as a result of specific experiences. Habituation is a very simple form of learning, in which there is a loss of responsiveness to stimuli that convey very little or no information.

29. (A) is correct. Classical conditioning is a form of associative learning (the ability of animals to learn to associate one stimulus with another). It specifically refers to an animal's ability to associate an arbitrary stimulus with a reward or a punishment.

30. (B) is correct. Imprinting is a form of learning that occurs during a sensitive period. Imprinting is generally irreversible, and the sensitive period is a limited phase in the animal's development when the learning of a particular behavior can take place.

31. (E) is correct. Altruism is thought to occur in populations because if parents sacrifice their own well-being for that of their offspring, this increases their fitness by better ensuring that the genes that they passed on will make it to the next generation. Likewise, helping other close relatives increases the chances that they will survive to pass on genes that are shared between them and the altruistic member.

32. (A) is correct. You should review the basic structure of an antibody, so that you can recognize the antigen-binding sites, heavy and light chains, and variable and constant regions.

33. (C) is correct. Reabsorption occurs when materials that are in the filtrate are taken back into the blood. The filtrate is found in the nephron tubules.

34. (D) is correct. You will want to review the role of each of these important cell types. Helper T cells are targeted by HIV infections, which is why AIDS patients have compromised immune systems.

35. (C) is correct. Gastrulation is an important event in embryonic development, as it establishes the three germ or tissue layers. Answer D describes the process of neurulation, which follows gastrulation in chordates.

36. (E) is correct. You should review the primary structures seen in this figure, and know the function of each region. The cerebellum coordinates motor activities.

37. (D) is correct. This is the medulla, which controls many involuntary activities.

38. (C) is correct. The autonomic nervous system is also referred to as the involuntary nervous system and has two components, the sympathetic branch and the parasympathetic branch. Nerves of the sympathetic nervous system activate organ systems to deal with stress, while the parasympathetic nervous system maintains the steady, relaxed situation.

39. (D) is correct. You need to know that depolarization can occur at a minimal level, and no impulse is transmitted. When the threshold level of depolarization occurs, the nerve impulse or action potential is transmitted. This is sometimes referred to as an "all-or-none" response.

▌ **40. (D) is correct.** You will want to review events at the synapse that trigger the release of neurotransmitter from the presynaptic cell. Vesicles filled with neurotransmitter migrate to the plasma membrane, fuse with it, and neurotransmitter is released by exocytosis into the synapse, where it will bind receptors on the next neuron or cell in the pathway, and generate a response.

Level 2: Application/Analysis/Synthesis Questions

▌ **1. (B) is correct.** Maintenance of glucose homeostasis by insulin and glucagon is the classic system for explaining negative feedback. The two hormones are antagonistic with insulin triggering the uptake of glucose by the cells, thus lowering blood glucose levels, and glucagon promoting the release of glucose into the blood from liver glycogen, raising blood glucose levels. Sensors monitor glucose levels while the two antagonistic hormones are used to maintain steady glucose levels in the blood. It is important to be able to explain both negative and positive feedback systems.

▌ **2. (B) is correct.** Glucagon raises blood glucose levels, as noted in question 1.

▌ **3. (C) is correct.** How the environment affects energy budgets and behavior and how free energy flows through organisms are the key points. In this case, if the organism is increasing in mass the inflow of energy must be greater than the outflow.

▌ **4. (B) is correct.** B cells are part of the humoral immune response which occurs in the blood and lymph. In the humoral response, antibodies help neutralize or eliminate toxins and pathogens in the blood and lymph. Cytotoxic T cells are part of the cell-mediated immune response, which destroys infected host cells.

▌ **5. (D) is correct.** Lymphocytes (B cells and T cells) have receptors for a single antigen. Antigens are *anti*body *gen*erating and generally stimulate more than one lymphocyte, but each lymphocyte is specific for the one antigen it is genetically equipped to recognize. Also note the true statements in this question to help review basic information about antibodies, antigens, and epitopes.

▌ **6. (C) is correct.** This figure depicts clonal selection, a central idea in immunology. The active plasma cell is the plasma cell producing antibodies. Use this figure to review clonal selection. This would be a good time to label antigen, plasma cells, memory cells, and antibody on the figure.

▌ **7. (A) is correct.** Note the specific nature of the interaction between antigen and antibody and think about how this interaction can affect the entire organism. The role of the immune system in homeostasis, as an example of molecular interactions, and how the environment can affect organisms is the context in which questions about the immune system will arise.

▌ **8. (C) is correct.** This question is at the center of understanding active immunity, vaccinations, and how B cells protect the body by producing antibodies and memory cells.

▌ **9. (B) is correct.** The secondary immune response is a heightened response to the same antigen that generated a primary immune response. Figure 43.15 in *Campbell Biology,* Ninth Edition, by Reece et al. has been on numerous AP exams, where it clearly shows primary and secondary immune responses.

10. (B) is correct. The receptor protein is marked as B with the water-soluble hormone clearly outside the cell membrane. It is not necessary to memorize the human hormones. The important thing is to understand the role hormones play in cell-to-cell communication, even over long distances.

11. (D) is correct. This diagram uses a steroid hormone that is lipid-soluble. The hormone functions in cell-to-cell communication by causing the production of transcription factors that in turn activate genes.

12. (D) is correct. An axon is a neural extension that conducts signals away from the cell body to another neuron or effector cell. Use the diagram to review parts of the neuron.

13. (C) is correct. The insecticide blocks the enzyme acetylcholinesterase, which normally removes acetylcholine from the synapse after the signal is received. Without this removal the neurotransmitter acetylcholine remains in the synapse, causing constant muscle contraction. Knowing specific neurotransmitters is not required, but understanding how neurotransmitters work is required.

14. (B) is correct. Since the neurotransmitter acetylcholine stimulates muscle contraction in both roaches and people, it is logical to conclude that the mechanism of stimulating skeletal muscle contraction must be similar in humans and roaches.

15. (C) is correct. If depolarization shifts the membrane potential enough, the result is a change in membrane potential called an action potential. Action potentials occur fully or not at all, they represent an all-or-none response to stimuli.

16. (C) is correct. Neurotransmitters are released from the presynaptic cell, but neurotransmitter receptors are located on the membrane of the postsynaptic cell.

17. (A) is correct. The action potential depolarizes the presynaptic membrane, which opens voltage-gated calcium channels. The key to this question is the emphasis on the work *directly*.

18. (B) is correct. Immediately behind depolarization caused by Na^+ inflow is a zone of repolarization caused by K^+ outflow. In the repolarized zone, the sodium channels remain inactivated. This prevents an action potential from moving backward. The interval between the onset on an action potential and end of the refractory period is only 1–2 milliseconds.

19. (D) is correct. The graph shows a clear peak at the 7-mms mark along the x-axis. This question is based on the idea of optimal foraging, the analyzing of feeding behavior as a compromise between feeding cost and feeding benefits. Understanding how organisms use energy, how energy use affects evolution, and how organisms interact with their environment are all themes to guide your review.

20. (C) is correct. Learned behavior occurs when behavior is modified based on specific experiences. If the salmon returned to Cold Bay, then they learned the new magnetic bearings during the experiment. Innate behaviors are the same in virtually all members of a population. The behavior is developmentally fixed and under strong genetic control.

21. (A) is correct. Imprinting occurs at a specific stage in life and results in a long-lasting behavioral response. Imprinting has a sensitive or critical period when the learning must occur. The pictures of young geese following Konrad

Lorenz or whooping cranes following an ultralight plane are both examples of imprinting in action as the young animals imprint during their sensitive period.

22. **(A) is correct.** Behavior is best understood as a consequence of natural selection. An organism's reproductive success depends in part on how the behavior is performed. Part of the behavior has a genetic basis, but the entire behavior does not have to be determined by genes.

Free-Response Questions

(a) Skeletal muscle is fibrous, and each fiber is a single long cell. The fibers are composed of myofibrils, which are, in turn, composed of two kinds of myofilaments. These are thin filaments—which are made up of two actin strands and one regulatory protein strand, coiled—and thick filaments made up of myosin molecules. The sarcomere is the basic contracting unit of the muscle. During muscle contraction, the length of the sarcomere decreases. The sliding-filament model of muscle contraction states that the thick and thin filaments slide past each other horizontally, due to the interactions of actin and myosin.

The myosin molecules look like golf clubs arranged horizontally in a group, with the heads of the golf clubs pointing up. This head region is the center of the reactions that take place during muscle contraction. Myosin binds ATP and hydrolyzes it into ADP, and its structure is changed in the process, which causes it to bind to a specific site on actin and form a cross-bridge. Myosin then releases the stored energy and relaxes to its normal conformation. This changes the angle of attachment of the myosin head relative to its tail. As myosin bends inward upon itself, tension increases on the actin filament, and the filament is pulled toward the middle of the sarcomere.

(b) When oxygen is scarce, as in situations in which a person is taking part in strenuous exercise, human muscle cells switch to lactic acid fermentation (which normally undergoes regular aerobic cellular respiration) to produce ATP. In lactic acid fermentation, pyruvate is reduced by NADH to form lactate with no release of CO_2.

This response uses the following key terms in context, showing the writer's knowledge of their meanings and relatedness:

skeletal muscle	*cross-bridge*
myofibrils	*action potential*
myofilaments	*thin filaments*
actin strands	*aerobic cellular respiration*
thick filaments	*lactic acid fermentation*
myosin	*pyruvate*
sarcomere	*lactate*
sliding-filament model	

It also shows knowledge of how these important biological processes take place: how muscle cells contract and ultimately cause bones to move, how tetanus is reached, and how and why strenuous exercise produces muscle pain.

Topic 10: Ecology

ANSWERS AND EXPLANATIONS

Multiple-Choice Questions

1. **(C) is correct.** The ozone layer is located in the stratosphere and surrounds Earth. It is composed of O_3, and it absorbs UV radiation, reducing the level of UV radiation reaching the organisms in the biosphere. Researchers have been observing the thinning of the ozone layer since about 1975. The destruction of the ozone layer has been attributed to the widespread use of chlorofluorocarbons.

2. **(C) is correct.** In a savanna, grass constitutes the primary producer. A primary producer traps the energy of sunlight and turns it into chemical energy through photosynthesis. Primary consumers (herbivores) consume primary producers, secondary consumers (carnivores) eat herbivores, and tertiary consumers eat carnivores.

3. **(B) is correct.** The carrying capacity of a population is defined as the maximum population size a particular environment can support at a particular time with no degradation of the habitat. It is fixed at certain times, but it varies over the course of time with the amount of resources that exist in an environment.

4. **(A) is correct.** Biomes are major types of ecosystems that occupy broad geographic regions. Some examples of biomes are coniferous forests, deserts, grasslands, and tropical forests.

5. **(D) is correct.** Lakes are classified according to how much organic matter they produce. Oligotrophic lakes are deep and generally poor in nutrients, and therefore, they have relatively little phytoplankton. Eutrophic lakes are usually shallower and have greater nutrient content, which allows the growth of more phytoplankton.

6. **(C) is correct.** An estuary is an area where a running freshwater source, such as a stream or river, meets the ocean. Often estuaries are bordered by large areas of coastal wetlands, and salinity varies with location within them, as well as with the rise and fall of the ocean tides. Estuaries are among the most biologically productive biomes, and they also are home to many of the fish and other animals that humans consume.

7. **(C) is correct.** Many aquatic biomes display vertical stratification, which forms two ecologically unique areas, the photic zone in which there is enough light for photosynthesis to occur and an aphotic zone, where very little light penetrates.

8. **(D) is correct.** Tundra is characterized by having permafrost (which is a permanently frozen layer of soil), very cold temperatures, and high winds. These factors prevent tall plants from growing in the tundra. Tundra generally does not receive much rainfall throughout the year, and what rain does fall cannot soak into the soil because of the permafrost.

9. **(B) is correct.** Tropical forests generally have thick canopies that prevent much sun from filtering through. This means that in breaks in the canopy,

other plants grow quickly to compete for sunlight. Tropical forests are home to epiphytes, and rainfall is frequent.

■ **10. (C) is correct.** Temperate broadleaf forests are characterized by dense populations of deciduous trees, which drop their leaves in the fall when the weather turns cold.

■ **11. (A) is correct.** Temperate grasslands are characterized by having thick grass, seasonal drought, occasional fires, and large grazing animals. Their soil is generally rich with nutrients, making them good areas for agriculture. Most of the temperate grassland in the United States is used today for agriculture.

■ **12. (E) is correct.** Deserts experience very little rainfall, so they are home to many plants and animals that have adaptations for storing and saving water. Deserts are marked by drastic temperature fluctuation; they can be very hot in the day but freezing at night. Many desert plants rely on CAM photosynthesis.

■ **13. (D) is correct.** A bacterial colony growing where it has limitless nutrients, and other ideal conditions, will experience what is called exponential growth. In exponential growth, all members are free to reproduce at their physiological capacity.

■ **14. (C) is correct.** A species' ecological niche is defined as the sum of its use of the abiotic and biotic factors in an environment. For instance, a particular bird's niche refers to many things, including the food it consumes, where it builds its nest, the time of day it is active, and what climate it lives in.

■ **15. (D) is correct.** In Batesian mimicry, a harmless or palatable animal evolves the same markings and/or colorings as a harmful or unpalatable animal and, in this way, may reduce predation.

■ **16. (C) is correct.** The dominant species in a community has the greatest biomass, or sum weight of all of the members of a population. Dominant species are also hypothesized to be the most competitive in exploiting the resources in an ecosystem.

■ **17. (C) is correct.** Almost all organisms use solar energy stored in organic molecules produced by green plants to power life processes. Autotrophs convert solar energy to a form useful to both autotrophs and heterotrophs. At each successive trophic level, less energy is available because so much is converted to a form not useful to the organisms. Energy cannot be recycled.

■ **18. (B) is correct.** Secondary succession refers to a situation in which a community has been cleared by a disturbance of some kind, but the soil is left intact. The area will begin to return to its original state through the process of plants invading the area and recolonizing.

■ **19. (C) is correct.** Most of Earth's nitrogen is in the form of N_2, which is unusable by plants. The major pathway for nitrogen to enter an ecosystem is nitrogen fixation, the conversion of N_2 by bacteria to forms that can be used by plants.

■ **20. (E) is correct.** The greenhouse effect is the process by which carbon dioxide and water vapor in the atmosphere intercept reflected infrared radiation from the sun and re-reflect it to Earth. Global warming is the process by which the amount of carbon dioxide in the atmosphere is increasing because of humans' combustion of fossil fuels, leading to higher temperatures on Earth.

■ **21. (B) is correct.** Type I survivorship curve is the pattern described in the question. Type II curves show an equal chance of death throughout the life span, while Type III shows heavy mortality in early stages of the life cycle.

▌ **22. (D) is correct.** An early fall frost is a density-independent factor—it occurs without regard to the density of the population. Density-dependent factors increasingly slow population growth as density increases.

▌ **23. (E) is correct.** All of the answers are contributing to increased human population growth, as much of the world is experiencing early demographic transition. Death rates have fallen, but birth rates are much slower to change.

▌ **24. (D) is correct.** Figure 10.5 shows the effect of the removal of a keystone species from a tide pool. In this case the removal of *Pisaster* allowed for the unrestrained growth of mussels, which eventually took over the rock faces of the tide pool and eliminated most other invertebrates.

▌ **25. (C) is correct.** A +/− notation indicates one species benefits (+), whereas the other species is harmed (−). Answer C does not fit this notation because both the honeybee and the flower benefit, an example of mutualism (+/+).

Level 2: Application/Analysis/Synthesis Questions

▌ **1. (C) is correct.** The answer gives the sequence of events leading to the environmental degradation of the ecosystem. Had the stem of this question been the basis for an essay question, you might have been asked to explain how the alteration of an abiotic factor impacted the ecosystem. The answer C would have been the backbone of your response.

▌ **2. (C) is correct.** Nitrogen and phosphorus are often limiting factors in both aquatic and terrestrial ecosystems as they are often in short supply for producers. This is why our lawn fertilizers are rich in nitrogen and phosphorus. When producers are limited, this limits energy flow throughout the ecosystem. Again note how the impact of an abiotic factor influenced the ecosystem.

▌ **3. (C) is correct.** The carrying capacity is the number of individuals that the environment can sustain. This is shown by a decrease of growth rate, shown by leveling off on a graph. When the graph line is horizontal, the rate of growth is zero. Population growth reaches that point in this graph very close to 1940.

▌ **4. (A) is correct.** When $K = N$, then the population estimate and the carrying capacity are equal. This causes the right side of the equation to equal zero as indicated by the flat line in 1945.

▌ **5. (B) is correct.** This is an example of a biotic factor impacting an ecosystem. Exotic species often reduce species diversity by increasing competition for resources and sometimes replacing native species that cannot compete. This is a good time for you to review the components of species diversity.

▌ **6. (B) is correct.** Option B is the best of the answers suggested. In practical terms, finding a predator that only eats snakeheads is problematic at best.

▌ **7. (C) is correct.** As with any population, if the number of individuals entering the reproductive years is much greater than the number of postreproductive individuals the population will expand. The large increase in the number of individuals entering the child-bearing years in Nigeria foretells rapid population growth.

▌ **8. (B) is correct.** The expanding population will place increased demands on resources. Although this question and question 7 deal with human populations, the underlying concepts of population ecology apply. The demands of a population that is above carrying capacity can result in the degradation of the ecosystem.

(a) The hawk, mouse, and plant in this particular ecosystem are related by the passage of energy through them. Together they comprise a food chain—the mouse is a primary consumer, and it consumes the plant, which is a primary producer (the plant is an autotroph—capable of trapping the energy of the sun and converting it into chemical energy in the form of carbohydrates). The hawk then is a predator of the mouse—and a secondary consumer. Secondary consumers eat herbivores. All of these animals are dependent on the soil in which the plant has grown because the soil provides the plant with nutrients. The plant needs a variety of organic elements to produce carbohydrates, but it also needs mineral nutrients, such as nitrogen, to make proteins and nucleic acids.

(b) One example in which the biotic factors of the biosphere impact the abiotic factors is seen in the case of global warming. We rely on the greenhouse effect (in which atmospheric carbon dioxide acts as an insulator, trapping infrared radiation from the sun and re-reflecting it) to help maintain the hospitable temperature of Earth. Yet, due to the burning of fossil fuels—beginning during the Industrial Revolution—the concentration of carbon dioxide in the atmosphere has increased significantly, and this has led to an increase in global temperatures.

The thinning of the ozone layer is another way in which humans (a biotic factor of the biosphere) impact abiotic processes. Organisms are protected from ultraviolet radiation from the sun by a protective layer of ozone that surrounds Earth. Decreased ozone levels are expected to increase human skin cancer rates as well as increase DNA damage in a wide variety of organisms. The ozone layer has been degraded by humans' use of chlorofluorocarbons (CFCs), which are chemicals used in refrigeration and other industrial processes. Many countries have stopped using these chemicals, but chlorine molecules already in the atmosphere continue to have an effect on ozone.

This response uses the following key terms in context, showing the writer's knowledge of their meanings and relatedness:

ecosystem	*nitrogen fixation*
primary producer	*herbivore*
primary consumer	*global warming*
secondary consumer	*greenhouse effect*
food chain	*ozone layer*
autotroph	*chlorofluorocarbons*
predator	

This response also shows knowledge of the following important biological processes: food chains and the interaction of organisms in a community, global warming, and the depletion of the ozone layer.

BIG IDEA 1: Investigation 2, Mathematical Modeling: Hardy-Weinberg

ANSWERS AND EXPLANATIONS

Multiple-Choice Questions

▌ **1. (C) is correct.** The question tells you that $p = 0.9$ and $q = 0.1$. From this, you can calculate the heterozygotes: $2pq = 2 \ (0.9) \ (0.1) - 0.18$. If you selected E as your response, you may have confused the allele frequency with genotypic frequency. This problem gives you the allele frequency of a, which is 10%.

▌ **2. (B) is correct.** The conditions described all contribute to genetic equilibrium, where it would be expected for initial gene frequencies to remain constant generation after generation. If you chose E, remember that genetic equilibrium does not mean that the frequency of A = the frequency of a.

▌ **3. (D) is correct.**

$$q^2 = 0.09, \text{ so } q = 0.3$$

$$p = 1 = q, \text{ so } p = 1 = 0.3 = 0.7$$

$$AA = q^2 = 0.49$$

BIG IDEA 2: Investigation 4, Diffusion and Osmosis

ANSWERS AND EXPLANATIONS

Multiple-Choice Questions

▌ **1. (B) is correct.** The water potential of distilled water in a container open to the environment is 0. The water potential of the beet core is –0.2. (Since water potential = solute potential (–0.4) + pressure potential (0.2), water potential of the beet core = –0.2.)

▌ **2. (A) is correct.** The water potential for both the distilled water and the beet core in beaker a is 0.

BIG IDEA 2: Investigation 5, Photosynthesis

ANSWERS AND EXPLANATIONS

Multiple-Choice Questions

▌ **1. (C) is correct.** DPIP is reduced in this experiment as it receives high-energy electrons from chlorophyll.

▌ **2. (B) is correct.** When DPIP becomes colorless very rapidly either there are too many chloroplasts or the concentration of DPIP is so dilute that the quantity can be reduced rapidly.

▌ **3. (C) is correct.** A flat curve indicates that no DPIP is being reduced. This must mean the chloroplasts are not functioning.

▌ **4. (B) is correct.** You should recall the equation for photosynthesis. Without a source of CO_2, photosynthesis cannot proceed.

BIG IDEA 2: Investigation 6, Cellular Respiration

ANSWERS AND EXPLANATIONS

Multiple-Choice Questions

▌ **1. (B) is correct.** To calculate this, it is easiest to find the change in y at 10 minutes (0.4 ml – 0 ml = 0.4 ml) and divide by the change in x (10 minutes – 0 minutes = 10 minutes). 0.4 ml/10 minutes = 0.04 ml/min.

▌ **2. (A) is correct.** Study Figure 5.2 carefully to see that at 10 minutes the 22°C germinating corn consumed 0.8 ml of oxygen, while the 12°C germinating corn consumed 0.04 ml of oxygen.

▌ **3. (D) is correct.** This is the only statement that is supported by information provided on the graph in Figure 5.2.

▌ **4. (D) is correct.** As carbon dioxide is released, it is removed from the air in the vial by this precipitation. Since oxygen is being consumed during cellular respiration, the total gas volume in the vial decreases. This causes pressure to decrease inside the vial, and water begins to enter the pipette.

BIG IDEA 3: Investigation 7, Cell Division: Mitosis and Meiosis

ANSWERS AND EXPLANATIONS

Multiple-Choice Questions

▌ **1. (D) is correct.** There are two nuclear divisions in meiosis, and only one in mitosis. Crossing over occurs only in meiosis; it does not occur at all in mitosis. Replication occurs only once in preparation for both mitosis and meiosis. The daughter cells of mitosis are identical to the parent cell, but in meiosis the daughter cells have only one of each homologous chromosome pair. Synapsis occurs only in prophase I of meiosis.

- **2. (D) is correct.** Remember that if crossing over does not occur, the arrangement of spores will be 4 light and 4 dark. All other combinations are the result of crossing over.
- **3. (B) is correct.** Map distance = number of crossovers divided by total number of asci × 100 divided by 2.

BIG IDEA 3: Investigation 8, Biotechnology: Bacterial Transformation

ANSWERS AND EXPLANATIONS

Multiple-Choice Questions

- **1. (B) is correct.** Since not every cell takes up the plasmid with resistance to kanamycin, the result will be scattered colonies of bacteria.
- **2. (D) is correct.** Plate IV has kanamycin but the bacteria cells were not exposed to the kanamycin plasmid, and so there was no transformation.
- **3. (D) is correct.** Plate II has colonies of transformed cells, so these should be used to verify transformation. If these cells are spread on agar with kanamycin, all cells will be able to grow and a lawn will result.
- **4. (C) is correct.** Since not every cell takes up the plasmid with resistance to kanamycin, the result will be scattered colonies of bacteria when the cells are spread on LB agar with kanamycin.

BIG IDEA 3: Investigation 9, Biotechnology: Restriction Enzyme Analysis of DNA

ANSWERS AND EXPLANATIONS

Multiple-Choice Questions

- **1. (D) is correct.** There are approximately 3,500 base pairs in the circled fragment. To determine this, measure the migration distance of the circled fragment. It is 22 mm. Locate 22 mm on the graph, and take a straight-edge up to the point of intersection with the curve. Read the number of bp from the *y*-axis (left side of graph).
- **2. (B) is correct.** Small fragments migrate farther than large fragments. All DNA is negatively charged and so moves toward the positive electrode.
- **3. (C) is correct.** Since *Bam*HI has three restriction sites on the plasmid, there will be three fragments.
- **4. (D) is correct.** There are four restriction sites for both enzymes, so there will be four fragments.

BIG IDEA 4: Investigation 11, Transpiration

ANSWERS AND EXPLANATIONS

Multiple-Choice and Other Questions

▌ **1. (D) is correct.** When K⁺ leave the guard cells, water follows, the guard cells become flaccid, and the stomata close. As a result, transpiration doesn't take place.

▌ **2. (D) is correct.** Water moves from a region of high water potential to a region of low water potential. The region of lowest water potential would be at the farthest end of the transpiration pathway.

▌ **3.** The correct tissue type: epidermis. The correct function: protection.

▌ **4.** The correct tissue type: phloem. The correct function: transport solutes.

▌ **5.** The correct tissue type: xylem. The correct function: transport water.

▌ **6.** The correct tissue type: parenchyma. The correct function: store food.

BIG IDEA 4: Investigation 12, Fruit Fly Behavior

ANSWERS AND EXPLANATIONS

Multiple-Choice Questions

▌ **1. (C) is correct.** The amount of fertilizer is the variable factor, and the use of two similar plants in the same location receiving equal amounts of water implies the concept of a control. However, with such a small sample size (two plants) and no repetition, this is a flawed experiment.

▌ **2. (E) is correct.** However, this experiment has no control, so any of the other choices are reasonable conclusions from the data. In order to reach a reliable single conclusion, you must design a new controlled experiment.

▌ **3. (B) is correct.** Much of the early movement is simply exploration, so data analysis should not begin until the pill bugs have acclimated. Choice C may seem reasonable, but it is not as good as B because it looks only at a single point in time.

▌ **4. (A) is correct.** It is the only statement that has the IF, THEN format, and gives both the conditions and a measurable predicted result.

BIG IDEA 4: Investigation 13, Enzyme Activity

ANSWERS AND EXPLANATIONS

Multiple-Choice Questions

▮ **1. (C) is correct.** The rate of the reaction decreases when the number of substrate molecules has been reduced by the enzyme. There are fewer and fewer substrate molecules available. Above the maximum substrate concentration, the rate will not be increased by adding more substrate; the enzyme is already working as fast as it can. An enzyme can catalyze a certain number of reactions per second, and if there is not sufficient substrate present for it to work at its maximum velocity, the rate will decrease. Therefore, to keep the enzyme working at its maximum, you must add more substrate.

▮ **2. (E) is correct.** H_2SO_4 lowers the pH so that the globular shape of the protein is altered. The active site is distorted to the point that the enzyme no longer functions.

Statistical Analysis: Chi-Square Analysis of Data

ANSWERS AND EXPLANATIONS

Multiple-Choice Questions

▮ **1. (D) is correct.** There is no evidence for any type of linkage, since both males and females show the traits in approximately equal proportions, and eye color and wings appear to sort independently. If the parents were homozygous for these traits, the offspring would not show different phenotypes from both parents.

▮ **2. (C) is correct.** This is obtained by using the formula x^2 the sum of $(o - e)^2/e$.

	o	e	$o-e$	$(o-e)^2/e$
Red eyes/normal wings	98	90	8	0.7111
Red eyes/no wings	22	30	8	2.1333
Sepia eyes/normal wings	26	30	4	0.5333
Sepia eyes/no wings	14	10	4	1.6
				$x^2 = 4.977$

▮ **3. (A) is correct.** The expected results are based on obtaining a 9:3:3:1 ratio from two heterozygous parents. There are three degrees of freedom. Since the Chi-square value is below 7.82, the results support the null hypothesis.

Sample Test

Part A

▮ **1. (D) is correct.** The carrying capacity of a population is defined as the maximum population size that a certain environment can support without being degraded. You should note that the population increases in number until it reaches about 900 members, and at that point it stabilizes. This is carrying capacity.

▮ **2. (A) is correct.** When $K = N$, then the population estimate and the carrying capacity are equal. This causes the right side of the equation to equal zero as indicated by the flat line at carrying capacity. The equation calculates change over time. When the population is at carrying capacity the change in population is zero.

▮ **3. (A) is correct.** This is a simple probability question. In order to calculate the chance that two or more independent events will occur together in a specific combination, you can use the multiplication rule. Take the probability that gene P will segregate into a gamete (1/4), and multiply it by the probability that gene Q will segregate into a gamete (1/4). Then multiply that by the probability that gene R will segregate into a gamete to get $1/4 \times 1/4 \times 1/4 = 1/64$.

▮ **4. (D) is correct.** The question asks you to look at gradual changes over time, and this is the only one where a series of gradual changes is indicated. The lower speciation event is occurring over time, sometimes termed "gradualism." The event that leads to the speciation of A and B was termed "punctuated equilibrium" by Eldridge and Gould—a rapid change, followed by periods of stasis.

▮ **5. (D) is correct.** If you consider that the homozygous long-haired deer is *HH*, and the homozygous short-haired deer is *hh*, then crossing them would give all offspring with the genotype *Hh* and medium-length hair phenotype. Crossing the heterozygotes would give you offspring in the ratio of 1:2:1—*HH*:*Hh*:*hh*. This means that 25% of the offspring would have long hair (*HH*), 25% of them would have short hair (*hh*), and 50% of them would have medium-length hair (*Hh*). This cross is an example of incomplete dominance.

▮ **6. (D) is correct.** Option A indicates incorrectly that chloroplasts function in cellular respiration, whereas option B indicates incorrectly that all eukaryotes have a cell wall, and C incorrectly states that all prokaryotes have a double membrane. Only answer D, which states that mitochondria, chloroplasts, and prokaryotes have similar DNA and chromosomes, is correct and supports the idea of symbiosis.

▮ **7. (D) is correct.** The only one of the statements not included in Darwin's theory of natural selection is answer D. Darwin's theory stated that the individuals that are least well suited will not leave behind as many offspring, but physical weakness does not necessarily lead to an individual's being unsuited for its environment. You should be prepared to take the correct statements in the question and develop them as a basis for answering a short essay question about natural selection. Asking about natural selection is a favorite topic on the essay portion of the exam.

■ **8. (C) is correct.** Asexual reproduction is a form of reproduction in which just one parent contributes genetic material to the offspring, which are clones of the parent and of each other. On the other hand sexual reproduction, in which two parents contribute genetic material to the offspring, results in offspring that differ from each other and their parents.

■ **9. (D) is correct.** Think of this problem as two monohybrid crosses. With tail length, all offspring show the dominant long tail. Answers A, B, and D satisfy this requirement. The coat color gene shows a 3 yellow to 1 white ratio. Answers C and D satisfy this requirement. Only answer D satisfies both requirements.

■ **10. (B) is correct.** Crossing guinea pigs with genotype *BbTt* gives a ratio of offspring of 9:3:3:1; this is a dihybrid cross between two independently assorting characters. Nine-sixteenths of the offspring will display both of the dominant traits; 3/16 of the offspring will display one of the dominant traits and one recessive one; 3/16 of the offspring will display the other dominant trait and the other recessive trait; 1/16 of the offspring will display both recessive traits. This question asks how many of the offspring will display one of the dominant traits (black fur) and one of the recessive traits (tail); the answer is 3/16.

■ **11. (C) is correct.** Genetic drift is defined as a change in the allele frequencies of a population due to chance. It is not an adaptive change. When a large part of a certain population is destroyed by a disaster such as an earthquake, drought, or fire, the surviving members may not be representative of the population's gene pool. Review types of genetic drift, the *bottleneck effect* and the *founder effect*.

■ **12. (D) is correct.** Regulatory genes are those that code for a protein that controls the transcription of another gene or a group of genes. In this case, allolactose is an inducer. When it is present in the cell, it binds to the repressor protein on the operator site of the *lac* operon. The binding of allolactose causes the repressor protein to change shape and fall from the operator site. The operon can now be transcribed, allowing mRNA to be formed. This represents an example of an inducible operon.

■ **13. (C) is correct.** Since the frequency of the recessive homozygote is 36%, we know that $q^2 = 0.36$, so $q = 0.6$ and $p = 0.4$. (because $p + q = 1.0$). The heterozygote condition, represented by $2pq$ in the Hardy-Weinberg equation, would be 2 (0.4)(0.6) in our problem or 0.48. If the answer is converted to a percent by multiplying by 100, the result is a final answer of 48%. If this answer is puzzling, review Chapter 23, Topic 6 paying particular attention to the information that is boxed.

■ **14. (C) is correct.** This is an autosomal recessive trait. If it were sex-linked, it would be expressed in only one of the sexes—usually the male, since males have only one X chromosome. We know it is recessive because it does not appear in every generation; only dominant traits appear in every generation. Also, the first generation does not show the trait, but they have children who do.

▌ **15. (A) is correct.** As the answer explanation from question 14 explains, this is an autosomal recessive trait. This means it is not on the sex chromosomes, eliminating answers C and D. Answer B indicates a dominant autosomal, not a recessive autosomal, as in the given answer, A.

▌ **16. (B) is correct.** In facilitated diffusion, the transport of certain substances across a membrane is aided by transport proteins that span the membrane. These transport proteins are specialized for the solute that they transport. In the case of chemiosmosis, an H^+ gradient is established across the inner mitochondrial membrane by the electron transport chain. This is an excellent time to review types of passive and active transport, a common area for essay and multiple choice questions.

▌ **17. (D) is correct.** The activation energy of a reaction is the initial energy investment required in order for the reaction to proceed. It is the energy required in order to break the bonds of the substrate enough for the substrate to reach the highly unstable transition state. You can tell that answer D is the reaction energy of the uncatalyzed reaction (because the presence of a catalyst would decrease the overall energy of the reaction), so the taller curve must be the uncatalyzed reaction.

▌ **18. (C) is correct.** The activation energy of the catalyzed reaction is represented by answer C. The overall energy of this reaction is significantly lower than that of the uncatalyzed reaction. This is because enzymes speed up the course of reactions by lowering the energy of activation so that the transition state is much easier to reach.

▌ **19. (B) is correct.** The transition state of a reaction is the highest-energy, most unstable form of the reactants in the reaction. The energy put into the reaction in order to make it "go"—also known as the activation energy—must be sufficient to enable the reactants to reach this transition state. These three questions should remind you to examine graphs closely and match potential answers carefully to specific questions. Expect questions on enzymes!

▌ **20. (D) is correct.** Hypotonic means having less solute, and therefore more water than the comparison tissue. Since water is moving out of the cells in this figure, they are hypotonic to the surrounding cells. Their turgor pressure is decreasing, and they are becoming flaccid. While the new AP Biology framework does not require much in the area of plants, this question is an example of taking an area that is required, osmosis, and applying it in one of many possible illustrative examples. Be prepared for this and do not quit on a question that seems to be about an area you might not have studied.

▌ **21. (D) is correct.** Stomata open when guard cells actively accumulate K^+ from neighboring epidermal cells, causing the guard cells to become hypertonic. This causes water to enter the guard cells, as shown in the figure. The guard cells are hypertonic due to the accumulation of K^+, a process of active transport that requires ATP. This question is about understanding Chapter 7, Membrane Structure and Function, not directly about plants.

▌ **22. (C) is correct.** In an inversion mutation a segment of a chromosome is reversed. This may occur during a crossing-over event if one of the chromosome segments being exchanged is inserted backwards. This reverses the order of the genes. Answer D describes a deletion mutation.

23. (C) is correct. An inversion of chromosome structure reverses a segment within a chromosome. In this case that would make the order GEWA. Other alterations of chromosome structure include deletions, duplications, and translocations.

24. (A) is correct. Species A is autotrophic and the primary producer of the ecosystem. This species is capable of capturing solar energy and converting it into the chemical energy contained in the bonds of organic compounds, which are used by the other organisms in this ecosystem. It also is the only one that has arrows flowing only from it, indicating that it is not a consumer.

25. (B) is correct. Species B and C represent primary consumers—they consume only the autotrophs in this ecosystem, which are the primary producers. They are presumably herbivores since their only food source is the primary producer.

26. (C) is correct. Species E consumes species A (presumably a plant), species C (presumably an animal), and species B (also presumably an animal). This means that species E is an omnivore—it eats both plants and animals. Species E is also both a primary consumer (because it consumes species A) and a secondary consumer (because it consumes species B and C).

27. (D) is correct. The DNA fragments migrated along the gel at rates according to their size—the smaller DNA fragments migrated more quickly through the dense gel and can be found near the bottom of the gel, whereas the larger fragments migrated more slowly and can be found closer to the top.

28. (D) is correct. Sample 2 must have been cut at more restriction sites than was sample 4 because more DNA fragments of different sizes were produced. This is shown by the greater number of bands on the gel in the lane of sample 2. The length of the original DNA molecule cannot be determined by the gel because multiple fragments of the same length (for example, three different fragments each 1,000 bp) will only form a single band.

29. (C) is correct. Smaller fragments move more easily through the gel and are found near the bottom of the gel; larger fragments migrate more slowly and are found near the top of the gel. A common error is to reverse this relationship.

30. (B) is correct. The solution on side B is hypertonic to the solution on side A. It is more concentrated (total molarity of 1.2) than the solution on side A (total molarity of 0.9) at the time this experiment began.

31. (B) is correct. Because side B is hypertonic to side A, water will cross the membrane increasing the level of fluid on side B.

32. (B) is correct. Each solute moves down its own concentration gradient. NaCl is more concentrated on side A than side B and the membrane is permeable to NaCl. This means we can expect NaCl to move down its concentration gradient from side A to side B.

33. (C) is correct. This question requires you to know that steroid hormones are lipids, and so can easily diffuse across the plasma membrane phospholipid bilayer. They are bound by intracellular receptors, not by cell-surface receptors.

34. (A) is correct. Because steroid hormones are hydrophobic and can cross the plasma membrane, they bond with intracellular receptors forming transcription factors that help to stimulate the transcription of the gene into mRNA. Answers B and C both involve plasma membrane receptors and D does not include the binding of the hormone to an intracellular receptor.

■ **35. (B) is correct.** The trait is sex-linked and recessive. Two sets of unaffected parents in the second generation have a child that is affected. Another indication that the trait is sex-linked is that five out of the six people affected by the trait are males.

■ **36. (C) is correct.** Females 2, 3, and 5 do not have the trait in question, but they have sons with the trait. The sons received their Y chromosome from their father and the X chromosome with the allele that causes the recessive sex-linked disorder from their mother. The female in 6 may have the recessive allele, but the pedigree does not provide any information about her genotype.

■ **37. (D) is correct.** Because both parents have brown fur, but they produce some white offspring, you have to conclude that both rabbits were heterozygous. Each parent must have a recessive allele to pass on. Additionally, the ratio of 3:1 that would result from a monohybrid, heterozygous cross fits the reported results.

■ **38. (C) is correct.** The movement of oxygen from the clusters of alveoli at the tips of the bronchioles in the lung, across the epithelial walls, and into the bloodstream is an example of passive diffusion. All of the answers except C involve transport mechanisms more complicated than simple diffusion. Transport is a favorite topic on the AP exam. Be sure to review transport before the exam.

■ **39. (D) is correct.** All of the answers listed are conditions that must be met for a population to be in Hardy-Weinberg equilibrium—except the criteria that there must be only two alleles present for each characteristic. Any population that is not in Hardy-Weinberg equilibrium will evolve, and natural populations rarely achieve this type of equilibrium for extended periods of time.

■ **40. (B) is correct.** The answer is 1/2. This answer can be obtained by working the dihybrid cross, but that is extremely time-consuming. Instead, note that one parent is *RR*, meaning all the offspring will have red hair. The second trait is *Ff* × *ff*, a cross that yields 1/2 freckles and 1/2 without freckles. Although all the children will have red hair, only 1/2 of the children will have red hair and no freckles.

■ **41. (B) is correct.** One of the two general types of speciation is allopatric speciation. In allopatric speciation, two populations are geographically separate, with no link between the two populations. The other type of speciation is sympatric speciation. In this case, two populations are in the same geographic area, but biological factors (such as chromosome changes and nonrandom mating) reduce gene flow.

■ **42. (A) is correct.** In reactions that are exergonic, energy is given off—often in the form of heat—during the course of the reaction. Conversely, endergonic reactions require the input of energy in order to proceed. Energy flow is a central theme in the new curriculum, so a fundamental understanding of exergonic and endergonic reactions is a starting place for thinking about energy flow. Expect AP exam questions to be more difficult, but this should serve as a valuable review of the basic information.

43. (B) is correct. The cell consumes O_2 as it receives electrons from the electron transport chain and forms water during the process of oxidative phosphorylation. The cell makes ATP through oxidative phosphorylation, so measuring the rate of consumption of O_2 by the cell is a good way to determine its metabolic rate.

44. (D) is correct. This population is exhibiting exponential growth. Exponential growth can occur when the conditions in an environment are ideal—when there is enough of, or an excess of, required resources in an environment. The population then grows at its maximum rate until it reaches its carrying capacity.

45. (C) is correct. The removal of introns requires spliceosomes. Spliceosomes are complexes of snRNA and proteins that interact with certain sites along an intron, excising the intron and joining together the two exons that flanked the intron.

46. (C) is correct. Although the figure may seem complicated, ultimately the question is asking what complex converts single-stranded DNA to double-stranded DNA. In the list of possible answers, only the enzyme DNA polymerase can accomplish this task.

47. (A) is correct. The carrier parents would each have the genotype *Aa*. This means their children would have a 25% chance of inheriting both of the recessive alleles. The fact that their first three children do not have cystic fibrosis does not affect the probability of the fourth, unrelated event.

48. (C) is correct. Punctuated equilibrium is the term used for the idea that evolutionary change in a species occurs in rapid bursts alternating with long periods of little or no change. Gradualism is the model of evolution in which species evolve gradually and diverge more and more as time passes.

49. (B) is correct. The first genetic material may have been short pieces of RNA that served as templates for aligning amino acids in polypeptide synthesis and for aligning nucleotides in a primitive form of self-replication. Early protobionts with self-replicating, catalytic RNA would have been more effective at using resources and would have increased in number through natural selection.

50. (A) is correct. The light reactions of photosynthesis convert solar energy to chemical energy in the form of ATP and NADPH. They do this when light is absorbed by various pigments in the thylakoid membrane of the chloroplasts; the pigments pass the energy down a chain of electron acceptors, and in the process, ATP and NADPH are produced.

51. (D) is correct. Enzyme 5 is the most efficient enzyme in this example—it has the highest rate of reaction. This means that its optimal rate is faster than that of any of the other enzymes depicted, assuming a standard measurement for the *y*-axes of both graphs.

52. (C) is correct. The pH scale is logarithmic, so the difference between scale numbers is a power of 10. Enzyme 4 has an optimal pH of 2, whereas enzyme 5 has an optimal pH of 8. The difference is 10^6 or 1,000,000 times more acidic!

53. (A) is correct. Temperature above the optimal causes thermal agitation of the enzyme resulting in disruption of hydrogen bonds, ionic bonds, and other weak interactions that normally stabilize the structure of the enzyme. With these disruptions the protein enzyme molecule denatures.

54. (D) is correct. The reason the action spectrum for photosynthesis doesn't match the absorption spectrum for chlorophyll *a* is because chlorophyll *a* is not the only photosynthetically important pigment in the chloroplast. Two other photosynthetically important pigments are chlorophyll *b* and the carotenoids.

55. (C) is correct. This cladogram is set up with the oldest common ancestor to the left, making branch point 1 the earliest common ancestor. Based on this, the most recent common ancestor *of the choices given* is that of lizard and ostrich.

56. (C) is correct. Branch point 5 indicates that crocodiles share a more recent common ancestor with ostriches than with lizards and snakes. The reptile clade consists of tuataras, lizards, snakes, turtles, crocodilians, and birds.

57. (D) is correct. Plant leaves are green because they reflect and refract green light, which is not utilized in photosynthesis. Red light is used in photosynthesis, meaning the plant would absorb CO_2 for photosynthesis. Check the action spectrum for photosynthesis in your text and be prepared to explain the peaks and valleys shown in the graph.

58. (A) is correct. The light reactions convert solar energy to the chemical energy of ATP and NADPH, which are utilized in the Calvin cycle to reduce CO_2 to sugar.

59. (D) is correct. A nucleotide insertion downstream and close to the start of the coding sequence creates a frameshift mutation, causing the regrouping of codons and a completely new list of amino acids, leading to a nonfunctional protein. If the amino acid sequence is changed after the insertion, a *missense mutation* occurs. If the regrouping leads to the formation of a stop codon, a *nonsense mutation* occurs, which terminates the forming protein. As you study different types of mutations, give emphasis to the *effect* of a mutation, and be able to predict and justify which type in a particular location would cause the greatest change in nucleotide sequence.

60. (A) is correct. Introns, not exons, are removed before mRNA leaves the nucleus. The rest of the choices in this question are true and worthy of your perusal and understanding. RNA editing is a fundamental concept in molecular genetics.

Part B

1. The Chi-squared value for this problem is **12.0.** If the parents were both heterozygous, you would predict a ratio of 3 red-eyed to 1 sepia-eyed. If there were 100 offspring, that would be 75 and 25.

Observed (*o*)	Expected (*e*)	(*o*–*e*)	(*o*–*e*)^2	(*o*–*e*)^2
60 red-eyed	75	15	225	225/75 = 3
40 sepia-eyed	25	15	225	225/25 = 9
				x^2 = 12.0

2. The answer is **0.47.** The question tells you that 24 birds show the recessive trait. Use this to calculate q^2 = # birds with defect/total # birds = 24/ 87 = 0.28 (28% of the birds show this defect). This is q^2. Take the square root of 0.28 = 0.53. But wait, this is q! Subtract this from 1 to get p = 0.47.

3. The answer is **1.89**. Refer to your formula sheet that may be used when taking the AP exam.

22/16 raised to the exponent of [10/(21 – 16)] = 1.89

Temperature Coefficient Q_{10}

$$Q_{10} = \left(\frac{k_2}{k_1}\right)^{\frac{10}{t_2 - t_1}}$$

Primary Productivity Calculation

mg O_2/L x 0.698 = mL O_2/L

mL O_2/L x 0.536 = mg carbon fixed/L

t_2 = higher temperature

t_1 = lower temperature

k_2 = metabolic rate at t_2

k_1 = metabolic rate at t_1

Q_{10} = the *factor* by which the reaction rate increases when the temperature is raised by ten degrees

Source: AP Biology—Course and Exam Description. © 2012. The College Board. www.collegeboard.org. Reproduced with permission.

This means that the rate of respiration is almost double (two times greater) for each 10° increase in temperature.

Free-Response Questions

1. (a) The cell membrane of a typical animal cell is composed of three main components: phospholipids, which are two fatty acids joined to two glycerol hydroxyl groups and a phosphate group connected to the third glycerol hydroxyl group; proteins, both integral (embedded in the cell membrane) and peripheral (associated with the outside of the membrane); and membrane carbohydrates, which are usually small carbohydrates associated with the outside of the membrane, forming glycolipids and glycoproteins. Animal membranes also contain cholesterol, a hydrophobic steroid embedded in the hydrophobic portion of the phospholipids.

Phospholipids form a hydrophobic foundation for the rest of the molecules in the cell membrane. Some very small nonpolar molecules, like O_2 and CO_2, can pass through the lipid membrane unaided. This type of movement across the membrane is called passive diffusion, because energy is not needed to move the substance across the membrane.

The function of the proteins is multifold, but one important function is to facilitate the passive transport of water and certain other solutes across the membrane. The proteins that serve this function are called transport proteins. Proteins can also participate in the active transport of certain substances across the membrane; they can act as pumps that use ATP energy to transport substances against their concentration gradient. Proteins can also act as important cell-surface receptors.

Membrane carbohydrates on the external surface of the cell membrane are important in cell-cell recognition; these carbohydrates vary from species to species, and from cell type to cell type, so they function in cell-cell recognition, an essential component of the immune system.

Cholesterol is also found in animal (but not plant) plasma membranes where it functions as a fluidity buffer. At high temperatures, such as human body temperature (37°C), cholesterol makes the membrane less fluid. At lower temperatures cholesterol prevents packing of phospholipids, helping to maintain fluidity.

(b) Glycoproteins found on viral envelopes bind to specific receptor molecules on the surface of a host cell. This promotes entry of the capsid and viral genome into the cytoplasm where cellular enzymes digest the capsid and release the genetic material. Viral reproduction follows, but the specific steps differ depending on the type of virus. This type of entry is a form of receptor-mediated endocytosis.

Hormones are the chemical messengers of the body. They travel through the bloodstream to their target cells. There are two ways by which hormones can gain entry into a target cell. Lipid-soluble hormones such as steroids diffuse through the cell membrane to bind to a specific receptor protein. Often, the protein receptor/hormone combination acts as a transcription factor, turning on specific genes. Water-soluble hormones bind to a specific receptor protein found on the target cell membrane. This binding initiates a signal transduction pathway that results in a specific cellular response.

Water enters the cell through a process called facilitated diffusion. This means that water crosses the cell membrane down its concentration gradient, but with the help of specific transport proteins termed "aquaporins." Transport proteins are specific for the molecules they assist across the membrane, but they do not require the input of energy.

This is a good free-response answer because it shows knowledge of the following important biological terms:

phospholipid	*passive diffusion*
fatty acid	*facilitated diffusion*
glycerol	*active transport*
integral protein	*hormones*
peripheral protein	*target cell*
carbohydrate	*signal transduction pathway*
glycolipid	*receptor*
capsid	*aquaporins*
transcription	*factors*

The response also shows knowledge of the following important biological processes: the importance and function of the cell membrane; how molecules get across the cell membrae; how viruses infect cells; and mechanisms of hormonal signaling. Remember to always read the questions carefully, answer all parts of the question, and organize your answer so it is easy for the person grading your response to follow your answer.

2. (a) There are three primary reasons the theory that glycolysis was the first ATP-producing metabolic pathway to evolve is true.

1. The first reason is that long ago, Earth's atmosphere contained almost no oxygen, and only relatively recently have the current atmospheric levels of gases come to be what they are. Glycolysis does not require oxygen, so it is possible that prokaryotes (which evolved before eukaryotes) used glycolysis for making ATP at a time when no free oxygen was present in the early atmosphere.

2. The second substantiating clue is that glycolysis is a very common method for making ATP conserved across organisms in the three domains: Bacteria, Archaea, and Eukarya. This commonality implies that it originated very early in the evolution of metabolic pathways.

3. The final reason has to do with the site of glycolysis—that is, it takes place in the cytosol, and not in an organelle. Prokaryotic cells, which evolved first, are much simpler than eukaryotic cells, and they contain no membrane-bound organelles (not even a nucleus). Therefore, if glycolysis were to take place in an early prokaryotic cell, it would have to evolve in the cytosol—for instance, it would have to evolve such that it did not rely on a specialized membrane in order to function.

(b) In the course of the citric acid cycle, acetyl CoA is first joined to oxaloacetate to form citrate, and then the molecule is manipulated extensively to finally re-form a molecule of oxaloacetate. In the course of these reactions, the citric acid cycle produces 1 ATP molecule, 3 NADH molecules, and 1 $FADH_2$ molecule (per turn).

The way that the citric acid cycle is related to oxidative phosphorylation is that the NADH and $FADH_2$ molecules produced during the citric acid cycle donate electrons to the electron transport chain, which is embedded in the wall of the inner mitochondrial membrane. This electron transport chain shuttles the electrons down its length, in an exergonic reaction. The energy produced in this series of electron transfers is used to pump hydrogen ions across the inner membrane. The electrochemical gradient formed by the hydrogen ions is a source of potential energy for the cell (much like water behind a dam). The cell utilizes this potential energy through the enzyme ATP synthase, which is also embedded in the wall of the inner mitochondrial membrane. When hydrogen ions flow down their electrochemical gradient through ATP synthase the enzyme catalyzes the phosphorylation of ADP to form ATP. In the last step of the electron transport chain the electrons from NADH and $FADH_2$ are combined with O_2 to form water. Thus, oxygen is the ultimate electron acceptor and reason for the term "oxidative phosphorylation."

This is a good free-response answer because it knowledgeably uses the following key terms:

glycolysis	*oxaloacetate*
prokaryotes	*citrate*
eukaryotes	*citric acid cycle*
ATP	*NADH*
cytosol	*FADH$_2$*
organelle	*inner mitochondrial membrane*
nucleus	*oxidative phosphorylation*
acetyl CoA	*electrochemical gradient*

The response also shows a working knowledge of the following important biological concepts: the origin of life and ancient Earth, relationships among living organisms, glycolysis, the citric acid cycle, energy transfers in cells, and the electron transport chain.

3. In the euchromatin form DNA is more accessible to proteins involved in transcription, like RNA polymerase. In chromatin packing, when the genetic material is in heterochromatin form, it is highly condensed and proteins involved in transcription do not have access to the DNA. In DNA methylation, methyl groups are attached to specific regions of DNA immediately after it is synthesized. In some cases, this is thought to be responsible for these genes' long-term inactivation. A second set of enzymes can also attach methyl groups to histone proteins, which also results in reduced access to the gene for transcription. Finally, in histone acetylation, acetyl groups are attached to certain amino acids of the tails of histone proteins. When the histones are acetylated, their shape alters so that they are less tightly bound to DNA; this enables the proteins involved in transcription to move in and begin work. When histones are deacetylated, DNA transcription is greatly reduced.

4. If this plant is rotated 180°, plant hormones will act to start its growth in the direction facing the sun. This is because a plant responds to light by an asymmetrical distribution of auxin going down from the tip of the plant, which causes the cells on the darker side of the plant to elongate more than the cells on the brighter side of the plant. Whether auxins are asymmetrically distributed or auxins on the light side are inactivated, the auxin hormone activates enzymes in the cell wall that pump hydrogen ions (protons) into the cell wall causing a drop in pH. The lowering pH activates other enzymes in the cell wall to break cross-links between cellulose fibrils, which weakens the cell wall. This weakening of the cell wall allows turgor pressure from the central vacuole to stretch the cell wall, making the cell larger and in the process turning the plant toward the light. Growth of a plant toward a light source is known as positive phototropism.

5. Recessive. All affected individuals (Arlene, Tom, Wilma, and Carla) are homozygous recessive *aa*. George is *Aa*, since some of his children with Arlene are affected. Sam, Ann, Daniel, and Alan are each *Aa*, since they are all unaffected children with one affected parent. Michael also is *Aa*, since he has an affected

child (Carla) with his heterozygous wife, Ann. Sandra, Tina, and Christopher can each have the *AA* or *Aa* genotype.

6. Template sequence from problem: 3'-TTCAGTCGT-5'
   Nontemplate sequence: 5'-AAGTCAGCA-3'
   mRNA sequence: 5'-AAGUCAGCA-3'
   The nontemplate and mRNA base sequences are the same, except there is T in the nontemplate strand of DNA wherever there is U in the mRNA.

7. Genetic drift results from chance events that cause allele frequencies to fluctuate at random from generation to generation; within a population, this process tends to decrease genetic variation over time. Gene flow is the exchange of alleles between populations; a process that can introduce new alleles to a population and hence may increase its genetic variation (albeit slightly, since rates of gene flow are often low). Much as with the case of mutation, gene flow may introduce new alleles but the alleles will increase in frequency only if they are selected for through natural selection mechanisms.

8. (a) Species richness, the number of species in the community, and relative abundance, the proportions of the community represented by the various species, both contribute to species diversity.

   (b) Community 1 has a distribution of species A: 25%, B: 25%, C: 25% and D: 25%. Community 2 has a species distribution of species A: 80%, B: 5%, C: 5% and D: 10%.

   (c) Compared to a community with a very high proportion of one species, a community with a more even proportion of species is considered more diverse. Community 1 is considered more diverse because although species diversity is equal between the two communities, the relative abundance is more equitable in community 1.

   (d) If the blight damaged species A, the community structure of community 2 would be most seriously impacted because species A makes up 80% of the individual trees. Loss of species A from community 2 would result in large light gaps, more erosion, and greater loss of nutrients from the community. Community 1 would be less seriously impacted, as its distribution of species is more equal. In general, more diverse communities are more stable.

Index

NOTE: The letter *f* following a page number indicates that the information appears within a figure; the letter *t* following a page number indicates that the information appears within a table. (The topic may also be discussed in text on that page.)

Bound ribosomes, 50
Bowman's capsule, **232**
Brain, 194, 248
Brainstem, **248**
Breathing, 226–227
Breeding, plant, 209
Broadleaf forest, 266*f*, **267**
Bronchi, **226**
Bronchiole, **226**, 226*f*
Bronchus, 226*f*
Bryophytes, 179
Budding, **238**
Buds, 180*f*, 199–200
Buffers, 37
Bulk flow, **204**
Bulk transport, 58
Bundle-sheath cells, 91, 202*f*

C

C$_4$ photosynthesis, 91–92, 91*f*
Calcium ions, 65, **250**
Calvin cycle, **84**, 89–90
Campbell Biology (Reece), 3, 31
CAM photosynthesis, **92**
CAM plants, 91
Cancer, 134
Cancer cells, 70
Canopy, **267**
Capillaries, **223**
Capillary beds, 225, 226
Capsid, **134**, 135*f*
Carbohydrates, **39, 221**
    on cell membrane, 55
    in plasma membrane, 48
Carbon, 37–38
Carbon compounds, 37
Carbon cycle, **277**, 277*f*
Carbon dioxide (CO$_2$)
    breathing and, 227
    Calvin cycle and, 89–90, 91
    transpiration and, 316
    transported in blood, 227
Carbon fixation, 84
Carbonic acid, **37**
Carbonic anhydrase, **227**
Carboxypeptidase, **221**
Cardiac muscle, 224
Carnivores, 276
Carnivorous plants, **208**
Carpel, **182**, 183*f*
Carrier protein, 57
Carrying capacity, **270**
Casparian strip, 205
Catabolic pathway, **59**
Catabolic pathways, **77**
Catalysts, **60**
Catalytic cycle of enzymes, 61*f*
Catastrophism, 151
cDNA library, **138**
Cecum, **222**

Cell
    animal cell structure, 48*f*
    cancer, 70
    chloroplasts, 51
    communication, 62–65
    cytoskeleton, 52–53
    endomembrane system, 50–51
    eukaryotic, 47–50
    extracellular components and
        connections between, 53
    membrane structure and function,
        53–58
    metabolism, 58–62
    mitochondria, 51
    peroxisomes, 51
    plant cell structure, 49*f*
    reviews questions and answers about,
        70–75, 354–360
Cell body, **243**
Cell-cell signals, **133**
Cell communication, 62–65
Cell cycle, 66–70, 134
Cell cycle control, 134
Cell cycle control system, 69
Cell differentiation, **132**, 133, 140
Cell division, **132**, 395–396
Cell-mediated immune response, **230**
Cell plate, 69, **103**
Cell signals, 62–65
Cell-surface receptors, **236**
Cellular drinking, 58
Cellular eating, 58
Cellular innate defenses, **228**
Cellular membranes, 54
    See also Plasma membrane
Cellular respiration, 77–82, 87–88
    laboratory, 300–302
    multiple choice answers, 395
Cellular signaling, 62–65
Cellulose, **39**, 185
Cell wall (plant cell), 49*f*, **53**
Central nervous system (CNS), **244, 248**
Central vacuoles, 49*f*, **51**
Centrioles, **52**
Centromere, 66
Centrosomes, 48*f*, 49*f*, **52**
Cephalization, **186**, 189
Cerebellum, **248**
Cerebral cortex, **249**
Cerebrospinal fluid, **247**
Cerebrum, **249**
Cervix, 239
Channel protein, 57
Chaparral, **266**
Chaperonins, **41**
Charophytes, 178–179
Chase, Martha, **119**
Checkpoints, 69
Chemical bonds, 34–35
Chemical cycling, 276

Chemical energy
    Calvin cycle and, 89–91
    chemiosmosis and, 88
    defined, **57, 59**
    glycolysis and, 79
    life energy and, 83–84
    light energy converted to, 83, 86
    solar energy converted to, 84, 85–88
Chemical equilibrium, **35**
Chemical force, 56
Chemical reaction, **35**
Chemical synapse, 246
Chemiosmosis, 81, 81*f*, **82**, 87, 88*f*
Chemistry of life, 33–46
    carbon and molecular diversity of life,
        37–38
    chemical context of life, 33–35
    review questions and answers about,
        42–46, 351–354
    structure/function of biological
        molecules, 38–42
    water and life, 35–37
Chemoautotrophs, **176**
Chemoheterotrophs, **176**
Chemoreceptors, **249**
Chiasmata, **103**
Chi-square analysis of data, 326–329, 398
Chitin, **39**, 185
Chlorophyll, 85
Chloroplasts, 49*f*, **51**, 84*f*, 88
Choanocytes, 189
Cholecystokinin (CCK), **222**
Cholesterol, **40**
Choline, 40*f*
Chondrichthyes, Class, **193**
Chordata, **191**
Chordata, Phylum, 190, 191
Chordates, 191, 192*f*
Chorion, 193
Chorionic villus sampling, **110**
Chromatids, 66, 67*f*, 68*f*
Chromatin, 48*f*, **50**, 123
Chromatin packing, 123, 124*f*
Chromosomal mutations, **155**
Chromosome duplication, 66, 67, 67*f*
Chromosomes, 99
    alterations of number
        or structure of, 113
    cell cycle and, 66–67, 67*f*, 68*f*, 69
    chromatin and, 50
    chromatin packing, 124*f*
    DNA and, 119
    DNA molecule with protein, 123
    eukaryotic cells and, 48
    genetic disorders and, 113
    homologous, 100
    independent assortment of, 105
    linked genes and, 112
    lost/rearranged, 113
    meiosis and, 101, 103–105

Linked genes, **112**
Linnaeus, Carolus, **151, 171**
Lipids, 39, 40, 54
Literature search, 321
Littoral zone, **267**
Liverworts, 179
Lobsters, 186
Locus, **99**
Logistic growth model, **270**
Long-term regulation of blood pressure, 224, 235
Looped domain, 124*f*
Loop of Henle, **232**, 233, 234
Lorenz, Konrad, 253
Lower epidermis, 201, 202*f*
Leukocytes, 228
Lungs, 226*f*
Lupus, **231**
Luteinizing hormone (LH), **236**
Lyell, Charles, **152**
Lymphocytes, 229
Lysis, **231**
Lysogenic cycle, **136**
Lysosomes, 48*f*, **50**
Lysozyme, **228**
Lytic cycle, **136**

M

Macroclimate, **265**
Macroclimate patterns, **265**
Macroevolution, **159**
Macromolecules, 38
Macrophages, **228**
Mad cow disease, 136
Major histocompatibility complex (MHC) molecules, **229, 231**
Malignant tumor, 70
Mammalian respiratory system, 225–226, 226*f*
Mammals, 192*f*, 193, 194, 266, 267
  characteristics, 194
Mammary glands, 194
Map unit, **112**, 305
Marine biomes, **267**, 268
Marker, 312
Marsupials, 194
Mass extinction, 163
Mass number, **34**
Mating systems, **253**
Matter, **33**, 58, 59
Mechanical isolation, **160**
Mechanoreceptors, **249**
Medusa form, 189
Megaspores, **182**
Meiosis, 52, 66, **101**–105, 179, 180*f*, 183*f*, 184*f*
  laboratory, 304–305
  multiple choice answers, 395–396
  reproduction and, 238–239

Meiosis I, 101, 102*f*, **103**
Meiosis II, 101, **103**
Membrane potential, **56, 244,** 245
Membranes
  eukaryotic cells and, 47–48
  plasma, 48, 49*f*, 54
  structure and function, 53–58
Memory cells, **229**
Mendelian genetics
  chromosomal basis of inheritance, 110–113
  gene idea, 105–110
  meiosis and sexual life cycles, 99–105
  review questions and answers on, 114–118, 364–367
Mendelian inheritance, 105–110, 111, 113
Menstrual cycles, **239**, 240*f*
Menstrual flow phase, **239**, 240*f*
Menstruation, 241
Meristems, 200
Mesoderm, 242
Mesophyll, **83,** 201
Messenger RNA (mRNA), 49, **125,** 129, 132
Metabolic adaptations in prokaryotes, 176
Metabolic pathways, glycolysis and citric acid, 82
Metabolic rate, **218**
Metabolism
  cellular, 58–62
  defined, **58**
  endomembrane system, 50
  enzyme regulation and, 61
Metamorphosis, 186
Metaphase, 68*f*, 69, 102*f*
Metaphase chromosome, 124*f*
Metaphase I, 102*f*, **103,** 104
Metaphase II, 102*f*, **104**
Metastasis, 70
Methanogens, **176**
Methylation, **112**
MHC molecules, **229, 231**
Microclimate, **265**
Microevolution, **155, 159**
Microfilaments, 48*f*, 49*f*, **52**
Micro RNAs (miRNA), 132
Microsporangium, 184*f*
Microspores, **182**
Microtubules, 48*f*, 49*f*, **52**
Microvilli, 48*f*, 221, **221**
Migration, **252**
Milk production, 194
Miller, 162
Miller, Stanley, 37
Mineral transport, 200, 203*f*, 204, 205
Minimum viable population, 279
Mismatch repair, **123**
Missense mutations, **130**
Mitochondria, **51,** 80, 87, 88
Mitochondrial matrix, 51

Mitochondrial membrane, 81, 82
Mitochondrin, 48*f*, 49*f*
Mitosis, 52, 66, 67–69, 101, 104
  laboratory, 304–305
  multiple choice answers, 395–396
Mitotic phase, 66–69
Mnemonic devices, 171
Molecular clocks, **173**
Molecular control system, 69–70
Molecular data, phylogenies inferred from, 172
Molecular diversity of life, 37–38
Molecular genetics
  biotechnology, 136–141
  gene expression, 130–134
  genomes and their evolution, 141–143
  molecular basis of inheritance, 119–123
  review questions and answers on, 143–150, 367–372
  transcription and translation, 125–130
  viruses, 134–136
Molecular homologies, **153**
Molecular systematics, **173,** 176
Molecules
  carbon atoms forming, 37
  formed by chemical bonding, 34
  structure and function of large biological, 38–46
Mollusca, Phylum, 189
Monocots, **184,** 211
Monocytes, **228**
Monogamous mating system, 253
Monohybrid cross, **107**
Monomers, 38
Monophyletic grouping, 173
Monosaccharides, **39**
Monosomic, 113
Monotremes, 194
Morgan, Thomas Hunt, 111
Morphogenesis, **132,** 133
Morphogens, **133**
Morphological data, phylogenies inferred from, 172
Morula, 241
Mosses, 179, 180*f*
Motile, **175**
Motor mechanisms, 249–251
Motor molecules, 52, 69
Motor (somatic) nervous system, 248
Motor neurons, **244,** 247, **249**
Movement corridors, 279
MPF (mitosis promoting factor), 69
M phase checkpoint, 69
MRNA (messenger RNA), **125,** 129, 132
Mucus, **221**
Mucus-producing cells, 226
Müllerian mimicry, **273**
Multicelled organisms, origin of, 162–163
Multicellular eukaryotes, 142, 163